Signals and Communication Technology

More information about this series at http://www.springer.com/series/4748

Samuel J. Davey · Han X. Gaetjens

Track-Before-Detect Using Expectation Maximisation

The Histogram Probabilistic Multi-hypothesis Tracker: Theory and Applications

Samuel J. Davey
Defence Science and Technology
Edinburgh, SA
Australia

Han X. Gaetjens
Defence Science and Technology
Edinburgh, SA
Australia

ISSN 1860-4862 ISSN 1860-4870 (electronic)
Signals and Communication Technology
ISBN 978-981-13-3971-4 ISBN 978-981-10-7593-3 (eBook)
https://doi.org/10.1007/978-981-10-7593-3

Softcover re-print of the Hardcover 1st edition 2018

Printed on acid-free paper

This Springer imprint is published by Springer Nature
The registered company is Springer Nature Singapore Pte Ltd.
The registered company address is: 152 Beach Road, #21-01/04 Gateway East, Singapore 189721, Singapore

soli deo gloria

Foreword

The best compliment that can be paid to any author is to see other researchers extend and improve their work and to apply it to new problems. We are thus extraordinarily honored by Sam Davey and Han Gaetjens for writing this book. A cursory review of the table of contents reveals that the histogram probabilistic multi-hypothesis tracking (H-PMHT) algorithm has undergone considerable development and application over the last nearly 20 years. While many researchers have worked in the area, Sam and his colleagues have made many important contributions. This book gathers all these results together under one cover and discusses them in a unified manner that will enable others to understand what is known and to take the work further still.

Tracking algorithms do not usually arise from Bayesian classification problems, but H-PMHT surely did. The richness of these methods may account for the fact that a sufficient body of H-PMHT material now exists to warrant writing a book. In the early 1990s, we worked together on a series of classification problems that required us to estimate class-specific probability density functions (PDFs) from sample data. Little training data existed and getting more was extremely laborious. We sought ways to make the most of what data we had. The data were time stamped, so we sought to estimate time-varying PDFs over a sliding time window. Aggregating sample data over a time window was a natural thing to do given our signal processing backgrounds. After much work and many false steps, we were ultimately successful. Only belatedly did we recognize that a very special case of what we had done was a multiple target tracking problem. This specialized version of the algorithm is the PMHT in H-PMHT. Tracking weak targets is a data starved problem even when data are aggregated over a time window, and we needed more data to push the envelope lower. We recognized that the sample data we were given were points (e.g., local "peaks") extracted from a signal processor output power spectra (or similar). We sought to use the entire power spectrum, and not just the peaks. We wanted to avoid what we termed peak-picking losses. At that critical juncture (December, 1991), we met Leonid Perlovsky. He introduced us to the idea of replacing spectrograms with integer counts, that is, by using histograms to approximate the spectrum. He used Gaussian mixtures to model the spectral histograms, as did we, but his algorithm for estimating mixture parameters was not rooted in the Expectation-Maximisation (EM) method of Dempster, Laird and Rubin [1]. We were already using EM in PMHT, and we naturally sought to use EM again with the Perlovsky-style histogram data. This was the very beginning of what would eventually become the H-PMHT algorithm.

Our progress was slow, however, and project demands took us in different directions. One of us exploited the histogram idea as the basis of his Ph.D. thesis to track the components of a modulated time-periodic function. The other investigated Poisson point process models of spectrograms and multi-sensor fusion problems. Both of us were frustrated by issues we attributed to data poverty. The H-PMHT algorithm finally began to emerge on a train ride from Paris to Rennes, France, in November 1998. The night before we had each presented a version of our current work at a workshop in Paris on probabilistic methods for multi-target tracking (organized by our dear friend and colleague, the late Jean Pierre LeCadre). On the train to Rennes we recognized that our approaches could be merged to address the PMHT algorithm's data poverty problem, which we had never satisfactorily resolved some years earlier. We spent the entire train ride discussing the histogram model, connections to Poisson point processes, and the many implications of the idea set, as well the possibility of removing the quantization step by taking limit as the histogram quantization became infinitely fine to recover the spectral data. A little over a year later the H-PMHT algorithm was finally complete.

With this book Sam and Han bring the H-PMHT story up to the current day. We thank them for that. We feel sure there are many stories yet undiscovered and chapters yet to be written.

Tod E. Luginbuhl
Naval Undersea Warfare Center
Newport, RI, USA

Roy L. Streit
Metron Inc., Reston, VA, USA

Reference

[1] Dempster, A.P., Laird, N.M., Rubin, D.B.: Maximum likelihood from incomplete data via the EM algorithm. Journal of the Royal Statistics Society **140**, 1–38 (1977)

Acknowledgements

Thanks to my teachers and mentors: Neil Gordon, Doug Gray and Roy Streit. This book is the culmination of research that began with a PhD many years ago; you have always been supportive and encouraging over those years. I have spent many years standing on the shoulders of Tod Luginbuhl: thanks Tod, none of this would have happened without your work. My colleague Monika Wieneke was the driving force behind the random matrix work, which I have come to use continually. My colleague Han Gaetjens made the Poisson H-PMHT a reality and has been an enormous help in transforming my distilled erudite prose into something partially comprehensible!

I am indebted to Mark Rutten for reviewing this book, it must have surely required an epic dose of caffeine, and now I know at least one person has read it. And of course I must thank Caprice and my family for putting up with many hours of me trapped behind a screen.

One of the aims of this book was to demonstrate the methods on a variety of different applications, and this has depended on the assistance of many people who have made data available: Michael McDonald, James Palmer, Ashley Summers, Mark Rutten, Joe Fabrizio, Richard Ellem, Dragana Carevic, Ben White, Nick Redding, Bittany Morreale and the folk in the Performance Evaluation of Tracking and Surveillance community.

May 2017 Samuel J. Davey

Contents

About the Authors

Samuel J. Davey studied engineering and mathematics at the University of Adelaide, culminating with a Ph.D. in signal processing in 2003.

He has worked for the Defence Science and Technology Group, Australia, since 1995 in the areas of target tracking, tracker performance assessment, and multi-sensor fusion; he is currently Group Leader, Geophysical Phenomenology and Performance Assessment. He is also a Visiting Research Fellow at the University of Adelaide, a senior member of the IEEE, and an associate editor of IEEE Signal Processing Letters. He is a co-author of the book *Bayesian Methods in the Search for MH370.*

He received the 2011 JP LeCadre award for the best paper at the International Conference on Information Fusion and the 2012 DST Science and Engineering Excellence award for work on H-PMHT that led to this book.

Han X. Gaetjens received her Honours degree in Mathematics from the University of South Australia in 2007 and her Ph.D. from the University of Adelaide in 2015.

She has worked for the Defence Science and Technology Group, Australia, since 2007.

Parlance

General

A	A matrix
z	A vector
A^T	matrix transpose
t	A time and frame index
T	The number of frames in a batch
$\mathcal{Q}(\cdot)$	An Expectation-Maximisation auxiliary function
$E_Y[X]$	The expected value of random variable X with respect to Y
$p_{comp}(\cdot)$	The Expectation-Maximisation complete data likelihood
$p_{miss}(\cdot)$	The Expectation-Maximisation missing data likelihood
$\mathcal{N}(x; \mu, \Sigma)$	A Gaussian probability density with mean μ and covariance matrix Σ evaluated at x

Measurements

i	An image pixel index
I	The number of pixels in an image
W^i	The part of measurement space spanned by pixel i
W	The whole measurement space
P_{FA}	The probability of false alarm
P_D	The probability of detection
n_t	The total number of point measurements in frame t
n_t^i	The number of point measurements inside pixel i in frame t
$\mathbf{N}_t$	The set of point measurement counts across the pixels in frame t
$\mathbb{N}$	The whole collection of counts across pixels and frames
$\mathbf{y}$	An arbitrary point in the measurement space
r	An index over point measurements
$\mathbf{y}_t^r$	The rth point measurement in frame t
$\mathbf{Y}_t$	The set of all of the point measurements in frame t
$\mathbb{Y}$	The whole collection of point measurements across frames

k_t^i	The true source of measurement $\mathbf{k}_t^r$ *and* $\mathbf{y}_t^r$
$\mathbf{K}_t$	The set of measurement sources at frame t
$\mathbb{K}$	The whole collection of measurement sources in every frame
$\mathbf{z}_t^i$	The measured power in pixel i in the image at frame t
$\mathbf{Z}_t$	The complete image at frame t
$\mathbb{Z}$	The whole image sequence across frames
$\|\|\mathbf{z}_t\|\|$	The total image power at frame t
A	The attribute space, for example colours or textures
$\mathbf{a}_t^i$	The measured attribute for pixel i in frame t
$\mathbf{A}_t$	The complete attribute image at frame t
$\mathbb{A}$	The attribute image sequence across frames

Targets

m	A target index
M	The number of targets
$\mathbf{x}_t^m$	The kinematic state of target m at frame t
$\mathbf{X}_t$	The set of states of all of the targets at frame t
$\mathbb{X}$	The whole collection of kinematic states over the batch
θ_t^m	The appearance parameters of target m
ϕ_t^m	The attribute state of target m at frame t
Φ_t	The set of attribute states of all of the targets at frame t
Φ	The whole collection of attribute states over the batch
π_t^m	The mixing proportion for target m
Π_t	The set of mixing proportions at frame t
Π	The set of all mixing proportions over the batch
$\lambda_t^m(y)$	The spatially dependent measurement rate of target m at frame t
$\Lambda_t^{m,i}$	The measurement rate in pixel i of target m at frame t
$\mathbf{L}_t$	The measurement rates across all the pixels in frame t
$\mathbb{L}$	The whole collection of measurement rates over the batch
$g^m(\mathbf{y}\|\mathbf{x},\theta)$	The target appearance function
$G_t^{m,i}$	The probability of a measurement from target m landing in pixel i
G_t^i	The marginal probability of a measurement from an unknown target landing in pixel i
$w_t^{m,i}$	The target-pixel association weight
$\mathfrak{z}_t^i$	The associated image power in pixel i
$\|\|\mathfrak{z}_t^m\|\|$	The total power in the associated image from target m

Acronyms

ADSS	Analysts Detection Support System
CFAR	Constant false alarm rate
DEAP	Defence Experimentation Airborne Platform
DRDC	Defence Research and Development Canada

DST Group	Defence Science and Technology Group, Australia
ECM	Expectation coupled maximisation
EM	Expectation maximisation
GEM	Generalised expectation maximisation
H-PMHT	Histogram probabilistic multi-hypothesis tracker
H-PMHT-RM	H-PMHT with random matrices
JPDA(F)	Joint probabilistic data association filter
KLT	Kanade Lucas and Tomasi (feature tracking)
MAP	Maximum a posteriori
ML-H-PMHT	Maximum likelihood H-PMHT
ML-PDA	Maximum likelihood probabilistic data association
ML-PMHT	Maximum likelihood PMHT
OTHR	Over the horizon radar
PDA(F)	Probabilistic data association filter
PETS	Performance evaluation of tracking and surveillance
PMHT	Probabilistic multi-hypothesis tracker
PPP	Poisson point process
RMS	Root mean square (error)
ROC	Receiver operating characteristic
SSA	Space situation assessment
TkBD	Track before detect
VMTI	Video moving target indication

Chapter 1
Introduction

This monograph is an exposition of the Histogram Probabilistic Multi-Hypothesis Tracker, H-PMHT. It aims to be both an accessible and intuitive acquaintance with the mathematical mechanics of the H-PMHT and a definitive reference for the existing literature on the method. As such, the first few chapters of the monograph begin rather gently before we build up momentum and address extensions of the method to a broad class of tracking problems. The latter chapters of the monograph present applications using recorded data from experimental radar, sonar and video sensor systems.

The H-PMHT algorithm can be categorised as a track-before-detect (TkBD) approach, which is not altogether helpful since this paradigm has a rather relaxed membership policy. What is unusual about the method is that it performs data association over the energy inside pixels of an image measurement. It can assign parts of the energy in a particular pixel to different targets and to background clutter. This is achieved by treating an image as a histogram of point measurements and then by using PMHT data association to fit a dynamic mixture to the histogram data. Within this monograph, the term *image* is not restricted to a two-dimensional matrix of 32-bit pixel values. Rather an image is a set of continuous-valued scalars simultaneously collected by a sensor. Each of these values corresponds to a different physical pixel, which can also be called a bin or a cell. The pixels are non-overlapping intervals in $\mathscr{R}^n$ where the dimension n is application dependent: except where noted the mathematics is agnostic to the image dimension. Examples are passive sonar angle of arrival maps, where $n = 1$; electro-optical video sequences, where $n = 2$; and active radar azimuth-range-Doppler maps, where $n = 3$. By dynamic mixture, we mean that the measurement probability density is the superposition of a number of component densities whose parameters change over time.

The novelty of H-PMHT is the transformation of a continuous-valued image into a histogram and the use of point measurement data association to exploit this data. These two steps are unique to H-PMHT and are not undertaken by any other TkBD approach. The process is not without its problems: the transformation creates a collection of assumed independent point measurements from a single sensor image,

S. J. Davey and H. Gaetjens, *Track-Before-Detect Using Expectation Maximisation*, Signals and Communication Technology,
https://doi.org/10.1007/978-981-10-7593-3_1

so care must be taken with the treatment of these measurements. In particular, the implied information content of the point measurement collection is not the same as the information content in the original image, which affects the implementation of an estimator for the mixture parameters and requires a special Bayesian prior. These matters will be discussed in detail in due course.

The remainder of this chapter provides background information for the development of H-PMHT. First, the history of the algorithm's evolution is reviewed highlighting the most important contributions and defining the scope of the technical literature that will be covered by the manuscript. Since the H-PMHT is fundamentally built using the mechanics of expectation–maximisation (EM) a brief primer on EM is provided next, mostly as an introduction for readers unfamiliar with the approach. Throughout the monograph, numerous variants of the H-PMHT algorithm are described and applied to simulated data. These simulations all use a common scenario with numerous crossing targets which will be referred to as the canonical multi-target scenario. This scenario is introduced and some example realisations are provided. Next the measures of performance used throughout the book are introduced. Finally, the chapter concludes with a spoiler of the rest of the monograph that gives an overview of the material within.

The examples presented throughout the monograph make extensive use of the MATLAB H-PMHT toolbox [12, 19]. This toolbox provides generic MATLAB implementations for all of the variants of H-PMHT presented here as well as functions to generate realisations of the canonical multi-target scenario and evaluate performance measures. Footnotes are used through the text to indicate the specific functions relevant to that section. The H-PMHT toolbox is available to download from the Mathworks MatlabCentral website; an overview is given in the appendix.

1.1 Historical Development of H-PMHT

The terminology Histogram Probabilistic Multi-Hypothesis Tracker was coined by Streit [43] but his work was the culmination of work begun by others. Table 1.1 lists the most significant contributors to the development of H-PMHT and summarises the differences between the methods published by each. The use of EM to fit parameters to histograms dates back to Dempster, Laird and Rubin's famous EM paper [24] where the authors showed how to fit the parameters of a general distribution to multi-dimensional histogram measurements. McLachlan and Jones [28, 33] considered the specific case of fitting a finite mixture distribution to one-dimensional histograms. These methods apply only to the static case where the model parameters are fixed and are based on treating the location of each measurement within a histogram bin as missing data, just as H-PMHT does.

Perlovsky addressed the problem of fitting finite multivariate mixture distributions to point measurement data [37, 38]. This work differed from the one-dimensional examples above because it did not marginalise over the missing measurement locations, but rather used a low threshold to produce a large number of point

Table 1.1 Significant developments in the history of H-PMHT

Authors	Dim.	Components	Appearance	Attribute data	Measurement model
Dempster, Laird and Rubin [24]	N-D	Static	Known Gaussian	None	Multinomial
McLachlan and Jones [28, 33]	1-D	Static mixture	Known Gaussian	None	Multinomial
Luginbuhl and Willett [31, 32]	1-D	Dynamic mixture	Static Gaussian	None	Multinomial
Perlovsky [37]	N-D	Deterministic mixture	Static Gaussian	None	Point multinomial
Streit et al. [43]	N-D	Dynamic mixture	Static Gaussian	None	Multinomial
Streit et al. [44]	N-D	Dynamic mixture	Static Gaussian	Superposition	Multinomial
Davey [8]	N-D	Dynamic mixture	Known non-Gaussian	None	Multinomial
Wieneke and Davey [57, 58]	N-D	Dynamic mixture	Stochastic Gaussian	None	Multinomial
Davey et al. [23]	N-D	Dynamic mixture	Arbitrary stochastic	None	Multinomial
Gaetjens [26, 52]	N-D	Dynamic mixture	Known Gaussian	None	Poisson
Davey et al. [18]	N-D	Dynamic mixture	Arbitrary stochastic	Occluding	Poisson
Dunham et al. [25, 59]	N-D	Deterministic mixture	Known Gaussian	None	Multinomial

measurements. Perlovsky's method grew from the classification community rather than tracking: it treats the target trajectories as deterministic and fits a maximum likelihood parameter estimate. In retrospect, Perlovsky's work can be viewed as an early version of Maximum Likelihood PMHT, which will be discussed in detail later in this book. This method is closely aligned with point measurement PMHT [49], which applies EM mixture models to point measurement data. Perlovsky was not aware of the earlier work at the time.

Streit's H-PMHT report [43] was actually the first to present the multivariate version of McLachlan and Jones' mixture distribution with missing measurement locations. However, this aspect is not dealt with in particular detail in [43]; Cadez later discussed the effects of multiple dimensions on aspects such as computation complexity and numerical integration [5]. An accessible summary of fitting stationary finite mixture models to multivariate histogram data was presented by Ainsleigh [1].

The discussion so far has considered the development of methods to fit static mixture distributions to multivariate histograms. The important difference between the earlier mixture fitting methods and the tracking problem is that tracking requires dynamic components that evolve according to a (usually random) process. Luginbuhl and Willett were first to apply a Markov prior to the parameters of the mixture distribution for the case of one-dimensional quantised data. This was first shown for a single signal component in noise [30, 32] and later extended to the more general case of multiple signals [29, 31].

Chronologically, we now arrive at Streit's H-PMHT report [43], which made a number of key contributions that brought the work of Luginbuhl and Willett, and their predecessors, into the context of TkBD. The first contribution was to frame the dynamic mixture parameter estimation problem in the usual target tracking nomenclature of hidden Markov state models. The earlier work of Luginbuhl and Willett has a hidden model and is mathematically equivalent to the model used by Streit, but is presented in the context of frequency spectrum estimation. Secondly, Streit presented a limiting argument that removed the requirement for quantised measurement data and allowed the algorithm to be applied directly to intensity maps. The third contribution was to recognise that the Markov prior was diluted by the process of measurement quantisation and infinitely diluted in the limit when intensity data is recovered. To avoid this, Streit introduced a data-dependent Markov prior. This prestidigitation will be explored in more detail within. In addition to the original technical report, Streit published this work in open literature with collaborators Walsh and Graham [46–48, 55].

A specialised version of H-PMHT was later presented by Streit in [44]. Spectral H-PMHT combines the target state-space model with a parametric frequency domain model for passive data where the frequency spectrum of emitters can be used as a classification feature for the target. The spectral H-PMHT was further reported in [27, 42, 56]. It describes the sensor response to the target using a fixed distribution across frequency bins and an evolving kinematic state that determines the spatial characteristics. The spectral H-PMHT literature presents the case of a Gaussian spatial spread in azimuth and a Gaussian mixture distribution in frequency. Other models are possible but no examples have been published.

For around a decade, H-PMHT then lay rather dormant in the open literature. Pakfiliz and Efe presented an example two-dimensional implementation of the algorithm [35, 36]. Although the earlier H-PMHT work had focussed on one-dimensional bearing-time maps, the development in [43] is general enough to encompass an arbitrary dimensioned sensor measurement image: the single indexing can easily label two dimensions using a raster scan, for example. Thus, the example in [35, 36] did not extend the H-PMHT theory, although it was the first two-dimensional application. Davey et al. began to explore the comparative performance of different track-before-detect algorithms in [13, 14]. This comparison assessed H-PMHT, particle filter TkBD and Viterbi TkBD on thumbnail images with Rayleigh noise. It was the first study to compare H-PMHT with alternative methods. It demonstrated that the output quality of the algorithm, that is the estimated target states and the confidence that there is a target at all, was comparable with a particle filter but that the computation cost was considerably lower. This comparison only considered the case of a single target in clutter, for which the particle filter output quality is asymptotically optimal as the number of particles increases. The breadth of the comparison was widened in [16] to include the Baum Welch and the Maximum Likelihood Probabilistic Data Association algorithms; the take-home story remained that H-PMHT provided high output quality for low computation cost. In a later comparison, Vo et al. attempted to compare H-PMHT with a multi-Bernouli TkBD algorithm under the random finite set framework [50]. However, Davey showed that the H-PMHT implementation used for this comparison was questionable [9] and the SNR used was too high for the results to be of interest. Davey subsequently described an approximate method for applying H-PMHT to complex images and demonstrated that again the performance was comparable with particle filtering for a fraction of the computation cost [15].

The H-PMHT algorithm described by Streit in [43] provided an implementation guide for multivariate Gaussian targets. The theoretical framework is general enough to accommodate an arbitrary target spreading function, but no practical guidance was provided for implementation in such a case. Davey addressed this gap by formulating a particle filter based implementation of H-PMHT [8]. This particle H-PMHT uses Monte Carlo integration to solve the various integrals required by the algorithm, which are intractable for non-Gaussian targets. It also demonstrated that the H-PMHT auxiliary function could be viewed as a general non-linear optimisation problem and used the particles to solve it. A grid-based alternative was described by Gaetjens et al. (published as Vu) that used the Viterbi algorithm to estimate the target states [54]. This implementation is interesting because it uses the H-PMHT data association as the glue between a bank of parallel independent Viterbi algorithms, one for each target. Because the H-PMHT data association localises the image data, the single-target Viterbi estimators are able to deploy a dynamic grid focussed around the target state [51].

Implementation considerations were directly discussed in [7], which presented the application of H-PMHT to a airborne maritime surveillance problem. Although [48] presented a whitening approach that proves critical in track initiation, H-PMHT was still thought to assume a known target count. Davey [7] discussed track management in detail and demonstrated the scalability of H-PMHT to problems of significant size. Implementation efficiency was further improved with the so-called single-target chip approach proposed by Davey [11], which was demonstrated to provide high quality track output for 500 targets in an image of 4,000 $\times$ 7,000 pixels.

The H-PMHT variants described above were limited to the assumption of a known target appearance function, that is the probabilistic footprint of the target within the sensor image. This was addressed by a series of works by Wieneke and Davey. They first considered the case of a randomly evolving Gaussian appearance function by applying a Wishart prior to the shape covariance matrix [57, 58]. This was then extended to the case of a mixture of Gaussians and used to track a cluster of people in video imagery [21, 22]. A randomly evolving template was applied in [20] and used to achieve super-resolution estimation of target shape. Davey et al. brought together this work on target appearance estimation and reconciled it with the numerical implementation methods described above in Gaetjens et al. [23].

Vu et al. recognised that the multinomial measurement function used for all histogram expectation–maximisation methods is consistent with a Poisson mixture. This was discussed by Ainsleigh [1] but it was not until the work of Gaetjens [26, 52] that this alternative representation was put to use. Davey et al. combined this Poisson H-PMHT with the shape estimating H-PMHT [57, 58] to arrive at an algorithm that estimates a dynamic clutter map in parallel with target tracking [17]. This clutter map was shown to vastly decrease the number of false tracks for an airborne surveillance scene with cloud obscuration.

The original H-PMHT work of Luginbuhl, Willett and Streit [29, 31, 43] applied H-PMHT to a simulated passive sonar problem. Since then it has been applied to airborne maritime surveillance with X-band radar [7]; video surveillance of people in both greyscale [21, 22, 58] and colour [18]; and active sonar [53].

At the time of publishing, the most recent variant of H-PMHT was the ML-HPMHT or Quanta [25, 59], a version of the algorithm that uses the histogram measurement model but not the Markov target state model. Instead, it applies a deterministic target model with a small number of fixed unknown parameters and uses direct numerical optimisation to estimate their values. The deterministic model used for Quanta is the same approach that Perlovsky [37, 38] used for point measurements. A very new development that is still enfolding during the writing of this book is the reversal of the expectation and maximisation parameters in [2]. This has the potential to arrive at a version of H-PMHT that estimates densities instead of single points.

At this point in time, the preceding discussion includes all of the published works on H-PMHT in the open literature. Most of the different variations of H-PMHT are discussed in detail in subsequent chapters of this monograph.

1.2 Preliminaries

We will make no attempt in this monograph to derive standard textbook algorithms in tracking, such as the Kalman filter. However, there are a few fundamental concepts that are pervasive through tracking mathematics. The first of these is conditional independence. The ubiquitous example is position and velocity over time: if an object is moving with constant velocity then its position tomorrow is conditionally independent from its position yesterday given that we know its position and velocity today. The current speed and location completely define the system and older information is superfluous. Note that this does not mean that yesterday's position has no bearing on tomorrow's: only that it adds no extra information given what we already know. Mathematically, we write A *is conditionally independent from* C *given* B as $p(A|B, C) = p(A|B)$. A sequence of random variables $\mathbf{x}_1, \mathbf{x}_2, \dots \mathbf{x}_N$ is said to be a Markov process if each variable is conditionally independent of its ancestors given its preceding variable, that is $p(\mathbf{x}_T|\mathbf{x}_1, \dots \mathbf{x}_{T-1}) = p(\mathbf{x}_T|\mathbf{x}_{T-1})$. In this case, we can write the probability of the whole sequence as

$$p(\mathbf{x}_1)p(\mathbf{x}_2|\mathbf{x}_1)\dots p(\mathbf{x}_T|\mathbf{x}_{T-1}) = p(\mathbf{x}_1)\prod_{t=2}^{T} p(\mathbf{x}_t|\mathbf{x}_{t-1}). \tag{1.1}$$

The second concept is marginalisation. For a pair of random variables with joint distribution $p(A, B)$, the marginal probability of A is given by $p(A) = \int p(A, B)\mathrm{d}B$. If A and B are unconditionally independent then $p(A, B) = p(A)p(B)$.

The final fundamental concept is Bayes' rule, which is often used as a method to express conditional probabilities in terms of marginal probabilities. Defining conditional probability as $p(A|B) = p(A, B)/p(B)$, then Bayes' rule can be written as

$$p(A|B) = \frac{p(A, B)}{p(B)} = p(B|A)\frac{p(A)}{p(B)} \tag{1.2}$$

Marginalisation, conditional independence and Bayes' rule combine to give what is often called the Bayesian filter. This filter is usually written as a two-step process consisting of a prediction based on previous data and then a correction, or update, based on a new measurement. The prediction equation is

$$\begin{aligned} p(\mathbf{x}_t|\mathbf{y}_1, \ldots \mathbf{y}_{t-1}) &= \int p(\mathbf{x}_t, \mathbf{x}_{t-1}|\mathbf{y}_1, \ldots \mathbf{y}_{t-1})\mathrm{d}\mathbf{x}_{t-1}, \\ &= \int p(\mathbf{x}_t|\mathbf{x}_{t-1})p(\mathbf{x}_{t-1}|\mathbf{y}_1, \ldots \mathbf{y}_{t-1})\mathrm{d}\mathbf{x}_{t-1}, \end{aligned} \tag{1.3}$$

where $\mathbf{x}_t$ is the state of the object to be estimated and $\mathbf{y}_t$ is a measurement. This prediction is often referred to as the Chapman–Kolmogorov equation. The new measurement is incorporated using Bayes' rule

$$\begin{aligned} p(\mathbf{x}_t|\mathbf{y}_1, \ldots \mathbf{y}_t) &= \frac{p(\mathbf{y}_t, \mathbf{x}_t|\mathbf{y}_1, \ldots \mathbf{y}_{t-1})}{p(\mathbf{y}_t|\mathbf{y}_1, \ldots \mathbf{y}_{t-1})} \\ &\propto p(\mathbf{y}_t|\mathbf{x}_t)p(\mathbf{x}_t|\mathbf{y}_1, \ldots \mathbf{y}_{t-1}), \end{aligned} \tag{1.4}$$

The states form a Markov process and the measurements are conditionally independent on previous measurements given the current state. The two equations lead to a recursion that moves the state conditional probability forwards through time.

If any of this is not well-trodden ground, then we suggest that some prior reading is in order before further exploration of this text. [41] gives a good overview of Bayesian estimation in general, and [3] provides detailed discussions of the standard approaches to target tracking.

1.3 Expectation–Maximisation

Both the histogram and point measurement PMHT algorithms are derived by applying expectation–maximisation [24] to multi-target data association. This next section provides a brief introduction to the EM algorithm. At a simple level, EM provides an optimisation approach for a two variable cost function $p(X, Y)$. The premise is that the aim is to find the value of X that maximises $p(X, Y)$. Joint optimisation would also give the value of Y at this maximum but is not computationally feasible. A direct approach would be to deal with the marginal $\int p(X, Y)\mathrm{d}Y$ but the maximising value of this function can be different to the maximising value of $p(X, Y)$. In EM terminology, Y is referred to as the *missing data*. EM operates by alternating two single parameter optimisations. Beginning with an estimate $\hat{X}$, the expectation step forms the conditional probability of the missing data, referred to in this book as the *missing data likelihood* $p_{\text{miss}}(Y|\hat{X})$ and then the maximisation step forms a new estimate by maximising a cost function denoted $\mathcal{Q}(X|\hat{X})$. In the EM context, the full set of random variables are the *complete data* and the *complete data likelihood* is written as $p_{\text{comp}}(X, Y)$. The EM cost function, usually referred to as the *auxiliary function*, is the conditional expectation of the logarithm of complete data likelihood over the missing data, that is

$$\mathcal{Q}(X|\hat{X}) = E_{Y|\hat{X}}\Big[\log\left\{p_{\text{comp}}(X, Y)\right\}\Big] \tag{1.5}$$

$$= \int \log\left\{p_{\text{comp}}(X, Y)\right\} p_{\text{miss}}(Y|\hat{X})\mathrm{d}Y. \tag{1.6}$$

To illustrate how this works, consider a simple example with X and Y scalars and $p_{\text{comp}}(X, Y)$ Gaussian with mean $\mu = [\mu_X, \mu_Y]^{\mathsf{T}}$ and covariance Σ

$$p_{\text{comp}}(X, Y) = |2\pi\Sigma|^{-\frac{1}{2}} \exp\left\{-\frac{1}{2}\left(\begin{bmatrix}X\\Y\end{bmatrix} - \begin{bmatrix}\mu_X\\\mu_Y\end{bmatrix}\right)^{\mathsf{T}} \Sigma^{-1} \left(\begin{bmatrix}X\\Y\end{bmatrix} - \begin{bmatrix}\mu_X\\\mu_Y\end{bmatrix}\right)\right\}.$$

Clearly, this is maximised when $X = \mu_X$ and $Y = \mu_Y$ but for illustrative purposes, we will explore how expectation–maximisation might be used to find the optimal estimate for X.

The expectation step (E-step) is to find the missing data likelihood $p_{\text{miss}}(Y|\hat{X})$ and use it to form the auxiliary function. The E-step essentially consists of evaluating the expectation that defines the EM auxiliary function and this can often be solved analytically to find a closed form. Let A_{ij} denote the elements of the inverse of Σ, namely $\begin{bmatrix}A_{11} & A_{12}\\A_{12} & A_{22}\end{bmatrix} = \Sigma^{-1}$ where we have asserted that $A_{12} = A_{21}$ because Σ is symmetric. Starting with an estimate $\hat{X}$, it is relatively easy to show that the pdf $p_{\text{miss}}(Y|\hat{X})$ is Gaussian with mean $\hat{Y} = \mu_Y + (\hat{X} - \mu_X)A_{12}/A_{22}$ and variance $1/A_{22}$, this is a standard result in linear systems, for example [39].

The logarithm of the complete data likelihood is

$$\log\left\{p_{\text{comp}}(X, Y)\right\} = \\ -\frac{1}{2}\Big[\log\left\{|2\pi\Sigma|\right\} + A_{11}(X - \mu_X)^2 + 2A_{12}(X - \mu_X)(Y - \mu_Y) + A_{22}(Y - \mu_Y)^2\Big].$$

The determinant term is constant so it will not influence the optimisation and we can safely ignore it. Combining this with $p_{\mathsf{miss}}(Y|\hat{X})$ gives

$$\begin{aligned}\mathscr{Q}(X|\hat{X}) = &-\frac{1}{2}\Big[A_{11}(X-\mu_X)^2 + 2A_{12}(X-\mu_X)\int(Y-\mu_Y)p_{\mathsf{miss}}(Y|\hat{X})\mathrm{d}Y \\ &+ A_{22}\int(Y-\mu_Y)^2 p_{\mathsf{miss}}(Y|\hat{X})\mathrm{d}Y\Big], \\ = &-\frac{1}{2}\Big[A_{11}(X-\mu_X)^2 + 2A_{12}(X-\mu_X)(\hat{Y}-\mu_Y) + A_{22}\left(1/A_{22} + (\hat{Y}-\mu_Y)^2\right)\Big].\end{aligned}$$

Having defined the auxiliary function, the E-step is now complete.

The maximisation step (M-step) finds a new estimate for X by maximising the auxiliary function $\mathscr{Q}(X|\hat{X})$. In this case, the maximisation can also be performed analytically. The last term in $\mathscr{Q}(X|\hat{X})$ is constant with respect to X so in practice one would usually discard it but it is useful to retain in this case because it is iteration dependent and maintains the proper offset between $\mathscr{Q}$ functions at different iterations. The new estimate for X is found by finding the stationary point of $\mathscr{Q}(X|\hat{X})$: it is quadratic in X and so has a single maximum at

$$\frac{\mathrm{d}}{\mathrm{d}X}\Big[A_{11}(X-\mu_X)^2 + 2A_{12}(X-\mu_X)(\hat{Y}-\mu_Y)\Big] = 0, \tag{1.7}$$

$$2A_{11}(X-\mu_X) + 2A_{12}(\hat{Y}-\mu_Y) = 0, \tag{1.8}$$

$$X = \mu_X - \frac{A_{12}}{A_{11}}(\hat{Y}-\mu_Y). \tag{1.9}$$

Substituting the mean $\hat{Y} = \mu_Y + (\hat{X}-\mu_X)A_{12}/A_{22}$ gives

$$X = \mu_X + \frac{(A_{12})^2}{A_{11}A_{22}}(\hat{X}-\mu_X). \tag{1.10}$$

It is clear that the adjustment above moves the estimate towards μ_X. Since the function $\mathscr{Q}$ is of the form $-1/2f(X)^2$, it cannot be positive and is bounded above by 0. Each iteration is guaranteed not to decrease $\mathscr{Q}$ so the estimate must converge [40]. Figure 1.1 shows an example of this toy problem for $\Sigma = \begin{bmatrix} 1 & 0.9 \\ 0.9 & 1 \end{bmatrix}$ with $\mu_x = \mu_y = 1$ and $\hat{X} = -1$ on initialisation. Figure 1.1a shows the first three $\mathscr{Q}$ functions and the estimates they produce; Fig. 1.1b shows the first thirty $\mathscr{Q}$ functions and the final estimate, which has converged to μ_X.

The figure illustrates that the EM auxiliary functions $\mathscr{Q}$ at each iteration provide a family of functions that bound the true joint likelihood. In this case, the algorithm takes quite a few iterations but it does converge to the correct optimum $\hat{X} = \mu_x = 1$. As a side product, the algorithm also produces the conditional likelihood of Y at the optimal X $p(Y|X = \mu_X)$.

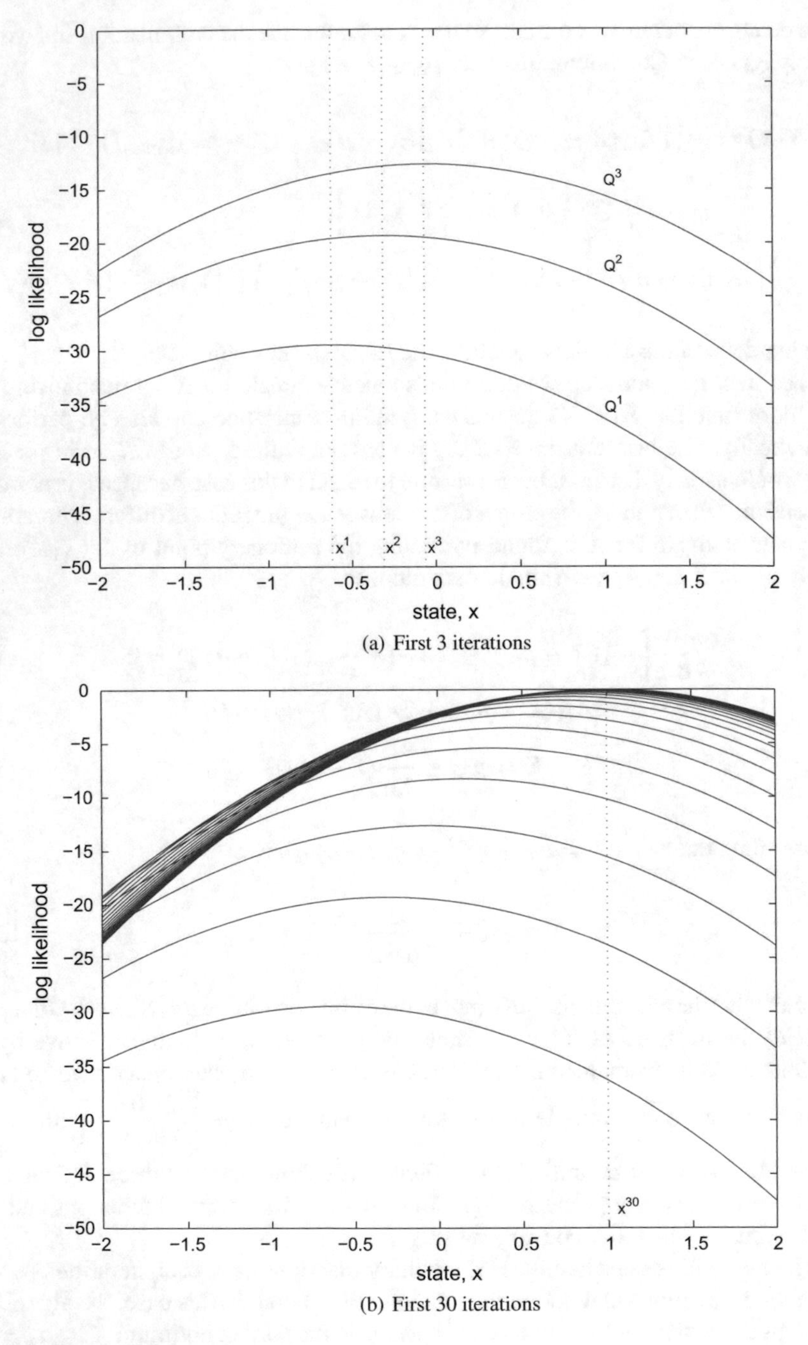

(a) First 3 iterations

(b) First 30 iterations

Fig. 1.1 Simple EM example

Of course, this simplistic example does not prove anything: what is really going on here? We now present a relaxed derivation of EM based on Bishop [4].

$$\begin{aligned}
\log\{P(X)\} &= \int q(Y)\log\{P(X)\}\,\mathrm{d}Y, \\
&= \int q(y)\log\left\{\frac{P(X,Y)}{P(X,Y)}P(X)\right\}\mathrm{d}Y, \\
&= \int q(y)\log\left\{\frac{P(X,Y)}{P(Y|X)}\frac{q(Y)}{q(Y)}\right\}\mathrm{d}Y, \\
&= \int q(y)\log\{P(X,Y)\}\,\mathrm{d}Y - \int q(y)\log\{q(Y)\}\,\mathrm{d}Y \\
&\quad - \int q(y)\log\left\{\frac{P(Y|X)}{q(Y)}\right\}\mathrm{d}Y,
\end{aligned}$$

which holds for any probability function $q(Y)$ under mild constraints such as $q(Y) > 0$. The second term above is the entropy of $q(Y)$ and the third term is the Kullback–Leibler divergence between $q(Y)$ and $p(Y|X)$, both of which are guaranteed to be non-negative. Denoting these as $\mathsf{Ent}(q)$ and $\mathsf{KL}\,(q, p(Y|X))$, the following inequality can be written

$$\begin{aligned}
\log\{P(X)\} &= \int q(y)\log\{P(X,Y)\}\,\mathrm{d}Y + \mathsf{Ent}(q) + \mathsf{KL}\,(q, p(Y|X)), \\
&\geq \int q(y)\log\{P(X,Y)\}\,\mathrm{d}Y + \mathsf{Ent}(q),
\end{aligned}$$

where equality holds when $\mathsf{KL}\,(q, p(Y|X)) = 0$, namely $q(Y) = p(Y|X)$. Choosing $q(Y) = p(Y|\hat{X})$ the first term in the inequality is the EM auxiliary function and the inequality is

$$\log\{P(X)\} \geq \mathcal{Q}(X|\hat{X}) + \mathsf{Ent}(q),$$

and the auxiliary function is clearly a lower bound for the log-likelihood. The entropy term is independent of X so the bound is maximised when $\mathcal{Q}(X|\hat{X})$ is maximised. The expectation step of EM can be viewed as forming a new approximation to $P(Y|X)$ based on the most recent estimate $\hat{X}$ and the maximisation step finds a new estimate by maximising a lower bound for the log-likelihood. If the estimate approaches the true X then $q(Y)$ approaches $P(Y|X)$ and the bound gets tighter. It can be shown that this process is guaranteed to converge to a local maximum of $P(X)$ under some mild conditions. An important feature of the process is that theory demands that the function $\mathcal{Q}(\hat{X})$ must increase on every iteration. In practice, we can use this as a handy health indicator for practical implementations: experience shows that implementation errors break the monotone properties of $\mathcal{Q}$ and this can be used to diagnose problems. It is advisable to keep track of $\mathcal{Q}$ during development.

Usually, the problem involves data Z as well as parameters and the generic EM auxiliary function is

$$\mathcal{Q}(X|\hat{X}) = E_{Y|\hat{X},Z}\Big[\log\{p_{\mathsf{comp}}(X, Y, Z)\}\Big]$$
$$= \int \log\{p_{\mathsf{comp}}(X, Y, Z)\}\, p_{\mathsf{miss}}(Y|\hat{X}, Z)\mathrm{d}Y. \quad (1.11)$$

As a coda, it is important to note that EM theory guarantees convergence of the peak of $\mathcal{Q}$, it does not guarantee that $\hat{X}$ will converge. In some cases, it is possible to have two (or more) states X^1 and X^2 that give $X^2 = \arg\max \mathcal{Q}(X|X^1)$ and $X^1 = \arg\max \mathcal{Q}(X|X^2)$, where $\mathcal{Q}(X|X^1) = \mathcal{Q}(X|X^2)$ and the distance between the two states is larger than an arbitrary constant, $||X^1 - X^2|| > \varepsilon$. If this happens then EM iterations swap between the two states X^1 and X^2 and the state estimate never converges even though the auxiliary function has. This may seem like an unlikely pathological example but it does occur in practice. When multiple states give the same maximum likelihood like this, then one can arbitrarily choose between them. Much more detailed discussions of EM can be found in [4, 24, 34, 45].

1.4 Notation

A full table of notation is given at the start of the monograph and a brief overview of conventions is presented here. Throughout, we use the following conventions: scalars use regular type, a; vectors use boldface, $\mathbf{x}$; and matrices are marked as, P. A subscript t is used for time indexing and superscripts for indexing within lists, for example $\mathbf{x}_t^m$ is the vector state of target m at time t. The full list of variables at a single time is written $\mathbf{x}_t^{1:M}$ or more compactly $\mathbf{X}_t$. The full list over a whole time sequence is written $\mathbf{x}_{1:T}^{1:M}$ or more compactly $\mathbb{X}$.

Two types of measurements are considered: point measurements are denoted $\mathbf{y}$ and each is a single point in a vector space. Typically, a sensor produces a random number of point measurements at each time instant. Image measurements are denoted $\mathbf{z}$ and consist of a scalar value $\mathbf{z}^i$ per pixel i over a (usually) fixed range of pixels $i = 1 \ldots I$.

It will often be convenient to deal with Gaussian distributions. The Gaussian distribution with mean μ and variance Σ evaluated at the point $\mathbf{y}$ is denoted as $\mathcal{N}(\mathbf{y}; \mu, \Sigma)$.

1.5 Canonical Multi-target Scenario

Throughout this monograph, numerous variations of the H-PMHT[1] will be described. The canonical multi-target scenario provides a benchmark with which these variants are compared with each other and alternative tracking approaches. The scenario

[1] *GenerateCanon*
The H-PMHT toolbox contains the function GenerateCanon that can be used to generate images, point measurements and truth trajectories for the canonical multi-target scenario. For details on how to use this function, refer to the H-PMHT toolbox documentation, or [12].

will be specialised for some situations, these details will be discussed as they are required. The canonical multi-target scenario contains 20 targets that move in a periodic fashion designed to ensure multiple crossing or near-crossing events. The states of the odd-numbered targets are defined by

$$\begin{bmatrix} X_t^m \\ Y_t^m \end{bmatrix} = \begin{bmatrix} 50 + 20\cos\left(\frac{2\pi t}{200} + \frac{(m-1)\pi}{20}\right) \\ 50 + 40\cos\left(\frac{2\pi t}{200} + \frac{(m-1)\pi}{20}\right) \\ +0.2\Delta_Y + \Delta_Y \frac{m-1}{2} \end{bmatrix},$$

and even numbered targets by

$$\begin{bmatrix} X_t^m \\ Y_t^m \end{bmatrix} = \begin{bmatrix} 50 - 20\cos\left(\frac{2\pi t}{200} + \frac{(m-2)\pi}{20}\right) \\ 50 + 40\cos\left(\frac{2\pi t}{200} + \frac{(m-2)\pi}{20}\right) + \Delta_Y \frac{m-2}{2} \end{bmatrix}.$$

Position is measured in units of pixels assuming a square grid for the sensor and the time step is the sensor update rate. The parameter Δ_Y controls how closely the targets are spaced in Y and also influences the total image size for a prescribed number of targets. It is probably not intuitive to examine the motion equations directly. However, notice that the X and Y components of each target have the same cosine term: this means that the positions move along straight lines. For odd-numbered targets, the positions move along lines of the form $Y_t^m = c^m + 2X_t^m$, and for the even-numbered targets $Y_t^m = c^m - 2X_t^m$. The targets are in mirrored pairs, so targets 1 and 2 have mirrored X positions and Y values offset by 0.2 Δ_Y and similarly for targets 3 and 4, 5 and 6, and so on. This mirroring means that the targets almost pass through the same position when the cosine term is zero. The phase delays between pairs of targets such as between 1–2 and 3–4 mean that the path of target 1 almost passes through the path of target 4 and similarly for 2 and 3. In fact, there are a multitude of near crossings, depending on the value of Δ_Y: if it is small the number of near crossings is very high, if it is large then there are no crossings. Because the space between pairs such as 1 and 2 is only 0.2 Δ_Y, the mirrored pairs are more likely to be close, whereas the space between pairs such as 1 and 3 is Δ_Y. Figure 1.2 shows the path of target 1 compared with targets 2, 4, 6 and 8 from frames 175 to 205. The figure shows the X-Y trajectories and the distance between target 1 and the even numbered targets. Each of the targets experiences similar interactions to this.

In addition to the crossing events, the targets continually manoeuvre: the velocity in X and Y follows a cosine function so each target follows a cycle: it starts moving quickly and then slows to a stop; it turns and accelerates in the other direction before slowing to stop again; the target accelerates and returns to its starting speed and velocity. The crossing events between mirrored pairs like 1 and 2 happen at the times of maximum speed, so these are the easiest to track through: a track on one of these has momentum leading up to the crossing and this helps it follow the true path. In contrast, some of the other crossing events happen close to a turn when the targets

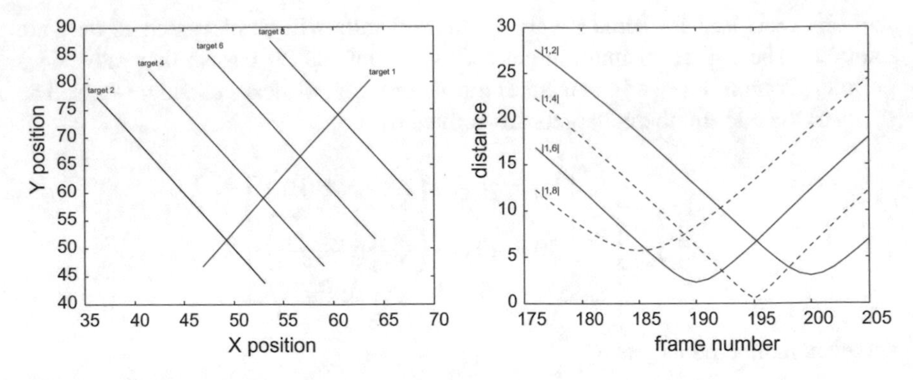

Fig. 1.2 Simulated scenario for $\Delta_Y = 10$, target 1

are manoeuvring and a track on that target can already be biased, making it more vulnerable to swaps during the crossing.

Figure 1.3 shows the trajectories for one cycle of the scenario, namely $t = 1 \ldots 200$, for a spacing of $\Delta_Y = 10$. The figure shows eight segments of the scenario. A square marks the position of each target at the most recent time step and a tail of 20 frames trails behind each target. Each target performs manoeuvres and makes several crossings or near crossings with other targets.

Figure 1.4 shows the distance to the nearest other target for three selected targets as a function of time. Three different spacing values are shown: $\Delta_Y = 5, 10, 20$. None of the targets occupy exactly the same location at any time. However, the closest approach is less than a pixel at several times for the closely spaced $\Delta_Y = 5$ case. Depending on the spread of the target energy, the targets are unlikely to be resolved at these times. For the more widely spaced $\Delta_Y = 20$ case, the targets never pass closer than a radial distance of five pixels and so are likely to always be resolved except for very big targets. The spread of each target will by default be assumed to be an isotropic Gaussian with unit variance, which is an appropriate model for low-resolution sensors such as some radars and sonars. For video imagery targets commonly occupy tens if not hundreds of pixels; broader spread targets will be specified when it is appropriate.

Unless otherwise discussed, the tracking filters used in this monograph will assume an almost constant velocity model, so the state is a four-element vector consisting of position and velocity in the plane $\mathbf{x}_t^m \equiv [X_t^m, \dot{X}_t^m, Y_t^m, \dot{Y}_t^m]^\mathsf{T}$. The state evolution is a random walk with Gaussian driving noise, $p\left(\mathbf{x}_t^m|\mathbf{x}_{t-1}^m\right) \sim \mathcal{N}\left(\mathbf{x}_t^m; \mathsf{F}_t\mathbf{x}_{t-1}^m, \mathsf{Q}_t\right)$, where

$$\mathsf{F}_t = \begin{bmatrix} 1 & \Delta_t & 0 & 0 \\ 0 & 1 & 0 & 0 \\ 0 & 0 & 1 & \Delta_t \\ 0 & 0 & 0 & 1 \end{bmatrix}, \qquad \mathsf{Q}_t = q \begin{bmatrix} \frac{1}{3}\Delta_t^3 & \frac{1}{2}\Delta_t^2 & 0 & 0 \\ \frac{1}{2}\Delta_t^2 & \Delta_t & 0 & 0 \\ 0 & 0 & \frac{1}{3}\Delta_t^3 & \frac{1}{2}\Delta_t^2 \\ 0 & 0 & \frac{1}{2}\Delta_t^2 & \Delta_t \end{bmatrix},$$

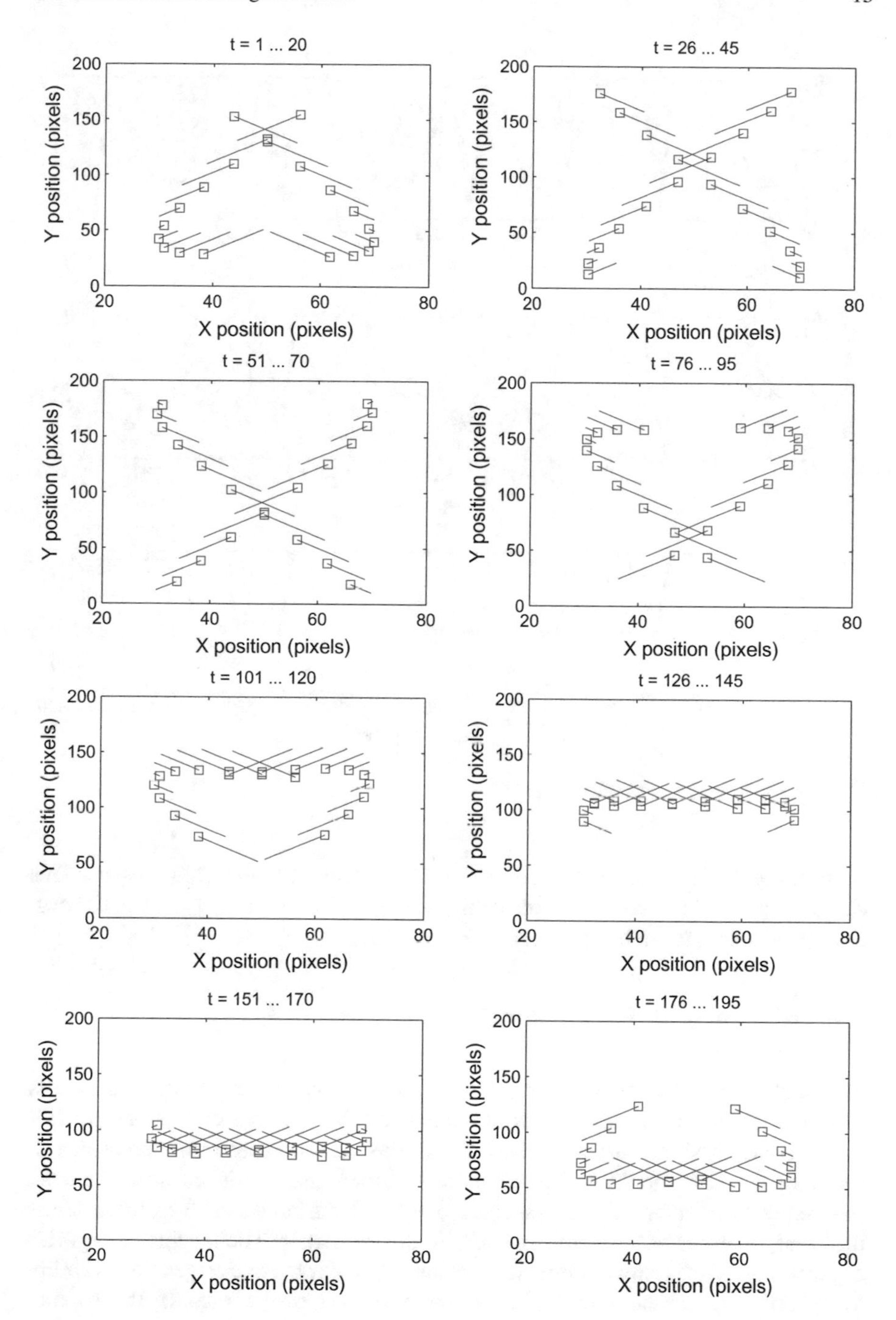

Fig. 1.3 Simulated scenario for $\Delta_Y = 10$

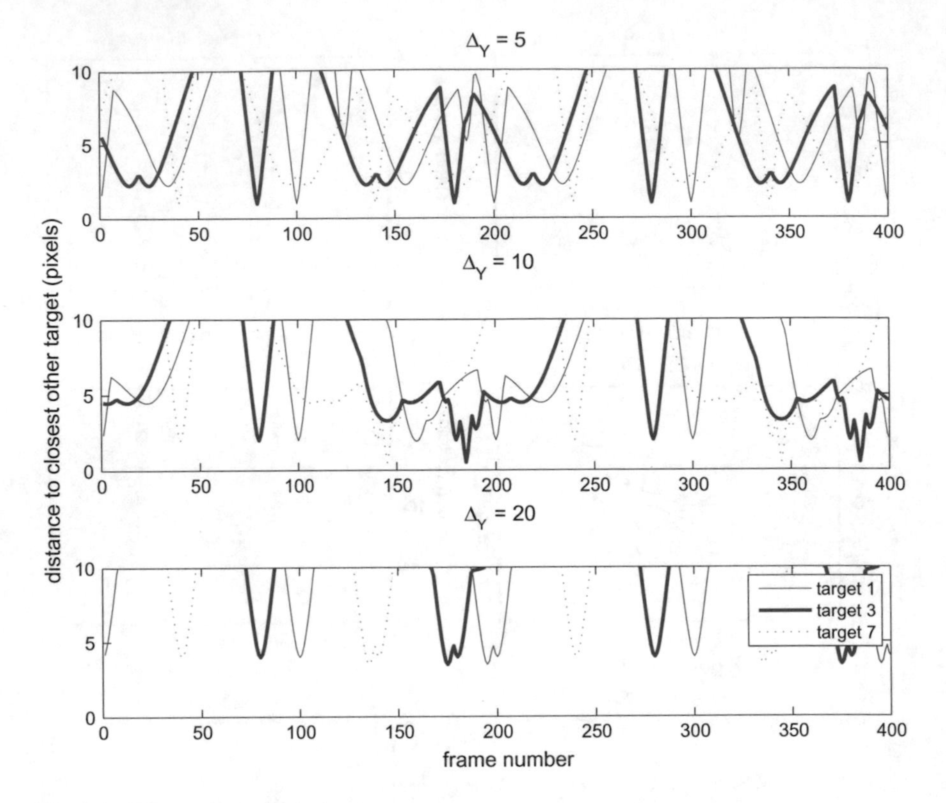

Fig. 1.4 Closest approach to other targets

with Δ_t the elapsed time between frames $t-1$ and t and q an adjustable constant. This elapsed time will be assumed to be a unit step $\Delta_t = 1$ for simplicity in simulations. For applications, it can be variable.

1.6 Measures of Performance

The multi-target tracking problem can be regarded as having three different aspects or sub-problems: the tracking system should determine the number of targets present in the scene; it should assign to each target the measurement information due to that target; and it should derive estimates of target characteristics from the data. Invariably the solutions to these aspects are intertwined, but it can be useful to define different performance measures for each. To this end, we will define three categories of performance measures: cardinality measures quantify the tracking system's ability to correctly estimate the number of targets; association measures quantify the tracking system's ability to discriminate one target from another; and accuracy measures quantify the statistical properties of the estimation errors. In order to evaluate tracking performance measures, it is inevitable that the tracks need to be ascribed to a particular target: processes for this are also defined. There are numerous other performance measures that are not considered in this monograph.

1.6.1 Cardinality Measures

The cardinality of a set is the number of elements contained within it; cardinality tracking measures quantify how well the tracker determines the number of targets in a scene. There are two types of cardinality measures used in this monograph: the Tracker Operating Characteristic curve and mean cardinality error counts. The first is used for the case where the overwhelming concern is target detection, such as for track-before-detect, whereas the second is used as part of a more holistic analysis.

For tracking problems where the target signal is very weak, it can be the case that the most important feature of a tracking system is its ability to correctly initiate tracks when targets are present and not when they are absent. In this case, the tracker can be viewed as a sophisticated detector and the target states as nuisance variables. A common performance measure for detectors is the Receiver Operating Characteristic (ROC) curve; the analogue for tracking systems is referred to as the Tracker Operating Characteristic (TOC) curve [2, 6, 10]. The ROC curve plots the locus of the probability of detection, the probability of correctly declaring a target when one is really present, as a function of the probability of false alarm, the probability of incorrectly declaring a target when there really is no target. For simple decision rules, such as a threshold, this is a curve but for composite decision rules, it is a manifold. A composite rule involves multiple chained tests, for example in Adelaide, South Australia, a *heatwave* is defined as 5 consecutive days where the maximum temperature is over 35C. In this case, there are two thresholds that could be varied, so the probability of a heatwave is a surface; a manifold is essentially a surface of arbitrary dimension. In such cases, we can project the manifold onto a curve by choosing the highest possible probability of detection for a given probability of false alarm. Figure 1.5 shows an example that compares a composite decision rule with incoherently integrating and applying a single threshold test. The ROC curve for two different target SNR values are plotted and the surface for the composite test is shown as a family of curves. In this example, the envelope of the composite curves is approximately the incoherent decision rule with a 1dB SNR penalty.

There are different ways to extend the ROC concept to tracking, two simple examples are to define the probability of detection over a whole batch (was there ever a track?) or to define an instantaneous probability of detection, which amounts to a proportion of frames containing a track. This monograph uses the second, which tends to be more discriminatory. So the TOC plots the instantaneous true track probability as a function of the instantaneous false track probability, where these are defined as the probability that a particular frame contains a track conditioned on a target being present or absent respectively.

Mean cardinality error counts are a simpler affair. There are two types of cardinality error: the tracker can fail to produce a track when a target is present, or the

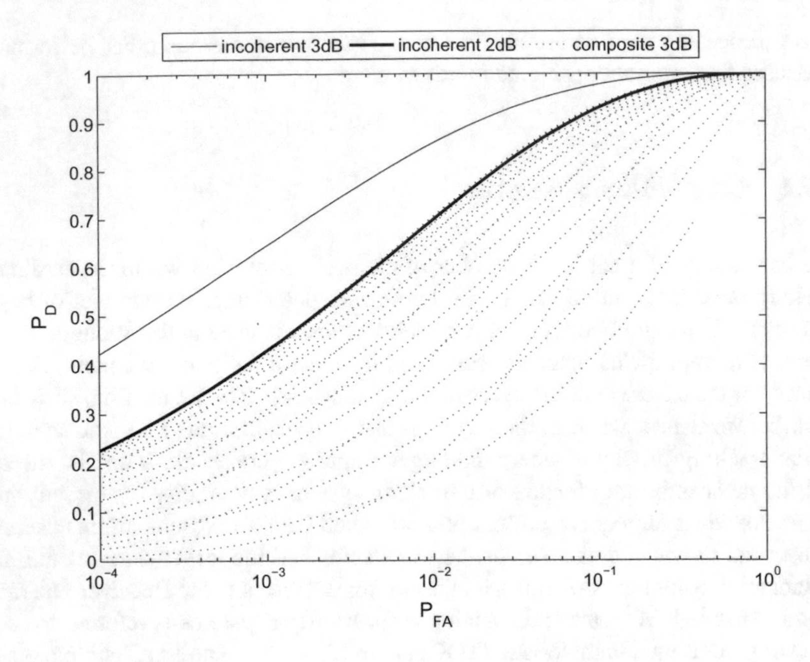

Fig. 1.5 Example ROC curves

tracker can produce more tracks than targets. This second type of error is sometimes subdivided into duplicate tracks, which are multiple tracks on a single target, and false tracks, for which no target can be blamed. However, in this monograph, we will simply lump them together and call them collectively redundant tracks. When the tracker fails to produce a track, this is referred to as a missed target. The count of missed targets and redundant tracks can be averaged over Monte Carlo trials and presented as a function of time.

1.6.2 Association Measures

The data association task for the tracking system is to assign data to a track that has been caused by the target it represents. This can go awry but we will not be interested in such micro management as to keep account of the number of poor assignments. Rather, what is important is the number of times such assignment errors cause a track to swap from following one target to another. As above, the number of swaps can be averaged over Monte Carlo trials and presented as a function of time. Usually, swaps occur when targets pass very close together or when one occludes another in an imaging sensor.

1.6.3 *Accuracy Measures*

By far the most common measure of estimation accuracy is the root mean square (RMS) error. If an estimator is unbiased then the RMS error is a sample approximation to the variance of the estimation error. If the estimator is biased then it couples bias and variance. In this monograph, RMS estimation error is calculated by averaging the squared Euclidian distance between the track estimated position and the true target position. It is generally averaged over Monte Carlo trials and sometimes also over time. Explicitly, the RMS error $\varepsilon_{\mathsf{RMS}}$ is given by

$$\varepsilon_{\mathsf{RMS}} = \sqrt{\frac{1}{J}\sum_{j=1}^{J}\left(\mathsf{H}\hat{\mathbf{x}}_t^j - \mathsf{H}\mathbf{x}_t^j\right)^{\mathsf{T}}\left(\mathsf{H}\hat{\mathbf{x}}_t^j - \mathsf{H}\mathbf{x}_t^j\right)}, \quad (1.12)$$

where j is an index over J Monte Carlo trials, $\hat{\mathbf{x}}_t^j$ is the state estimate on the jth trial, $\mathbf{x}_t^j$ is the true state, and H selects the position elements of the state vector.

1.6.4 *Track to Truth Association*

An important component of the tracker assessment is the algorithm for determining which target to compare a track with. Inconsistencies in this process can cause a track to be assigned to different targets from one frame to the next, leading to increased track swaps and potentially incorrect cardinality counts. These issues tend to arise at times when the true target locations are quite close together. The track to truth association can be formulated as an optimisation problem and when the targets are close together the objective function can be relatively flat around the local maximum. In order to minimise the impact of the association decisions on the performance measures, one could choose to make batch decisions. However, this is essentially formulating the association as a multi-hypothesis tracker and will generally not be feasible. As an approximation to this that is low in complexity, association is implemented by comparing the single-frame global nearest neighbour association with the association from the previous frame: the previous frame's association is only changed if its probability is lower by more than a threshold amount. In this way, the association decisions tend to stick unless the best alternative is significantly better. This amounts to an approximation to the recursive MAP association where there is a transition cost for changing the association vector.

1.7 Monograph Synopsis

The remainder of this monograph is loosely grouped into three parts, dealing with the background of H-PMHT, extensions to the method, and applications. The background material is presented in Chaps. 2–4, which develop the core H-PMHT algorithm and demonstrate its application to an academic TkBD problem. Within this monograph, the H-PMHT method described by Streit in [43] is referred to as the *core H-PMHT* since the other variants build on this algorithm. Chapter 2 motivates the study of H-PMHT by presenting an analysis of the core H-PMHT applied to an academic TkBD problem. For the single target case it is compared with existing standards in the TkBD literature, namely the particle filter and Viterbi algorithm. From here, the mathematical development of the core H-PMHT is built up. Chapter 3 reviews the point measurement Probabilistic Multi-Hypothesis Tracker. This chapter provides the development of the PMHT data association method that is used in the derivation of the core H-PMHT. Performance examples of PMHT tracking on the canonical multi-target scenario are provided. Chapter 4 reviews the derivation of the core H-PMHT algorithm and demonstrates it on the canonical multi-target scenario.

The second part of the monograph details extensions to the core H-PMHT algorithm. As presented in [43], the core H-PMHT is a method for updating the states of a fixed and known number of targets with known priors. Implementation also requires the evaluation of several integrals that are not necessarily trivial to evaluate. These issues are addressed in Chap. 5, which deals with the mundane implementation issues that are the difference between a filtering engine and a multi-target tracking system. It provides a method for automated track management, namely the creation of new tracks and the termination of old tracks when required. It also offers advice on efficient implementation strategies and demonstrates how these can be used to achieve high frame rate performance on scenes with hundreds of targets embedded in images with millions of pixels. Chapter 6 describes how the H-PMHT measurement model can be viewed as a Poisson target process and uses this perspective to create a dynamic model for target amplitude. This allows for fluctuating targets and improves track management decisions by providing an estimate of the mean target strength rather than the instantaneous power in a particular frame. Chapter 7 discusses implementation methods for applications where the target spread function is not Gaussian or the target dynamics are nonlinear. Simulated examples demonstrate how significant improvements in output quality can be achieved over Gaussian approximations. The spread of target energy in the sensor image frame is referred to as the target appearance function. Chapter 8 addresses the problem of tracking where this appearance function is unknown. It derives an extended H-PMHT for estimating the parameters of an appearance function and provides example implementations for the important classes of appearance function: Gaussians, Gaussian mixtures, templates and appearance libraries. Chapter 9 is motivated by the problem of tracking in imagery where non-kinematic attribute data is often available. Such attributes include colour and texture and are different from intensity in that they do not obey a superposition law. Chapter 9 derives an extended H-PMHT for the case where such attribute data is available and demonstrates how exploitation of attribute data can greatly improve output quality in scenes involving complex interactions and obscuration. Chapter 8 also deals with the problem of spatially and temporally correlated non-uniform clutter. It provides both mixture-based and grid-based approaches to map the evolving clutter distribution and demonstrates how this clutter mapping dramatically reduces false track rates when the clutter distribution is no longer uniform.

Chapter 10 reviews the maximum likelihood formulations of PMHT and H-PMHT that make use of a deterministic target model.

The third part of the monograph deals with applications of H-PMHT to experimental sensor systems rather than the simulations described in the earlier two parts. Chapter 11 presents examples with radar and sonar sensors (details to be confirmed). Chapter 12 demonstrates the application of H-PMHT to full motion video data and provides performance examples from the PETS archive and from airborne imagery collected by the DST Group DEAP sensor testbed.

Chapter 13 reviews other methods in the literature that have a close relationship with H-PMHT and discusses open areas for further research. The appendix reviews some alternative algorithms that are used from time to time in the book for comparison purposes. Chapter 14 summarises the key messages of the book.

References

1. Ainsleigh, P.L.: A tutorial on EM-based estimation with histogram data. Technical report 11807, NUWC, Newport, Rhode Island, USA (2008)
2. Bar-Shalom, Y., Campo, L.J., Luh, P.B.: From receiver operating characteristic to system operating characteristic: evaluation of a track formation system. IEEE Trans. Autom. Control **35**, 172–179 (1990)
3. Bar-Shalom, Y., Willett, P.K., Tian, X.: Tracking and Data Fusion: a Handbook of Algorithms. YBS Publishing, England (2011)
4. Bishop, C.M.: Pattern Recognition and Machine Learning. Springer, New York (2006)
5. Cadez, I.V., Smyth, P., McLachlan, G.J., McLaren, C.E.: Maximum likelihood estimation of mixture densities for binned and truncated multivariate data. Mach. Learn. **47**, 7–34 (2002)
6. Davey, S.J.: Extensions to the Probabilistic Multi-hypothesis Tracker For Improved Data Association. School of Electrical and Electronic Engineering, the University of Adelaide (2003)
7. Davey, S.J.: Detecting a small boat with histogram PMHT. ISIF J. Adv. Inf. Fusion **6**, 167–186 (2011)
8. Davey, S.J.: Histogram PMHT with particles. In: Proceedings of the 14th International Conference on Information Fusion, Chicago, USA (2011)
9. Davey, S.J.: Comments on joint detection and estimation of multiple objects from image observations. IEEE Trans. Signal Process. **60**(3), 1539–1540 (2012)
10. Davey, S.J.: SNR limits on Kalman filter detect-then-track. IEEE Signal Process. Lett. **20**, 767–770 (2013)
11. Davey, S.J.: Efficient Histogram PMHT via single target chip processing. IEEE Signal Processing Letters (In press)
12. Davey, S.J.: H-PMHT Toolbox: User Manual arxiv TBD (2017) (In Preparation)
13. Davey, S.J., Rutten, M.G.: A comparison of three algorithms for tracking dim targets. In: Proceedings of Information, Decision and Control 2007. Adelaide, Australia (2007)
14. Davey, S.J., Rutten, M.G., Cheung, B.: A comparison of detection performance for several track-before-detect algorithms. EURASIP J. Adv. Signal Process. **2008**(6), 1–11 (2008)
15. Davey, S.J., Rutten, M.G., Cheung, B.: Using phase to improve track-before-detect. IEEE Trans. Aerosp. Electron. Syst. **48**(1), 832–849 (2012)
16. Davey, S.J., Rutten, M.G., Gordon, N.J.: Track-before-detect techniques, in Integrated Tracking, Classification and Sensor Management: theory and applications, Wiley, New York (2012)
17. Davey, S.J., Vu, H.X., Arulampalam, S., Fletcher, F., Lim, C.C.: Clutter mapping for histogram PMHT. In: Statistical Signal Processing, 153–156. Gold Coast, Queensland (2014)

18. Davey, S.J., Vu, H.X., Fletcher, F., Arulampalam, S., Lim, C.C.: Histogram probabilistic multi-hypothesis tracker with color attributes. IET Radar Sonar Navig. **9**(8), 999–1008 (2015)
19. Davey, S.J., Gaetjens, H.X.: The H-PMHT Toolbox. (submitted to Fusion 2018)
20. Davey, S.J., Wieneke, M.: H-PMHT with an unknown arbitrary target. In: Proceedings of ISSNIP (2011)
21. Davey, S.J., Wieneke, M.: Tracking groups of people in video with histogram-PMHT. Def. Appl. Signal Process.**47**(1) (2011)
22. Davey, S.J., Wieneke, M., Gordon, N.J.: H-PMHT for correlated targets. In: SPIE Signal and Data Processing of Small Targets, Proceedings vol. SPIE 8393, 83930R. Baltimore, USA (2012)
23. Davey, S.J., Wieneke, M., Vu, H.X.: Histogram-PMHT unfettered. IEEE J. Sel. Top. Signal Process. **7**(3), 435–447 (2013)
24. Dempster, A.P., Laird, N.M., Rubin, D.B.: Maximum likelihood from incomplete data via the EM algorithm. J. R. Stat. Soc. **140**, 1–38 (1977)
25. Dunham, D.T., Ogle, T.L., Willett, P.K., Balasingam, B.: Advancement of an algorithm. In: SPIE Signal and Data Processing of Small Targets, vol. SPIE 9092 (2014)
26. Gaetjens, H. X., Davey, S.J., Arulampalam, S., Fletcher, F.K., Lim, C.C.: Histogram - PMHT for fluctuating target models. IET Radar Sonar and Navig, **11**(8), 1292–1301 (2017)
27. Graham, M., Luginbuhi, T., Streit, R., Walsh, M.: Method for tracking targets with hyper-spectral data (2007). http://www.google.com/patents/US7212652. US Patent 7,212,652
28. Jones, P.N., McLachlan, G.J.: Algorithm AS 254: maximum likelihood estimation from grouped and trncated data with finite normal mixture models. Appl. Stat. **39**(2), 273–282 (1990)
29. Luginbuhl, T.E.: Estimation of general discrete-time FM processes. Ph.D. thesis, University of Connecticut (1999)
30. Luginbuhl, T.E., Willett, P.: Tracking a general, frequency modulated signal in noise. In: IEEE Conference on Decision and Control, vol. 5, pp. 5076–5081 (1999)
31. Luginbuhl, T.E., Willett, P.: Estimating the parameters of general frequency modulated signals. IEEE Trans. Signal Process. **52**(1), 117–131 (2004)
32. Luginbuhl, T.E., Willett, P.K.: Tracking a single general frequency modulated signal in noise. In: Workshop on Probabilistic Approaches for Multitarget Tracking, Paris, France (1998)
33. McLachlan, G.J., Jones, P.N.: Fitting mixture models to grouped and truncated data via the EM algorithm. Biometrics **44**, 571–578 (1988)
34. McLachlan, G.J., Krishnan, T.: The EM Algorithm and Extensions. Wiley, New York (1997)
35. Pakfiliz, A.G.: Development of a probabilistic tracking algorithm for maneuvering targets in cluttered environment. Ph.D. thesis, Ankara University (2004)
36. Pakfiliz, A.G., Efe, M.: Multi-target tracking in clutter with histogram probabilistic multi-hypothesis tracker. In: IEEE Conference on Systems Engineering, pp. 137–142 (2005)
37. Perlovsky, L.I., McManus, M.: Maximum likelihood neural networks for sensor fusion and adaptive classification. Neural Netw. **4**, 89–102 (1991)
38. Perlovsky, L.I.: MLANS neural network for track before detect. In: IEEE National Aerospace and Electronics Conference (1993)
39. Poor, H.V.: An Introduction to Signal Detection and Estimation. Springer, Berlin (1988)
40. Rudin, W.: Principles of Mathematical Analysis. MCGraw Hill, New York (1976)
41. Särkkä, S.: Bayesian Filtering and Smoothing. Cambridge University Press, USA (2013)
42. Streit, R., Graham, M., Walsh, M.: Tracking in hyper-spectral data. In: Proceedings of the Fifth International Conference on Information Fusion, 2002. vol. 2, 852–859 (2002)
43. Streit, R.L.: Tracking on intensity-modulated data streams. Technical Report 11221, NUWC, Newport, Rhode Island, USA (2000)
44. Streit, R.L.: Tracking targets with specified spectra using the H-PMHT algorithm. Technical Report 11291, NUWC, Newport, Rhode Island, USA (2001)
45. Streit, R.L.: Poisson Point Processes: imaging, Tracking, and Sensing. Springer, New York (2010)
46. Streit, R.L., Graham, M.L., Walsh, M.J.: Multiple target tracking of distributed targets using Histogram-PMHT. In: Proceedings of the 4th International Conference on Information Fusion, Montreal, Canada (2001)

47. Streit, R.L., Graham, M.L., Walsh, M.J.: Multitarget tracking on intensity-modulated sensor data. In: SPIE Signal Processing, Sensor Fusion, and Target Recognition, vol. SPIE 4380 (2001)
48. Streit, R.L., Graham, M.L., Walsh, M.J.: Multitarget tracking of distributed targets using histogram-PMHT. Digit. Signal Process. **12**(2), 394–404 (2002)
49. Streit, R.L., Luginbuhl, T.E.: Probabilistic multi-hypothesis tracking. Technical Report 10428, NUWC, Newport, Rhode Island, USA (1995)
50. Vo, B.N., Vo, B.T., Pham, N.T., Suter, D.: Joint detection and estimation of multiple objects from image observations. IEEE Trans. Signal Process. **58**(10), 5129–5141 (2010)
51. Vu, H.X.: Track-before-detect for active sonar. Ph.D. thesis, The University of Adelaide (2015)
52. Vu, H.X., Davey, S.J., Arulampalam, S., Fletcher, F., Lim, C.C.: H-PMHT with a Poisson measurement model. In: Radar, pp. 446–451. Adelaide, South Australia (2013)
53. Vu, H.X., Davey, S.J., Fletcher, F., Arulampalam, S., Ellem, R., Lim, C.C.: Track-before-detect for an active towed array sonar. In: Acoustics. Victor Harbor, South Australia (2013)
54. Vu, H.X., Davey, S.J., Fletcher, F., Arulampalam, S., Lim, C.C.: Track-before-detect using histogram PMHT and dynamic programming. In: Digital Image Computing: Techniques and Applications, pp. 1–8. Fremantle, Western Australia (2012)
55. Walsh, M., Graham, M., Streit, R., Luginbuhl, T., Mathews, L.: Tracking on intensity-modulated sensor data streams. In: IEEE Proceedings Aerospace Conference, 2001. vol. 4, pp. 1901–1909 (2001)
56. Walsh, M.J., Graham, M.L., Streit: Adding power spectra to pmht. In: Proceedings of the 6th International Conference on Information Fusion (2003)
57. Wieneke, M., Davey, S.J.: Histogram PMHT with target extent estimates based on random matrices. In: Proceedings of the 14th International Conference on Information Fusion, Chicago, USA (2011)
58. Wieneke, M., Davey, S.J.: Histogram-PMHT for extended targets and target groups in images. IEEE Trans. Aerosp. Electron. Syst. **50**(3) (2014)
59. Willett, P.K., Balasingam, B., Dunham, D.T., Ogle, T.L.: Multiple target tracking from images using the maximum likelihood hpmht. In: SPIE Signal and Data Processing of Small Targets, vol. SPIE 8857 (2013)

Chapter 2
Idealised Track-Before-Detect

The H-PMHT is a multi-target tracking algorithm that uses images as the measurement input; we refer to tracking over images as track-before-detect (TkBD). What makes the method worth attention is that its execution cost grows linearly with the number of targets. Examples of other multiple target TkBD occasionally demonstrate as many as six concurrent targets. Later in this book, we will show you an application with almost 30,000! The mathematics under the bonnet assumes numerous targets and can handle significantly big problems. However, this mathematics and bookkeeping its notation can be a little daunting. Before we embark upon that sanative expedition, this chapter presents a simple example of the algorithm's performance, which will perhaps fortify the reader's resolve.

This chapter presents an academic case study of track initiation for a single target in a relatively small image and compares H-PMHT with a common point measurement tracker and three alternative track-before-detect (TkBD) algorithms. These alternatives are historically the most popular methods for TkBD. Rather than slow the reader down with the details of all of these methods, we here present only a numerical comparison for a idealised scene. The next two chapters are devoted to the detailed derivation of H-PMHT and the comparison algorithms are reviewed in Appendix A.

Track-before-detect is a processing paradigm that seeks to defer the use of hard decisions until as late as possible in the data processing chain. In conventional detect-then-track systems, the tracker assumes that the sensor creates point measurements. Its task is to determine how many targets are present; to associate across time point measurements arising from each particular target; and to use the measurement data and target dynamic models to infer parameters of interest. However, the assumed point measurements are not necessarily intrinsic to the sensor. Frequently, the sensor returns something more like a multidimensional map of the collected power across a physical region. Point measurements are then extracted by searching for local maxima within this power surface and by applying a threshold. We call this process a single-frame detector and its limitation is that it makes hard decisions based on data

S. J. Davey and H. Gaetjens, *Track-Before-Detect Using Expectation Maximisation*, Signals and Communication Technology,
https://doi.org/10.1007/978-981-10-7593-3_2

from only one frame. The task of associating together point measurements from a common target can be the most computationally demanding part of the detect-then-track process because there is usually an assertion that each target can make only 1 or 0 measurements. Under that assertion, association requires the treatment of a combinatorial number of measurement-to-target assignment hypotheses and much of the body of tracking literature deals with efficient ways to approximate or implicitly marginalise these hypotheses without explicitly enumerating them.

In contrast, TkBD avoids this early hard decision by supplying the sensor intensity image directly to the tracker. The TkBD tracker still has the same three tasks as the point measurement tracker but over a different kind of measurement. It must determine how many targets are present; associate sensor energy across time and space belonging to each particular target; and use this measurement data along with target dynamic models to infer parameters of interest. What makes this different to point measurement tracking is that point measurement trackers can often approximate the measurement function as a linear process with Gaussian noise. In essence, the mapping from a point representation of the target to a point measurement is a projection and the noise can usually be assumed to be additive. In TkBD, the point representation of the target determines where in the sensor image the target 'blob' is centred but the relationship between the pixel values and the target state is highly nonlinear. Frequently, the pixel noise is also non-Gaussian.

Track-before-detect is hence a nonlinear non-Gaussian estimation problem and many of the methods used are based on numerical approximations. A detailed history of methods for TkBD is presented in [5] along with summaries of the main methods used in the area.

We now present a simulation-based comparison of H-PMHT with the following algorithms:

- Probabilistic data association is a common point measurement tracking algorithm. It first applies a threshold detector to extract points from the sensor images. We use a version that exploits amplitude information and models target existence [2].
- The general Bayesian filter has no closed form solution for TkBD but can be numerically approximated over a grid in the state space using the forwards–backwards smoother [7].
- An alternative numerical approach is to use sequential Monte Carlo sampling which is referred to as the particle filter [6]. This is the most common TkBD approach in the current literature.
- The Viterbi algorithm is an efficient optimal sequence estimator for discrete states and has also been extensively applied to TkBD [1].

All of these methods are described in more detail in Appendix A. The three TkBD algorithms are all based on numerical sampling approximations. The purpose of this chapter is to compare the methods in terms of output quality and computation cost on an idealised TkBD problem. These comparisons were first presented in [3] and later extended to complex-valued data in [4].

Table 2.1 Scenario parameters

Scenario index	1	2	3	4	5	6	7	8	9	10	11	12
Target speed (Pixels/frame)	0.25	0.5	1	2	0.25	0.5	1	2	0.25	0.5	1	2
SNR (dB)	12	12	12	12	6	6	6	6	3	3	3	3

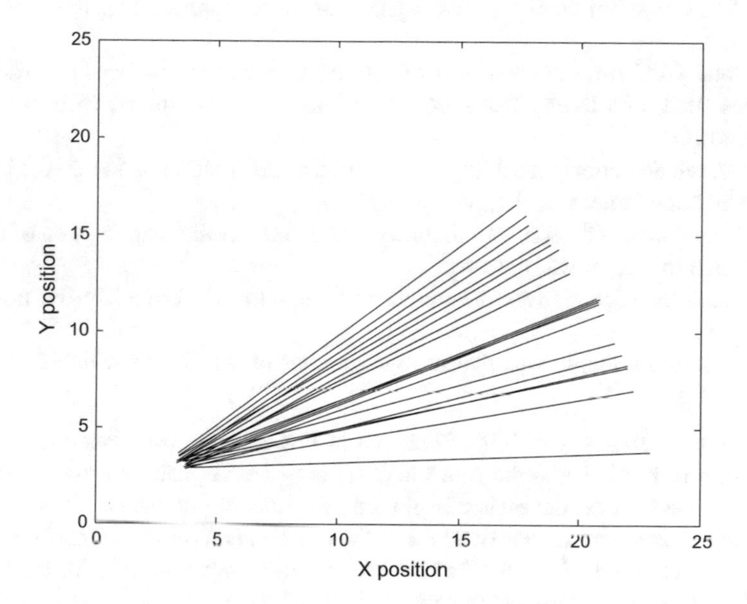

Fig. 2.1 Example scenario, target speed = 1

2.1 Single Target Comparison

The performance of the various algorithms was investigated by simulating a scenario with a single target. To keep things simple, we did not follow the canonical multi-target paths and instead used a straight line target for a total of 20 frames. The scenario was varied by selecting a particular target speed and SNR. In each case, Rayleigh pixel noise with unit variance was added. Table 2.1 summarises the different parameters considered. For each scenario, 100 Monte Carlo trials were performed.

The algorithms which sample the state space on a fixed grid may be affected by the position of the target relative to the grid, that is, whether the target is close to a grid point or midway between them. In order to average over this potential variation, the initial target position for each Monte Carlo trial was randomly sampled from the range 2.5 . . . 3 independently in X and Y. The target heading was also randomly sampled from East to North-East. Figure 2.1 shows an example of 20 Monte Carlo trials with a speed of 1.

False track performance was quantified using a single realisation of a scenario with no target present and 2000 scans. The long duration was chosen to test whether the particle filter algorithm suffered from degeneracy.

We illustrate output quality using metrics for accuracy, cardinality and computation cost. Six metrics were used:

- The RMS position error was averaged over those frames when the target was detected.
- The total CPU time required to evaluate all of the scenarios was recorded and is presented both in absolute seconds, and as a ratio compared with the fastest algorithm.
- The overall detection probability was defined as the fraction of Monte Carlo trials for which the target was detected *at any time*.
- The instantaneous detection probability was defined as the total fraction of frames for which the target was detected.
- The false track count was the number of false tracks formed in the no-target scenario.
- The false track length was the average number of frames for which these false tracks persist.

The overall detection probability is shown in Fig. 2.2, the instantaneous detection probability in Fig. 2.3 and the root mean square (RMS) position error in Fig. 2.4. In each of the figures, the metric is plotted as a function of scenario number. The horizontal (scenario) axis has two labels: the target speed for the scenario is shown below the axis, and the SNR for the scenario is shown above the axis. Vertical dotted lines delineate the scenarios with a particular SNR value.

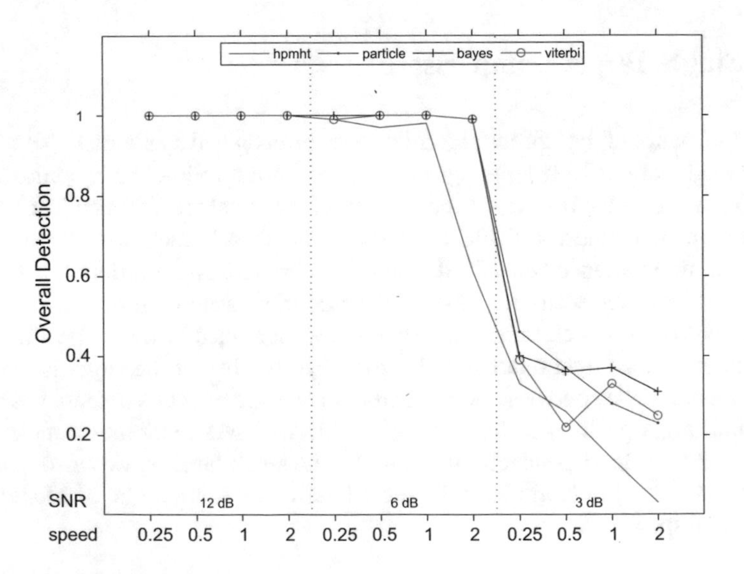

Fig. 2.2 Overall detection probability

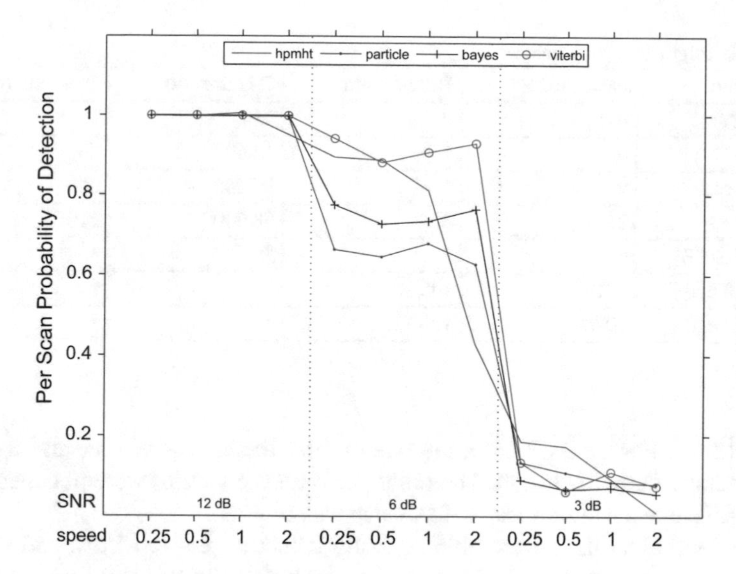

Fig. 2.3 Average instantaneous detection probability

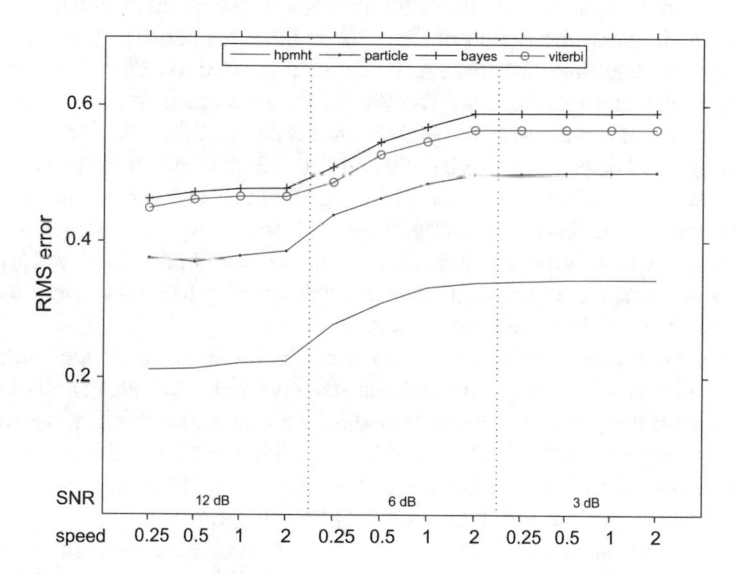

Fig. 2.4 RMS position estimation error

Table 2.2 shows the false track count and the computation resource. For comparison, a probabilistic data association filter (PDAF) was also run on the false track scenario. The PDAF used point measurements and included amplitude as a non-kinematic measurement feature. The PDAF was run assuming a known target SNR and a detection threshold to give 90track performance at the SNR values of interest is shown in Table 2.2. The false track performance of the PDAF is clearly unacceptable

Table 2.2 Algorithm performance

Algorithm	False number	False length	Cpu time (s)	Cpu time (ratio)
H-PMHT	0	*	420	1
Particle	0	*	16,600	39
Grid-bayes	0	*	85,600	204
Viterbi	11	2.27	88,000	210
PDAF (12 dB)	0	*	*	*
PDAF (6 dB)	1412	145	*	*
PDAF (3 dB)	1028	185	*	*

below 12 dB. For the 3 dB case, the rate of false tracks is lower because the false tracks persist for much longer. The other performance metrics were not considered for PDAF since the false track performance was so poor.

All of the algorithms were able to easily detect targets at 12 dB, and they all detected every target. The H-PMHT had a small delay in detecting some of the 2 pixels per frame targets due to initialisation: we will discuss this much later. For the 6 dB targets, in the centre region of Figs. 2.2 and 2.3, the numerical approximation techniques continued to detect almost all of the targets, but the H-PMHT only detected about half of the high-speed targets. The Viterbi algorithm gave the best instantaneous detection result because it was allowed to back-track. At 3dB, all of the algorithms showed degraded detection performance and found less than half of the targets.

The false track performance of all of the algorithms is comparable since this was a requirement of the algorithm tuning. This leads to the overall conclusion that the numerical approximation techniques considered give similar detection performance. If batch processing is acceptable, then back-tracking can provide some improvement in the percentage of time that a target is tracked.

The RMS estimation error curves shown in Fig. 2.4 show an obvious trend. The error increases as the target speed increases and amplitude reduces. For all cases, the H-PMHT error is significantly lower than particle filter error, which is in turn better than the grid approximations. The smoothing used in the Viterbi algorithm reduced the RMS error a little over the Bayesian filter. As may be expected, the error from the grid-based algorithms was approximately half the grid size.

The computation resource required by the different algorithms shows a more marked difference than the detection performance. As Table 2.2 shows, the H-PMHT was more than an order of magnitude faster than the particle filter, and more than two orders of magnitude faster than the grid-based algorithms. Significant effort was spent in optimising all of the algorithms. For the H-PMHT, there was no specific computation bottleneck: both the association and filtering code took similar resource. In contrast, the numerical approximation algorithms were all limited by the likelihood calculations. These incurred the vast majority of the processing cost, even though they used external library code. The H-PMHT was much faster because it does not perform likelihood computations.

2.2 Summary

This chapter has presented a very brief comparison of different methods for track-before-detect. The take-home message is that H-PMHT gave only slightly worse detection results than numerical approximations to theoretically optimal methods, but at a fraction of the computation cost. By its nature, H-PMHT is inherently a multi-target algorithm, whereas all the other methods assume a single target. The main weakness of H-PMHT is initialisation because it uses a hill-climbing optimisation approach.

The quality of the output tracks combined with a simple and low-complexity strategy for multiple targets makes H-PMHT the preferred method for imagery-based tracking.

References

1. Barniv, Y.: Dynamic programming algorithm for detecting dim moving targets In: Multitarget-Multisensor Tracking: Advanced Applications. Artech House, USA (1990)
2. Colegrove, S.B., Davis, A.W., Ayliffe, J.K.: Track initiation and nearest neighbours incorporated into probabilistic data association Journal of Electrical and Electronics Engineers. Australia **6**, 191–198 (1986)
3. Davey, S.J., Rutten, M.G., Cheung, B.: A comparison of detection performance for several track-before-detect algorithms. EURASIP J. Adv. Signal Process. **2008** (2008)
4. Davey, S.J., Rutten, M.G., Cheung, B.: Using phase to improve track-before-detect. IEEE Trans. Aerosp. Electron. Syst. **48**(1), 832–849 (2012)
5. Davey, S.J., Rutten, M.G., Gordon, N.J.: Track-before-detect techniques, in Integrated Tracking, Classification and Sensor Management: theory and applications, Wiley, New York (2012)
6. Ristic, B., Arulampalam, S., Gordon, N.J.: Beyond the Kalman Filter: Particle Filters for Tracking Applications. Artech House, USA (2004)
7. Stone, L.D., Streit, R.L., Corwin, T.L., Bell, K.L.: Bayesian Multiple Target Tracking. Artech House, USA (2013)

Chapter 3
Point Measurement Probabilistic Multi-hypothesis Tracking

The H-PMHT is fundamentally the point measurement PMHT applied to a histogram interpretation of image measurements. The PMHT algorithm is derived by applying Expectation Maximisation (EM) to point measurement tracking. Each point measurement is assumed to be caused by exactly one source, either one of the targets or the background clutter, and the EM missing data is an indicator variable that links each measurement to its source. This chapter reviews the derivation of PMHT and acts as warm up for the remainder of the book. An understanding of the mechanics of PMHT is very helpful in the development of H-PMHT, so don't be tempted to skip this chapter unless you're sure you can do without it. Before tackling the full complexity of the H-PMHT machinery, we will gradually build momentum by first applying the Expectation Maximisation algorithm to a static mixture model. The mixture components are then allowed to randomly evolve according to a known process model and the solution to this problem is PMHT. For the case that the measurement function is a linear function of the target state and the measurement noise is Gaussian, the PMHT can be implemented as a Kalman Smoother over equivalent measurements: this equivalence is demonstrated. For non-linear non-Gaussian applications the implementation is more complicated: an analytic expression is developed for the generic E-step, but in the general case the M-step must be tackled numerically. Finally the chapter concludes with an illustration of the algorithm applied to the canonical multi-target scenario, introduced in Chap. 1.

3.1 Gaussian Mixture Models

Consider a mixture of M Gaussian components with means $\mu^{1:M}$ and variances $\Sigma^{1:M}$. The mixing proportion vector $\pi^{1:M}$ is a probability vector, so $\sum_{m=1}^{M} \pi^m = 1$ and $\pi^m \geq 0$. When a measurement is drawn from the mixture, the probability of it

S. J. Davey and H. Gaetjens, *Track-Before-Detect Using Expectation Maximisation*, Signals and Communication Technology,
https://doi.org/10.1007/978-981-10-7593-3_3

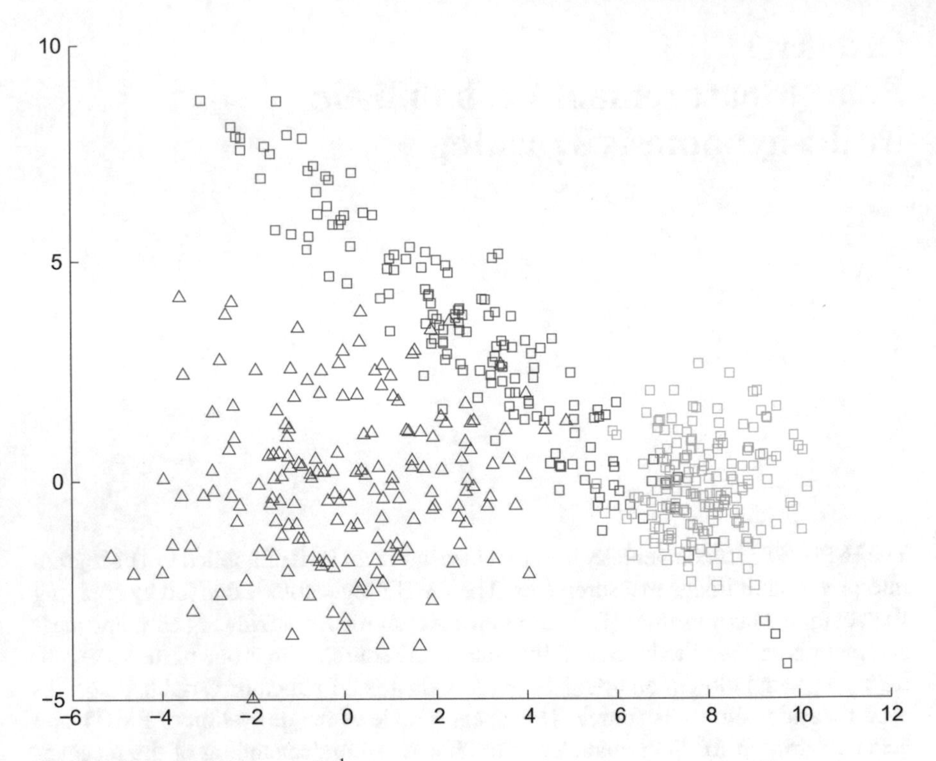

Fig. 3.1 Gaussian mixture model[1]

originating from component m is π^m and in this case, the spatial distribution of the measurement is Gaussian with mean μ^m and variance Σ^m. To simplify notation, μ, Σ and Π without superscripts are used in this section to denote the sets $\mu \equiv \mu^{1:M}$, $\Sigma \equiv \Sigma^{1:M}$ and $\Pi \equiv \pi^{1:M}$.

Figure 3.1 shows an example realization of the mixture with $n = 100$ measurements. In the figure there are $M = 3$ components and the source of each measurement is shown by the marker symbol, circle triangle or square. When these sources are available, then it is straightforward to estimate the component means and variances: the sample mean and variance is the Maximum Likelihood Estimate. Much more interesting is the case when they are not available.

When the measurement sources are unknown, one way to address the component parameter estimation problem is to treat the sources as missing data. Denote the rth measurement as $\mathbf{y}^r$ and its source as k^r; $k^r \in 1 \ldots M$. Following our usual convention $\mathbf{Y} \equiv \mathbf{y}^{1:n}$ and $\mathbf{K} \equiv k^{1:n}$ where n is the number of measurements. At this stage the component parameters μ, Σ and Π are assumed to be unknown constants, not random variables, so we write the complete data likelihood as $p_{\text{comp}}(\mathbf{K}, \mathbf{Y}; \mu, \Sigma, \Pi)$ not $p_{\text{comp}}(\mathbf{K}, \mathbf{Y}, \mu, \Sigma, \Pi)$. The EM auxiliary function is the expectation of the logarithm

[1] *GaussMixDemo*
The points in Fig. 3.1 can be reproduced using the H-PMHT toolbox function GaussMixDemo.

of the measurement likelihood over the conditional assignment probabilities. Starting with (1.11) this is

$$\begin{aligned}\mathcal{Q}(\mu, \Sigma, \Pi|\hat{\mu}, \hat{\Sigma}, \hat{\Pi}) &= E_{\mathbf{K}|\hat{\mu},\hat{\Sigma},\hat{\Pi},\mathbf{Y}}\Big[\log\left\{p_{\text{comp}}(\mathbf{K}, \mathbf{Y}; \mu, \Sigma, \Pi)\right\}\Big]\\ &= \sum_{\mathbf{K}} \log\left\{p_{\text{comp}}(\mathbf{K}, \mathbf{Y}|\mu, \Sigma, \Pi)\right\} p_{\text{miss}}(\mathbf{K}|\mathbf{Y}; \hat{\mu}, \hat{\Sigma}, \hat{\Pi}),\end{aligned} \tag{3.1}$$

where the integral in (1.11) is here a sum because the missing data is discrete. The E-step is now to determine the form of this auxiliary function and the M-step finds new estimates for μ, Σ, an Π by maximising it. A useful byproduct of the E-step is the conditional probability $p_{\text{miss}}(\mathbf{K}|\hat{\mu}, \hat{\Sigma}, \hat{\Pi}, \mathbf{Y})$

We start with the E-step, assuming that estimates already exist for the mixture parameters. In practice the sensitivity of the method to these initial points is of interest and depends on whether the auxiliary function is accommodatingly convex or more likely multi-modal. The measurement source variables are independent and identically distributed (IID) and the measurements from a single component are also IID, so the complete data likelihood is

$$p_{\text{comp}}\left(\mathbf{Y}, \mathbf{K}; \mu, \Sigma, \Pi\right) = \prod_{r=1}^{n} \pi^{k^r} \mathcal{N}\left(\mathbf{y}^r; \mu^{k^r}, \Sigma^{k^r}\right), \tag{3.2}$$

where recall that $\mathcal{N}(\mathbf{y}; \mu, \Sigma)$ is a Gaussian distribution with mean μ and variance Σ evaluated at the point $\mathbf{y}$. The logarithm of the complete data likelihood is

$$\begin{aligned}&\log\left\{p_{\text{comp}}\left(\mathbf{Y}, \mathbf{K}; \mu, \Sigma, \pi\right)\right\}\\ &= \sum_{r=1}^{n}\left[\log\left\{\pi^{k^r}\right\} - \frac{1}{2}\log\left\{|2\pi\Sigma^{k^r}|\right\} - \frac{1}{2}(\mathbf{y}^r - \mu^{k^r})^{\mathsf{T}}\left[\Sigma^{k^r}\right]^{-1}(\mathbf{y}^r - \mu^{k^r})\right].\end{aligned} \tag{3.3}$$

The conditional probability of the missing data is

$$\begin{aligned}p_{\text{miss}}\left(\mathbf{K}|\mathbf{Y}; \hat{\mu}, \hat{\Sigma}, \hat{\Pi}\right) &= \frac{p_{\text{comp}}\left(\mathbf{K}, \mathbf{Y}; \hat{\mu}, \hat{\Sigma}, \hat{\Pi}\right)}{p\left(\mathbf{Y}; \hat{\mu}, \hat{\Sigma}, \hat{\Pi}\right)}\\ &= \frac{p_{\text{comp}}\left(\mathbf{K}, \mathbf{Y}; \hat{\mu}, \hat{\Sigma}, \hat{\Pi}\right)}{\sum_{\acute{\mathbf{K}}} p_{\text{comp}}\left(\acute{\mathbf{K}}, \mathbf{Y}; \hat{\mu}, \hat{\Sigma}, \hat{\Pi}\right)}.\end{aligned} \tag{3.4}$$

Again, applying the independence of measurements,

$$p_{\text{miss}}\left(\mathbf{K}|\mathbf{Y};\hat{\mu},\hat{\Sigma},\hat{\Pi}\right)=\frac{\prod_{r=1}^{n}\hat{\pi}^{k^r}\mathcal{N}\left(\mathbf{y}^r;\mu^{k^r},\Sigma^{k^r}\right)}{\sum_{\acute{\mathbf{K}}}\prod_{r=1}^{n}\hat{\pi}^{\acute{k}^r}\mathcal{N}\left(\mathbf{y}^r;\mu^{\acute{k}^r},\Sigma^{\acute{k}^r}\right)} \tag{3.5}$$

The notation $\acute{k}^r$ explicitly recognises that the dummy variable in the denominator summation is not the same as the assignment k^r in the numerator. The sum of products in the denominator is the sum over all possible arrangements of the assignments. Each term inside the product only depends on the assignment for one measurement, so we can factorise this sum of products measurement by measurement and the result is a sum-product form, in simpler notation

$$\sum_{a=a_1}^{a_N}\sum_{b=b_1}^{b_N}ab\equiv\left(\sum_{a=a_1}^{a_N}a\right)\left(\sum_{b=b_1}^{b_N}b\right). \tag{3.6}$$

The conditional probability of the missing data is then

$$\begin{aligned}p_{\text{miss}}\left(\mathbf{K}|\mathbf{Y};\hat{\mu},\hat{\Sigma},\hat{\Pi}\right)&=\frac{\prod_{r=1}^{n}\hat{\pi}^{k^r}\mathcal{N}\left(\mathbf{y}^r;\mu^{k^r},\Sigma^{k^r}\right)}{\prod_{r=1}^{n}\sum_{m=1}^{M}\hat{\pi}^{m}\mathcal{N}\left(\mathbf{y}^r;\mu^{m},\Sigma^{m}\right)}\\&=\prod_{r=1}^{n}\frac{\hat{\pi}^{k^r}\mathcal{N}\left(\mathbf{y}^r;\mu^{k^r},\Sigma^{k^r}\right)}{\sum_{m=1}^{M}\hat{\pi}^{m}\mathcal{N}\left(\mathbf{y}^r;\mu^{m},\Sigma^{m}\right)}\\&=\prod_{r=1}^{n}p\left(k^r|\mathbf{Y};\hat{\mu},\hat{\Sigma},\hat{\Pi}\right).\end{aligned} \tag{3.7}$$

So the missing data probability is given by the product of the marginal conditional probabilities of each measurement. Define the data association *weight* $\tilde{w}^{m,r}$ as

$$\tilde{w}^{m,r}=p\left(k^r=m|\mathbf{Y};\hat{\mu},\hat{\Sigma},\hat{\Pi}\right)=\frac{\hat{\pi}^{m}\mathcal{N}\left(\mathbf{y}^r;\mu^{m},\Sigma^{m}\right)}{\sum_{s=1}^{M}\hat{\pi}^{s}\mathcal{N}\left(\mathbf{y}^r;\mu^{s},\Sigma^{s}\right)}. \tag{3.8}$$

Substituting the three terms in (3.3) into (3.1) leads to three terms in the auxiliary function $\mathcal{Q}$. The first of these is

$$\sum_{\mathbf{K}}\left(\sum_{r=1}^{n}\log\left\{\pi^{k^r}\right\}\right)p_{\text{miss}}\left(\mathbf{K}|\mathbf{Y};\hat{\mu},\hat{\Sigma},\hat{\Pi}\right). \tag{3.9}$$

It is useful to break the missing data likelihood into

$$p_{\text{miss}}\left(\mathbf{K}|\mathbf{Y},\hat{\mu},\hat{\Sigma},\hat{\Pi}\right)=p\left(k^r|\mathbf{Y},\hat{\mu},\hat{\Sigma},\hat{\Pi}\right)p\left(\mathbf{K}\setminus k^r|\mathbf{Y},\hat{\mu},\hat{\Sigma},\hat{\Pi}\right), \tag{3.10}$$

where $\mathbf{K}\setminus k^r$ denotes the collection of all the assignments except the rth one and the simple product occurs because all of the assignments are conditionally independent.

Each term in the internal sum in (3.9) depends on only one measurement assignment so swap the order of the sums to give

$$\sum_{r=1}^{n}\sum_{k^r=1}^{M}\left[\log\left\{\pi^{k^r}\right\}p\left(k^r|\mathbf{Y},\hat{\mu},\hat{\Sigma},\hat{\Pi}\right)\sum_{\mathbf{K}\setminus k^r}p\left(\mathbf{K}\setminus k^r|\mathbf{Y},\hat{\mu},\hat{\Sigma},\hat{\Pi}\right)\right]. \quad (3.11)$$

The inner sum is of the form $\sum_k p(k|\cdot)$ and covers the whole support of $p(k|\cdot)$ so it sums to unity. In other words, the rth term only depends on the marginal $p\left(k^r|\mathbf{Y},\hat{\mu},\hat{\Sigma},\hat{\Pi}\right)$

$$\sum_{r=1}^{n}\sum_{k^r=1}^{M}\log\left\{\pi^{k^r}\right\}p\left(k^r|\mathbf{Y},\hat{\mu},\hat{\Sigma},\hat{\Pi}\right), \quad (3.12)$$

which can be expressed in terms of data association weights as

$$\sum_{r=1}^{n}\sum_{m=1}^{M}\log\left\{\pi^m\right\}\tilde{w}^{m,r}. \quad (3.13)$$

The other two terms simplify in a similar way, giving the auxiliary function

$$\begin{aligned}
&\mathcal{Q}\left(\mu,\Sigma,\Pi|\hat{\mu},\hat{\Sigma},\hat{\Pi}\right)\\
&=\sum_{r=1}^{n}\sum_{m=1}^{M}\log\left\{\pi^m\right\}\tilde{w}^{m,r}-\frac{1}{2}\sum_{r=1}^{n}\sum_{m=1}^{M}\log\left\{|2\pi\,\Sigma^m|\right\}\tilde{w}^{m,r} \quad (3.14)\\
&\quad-\frac{1}{2}\sum_{r=1}^{n}\sum_{m=1}^{M}\left(\mathbf{y}^r-\mu^m\right)^{\mathsf{T}}\left[\Sigma^m\right]^{-1}\left(\mathbf{y}^r-\mu^m\right)\tilde{w}^{m,r}, \quad (3.15)\\
&=\mathcal{Q}_\pi-\frac{1}{2}\sum_{m=1}^{M}\mathcal{Q}^m_{\mu,\Sigma}, \quad (3.16)
\end{aligned}$$

where

$$\mathcal{Q}_\pi=\sum_{m=1}^{M}\log\left\{\pi^m\right\}\sum_{r=1}^{n}\tilde{w}^{m,r}, \quad (3.17)$$

and

$$\mathcal{Q}^m_{\mu,\Sigma}=\log\left\{|2\pi\,\Sigma^m|\right\}\sum_{r=1}^{n}\tilde{w}^{m,r}+\sum_{r=1}^{n}\left(\mathbf{y}^r-\mu^m\right)^{\mathsf{T}}\left[\Sigma^m\right]^{-1}\left(\mathbf{y}^r-\mu^m\right)\tilde{w}^{m,r}. \quad (3.18)$$

This concludes the E-step.

The M-step finds new estimates by maximising the auxiliary function. The mixing proportion is contained only in the $\mathscr{Q}_\pi$ and this term is independent of the other parameters, so the auxiliary sub-function $\mathscr{Q}_\pi$ can be separately maximised. The mixing proportion estimate is found by maximising $\mathscr{Q}_\pi$ under the constraint that $\sum_{m=1}^{M} \pi^m = 1$. This is achieved using the Lagrangian

$$L_\pi = \mathscr{Q}_\pi + \xi \left(1 - \sum_{m=1}^{M} \pi^m \right). \tag{3.19}$$

The derivative of the Lagrangian is

$$\frac{\mathrm{d}L_\pi}{\mathrm{d}\pi^m} = \left(\pi^m\right)^{-1} \sum_{r=1}^{n} \tilde{w}^{m,r} - \xi, \tag{3.20}$$

which is zero when $\pi^m = \sum_{r=1}^{n} \tilde{w}^{m,r} / \xi$. Reapplying the constraint gives

$$\sum_{m=1}^{M} \pi^m = \sum_{m=1}^{M} \frac{\sum_{r=1}^{n} \tilde{w}^{m,r}}{\xi} = 1, \tag{3.21}$$

which implies

$$\xi = \sum_{m=1}^{M} \sum_{r=1}^{n} \tilde{w}^{m,r} = \sum_{r=1}^{n} \left(\sum_{m=1}^{M} \tilde{w}^{m,r} \right) = n, \tag{3.22}$$

so the mixing proportion estimate is

$$\hat{\pi}^m = \frac{1}{n} \sum_{r=1}^{n} \tilde{w}^{m,r}. \tag{3.23}$$

This is essentially a relative frequency estimate except that the count of the number of m originated measurements is replaced with the expected value of that count under the conditional missing data probability distribution.

The $\mathscr{Q}^m_{\mu,\Sigma}$ term can be optimised more directly. Differentiating with respect to μ^m gives

$$\frac{\mathrm{d}\mathscr{Q}^m_{\mu,\Sigma}}{\mathrm{d}\mu^m} = -\sum_{r=1}^{n} \left[\Sigma^m\right]^{-1} \left(\mathbf{y}^r - \mu^m\right) \tilde{w}^{m,r}, \tag{3.24}$$

which is zero when

$$\hat{\mu}^m = \frac{\sum_{r=1}^{n} \mathbf{y}^r \tilde{w}^{m,r}}{\sum_{r=1}^{n} \tilde{w}^{m,r}}, \tag{3.25}$$

namely the weighted sample mean.

Using the matrix identity $\boldsymbol{v}^{\mathsf{T}}\mathsf{A}^{-1}\boldsymbol{v} = \text{trace}\left(\boldsymbol{v}\boldsymbol{v}^{\mathsf{T}}\mathsf{A}^{-1}\right)$ [10] we can rewrite $\mathcal{Q}^m_{\mu,\Sigma}$ as

$$\mathcal{Q}^m_{\mu,\Sigma} = \log\left\{|2\pi\,\Sigma^m|\right\}\sum_{r=1}^{n}\tilde{w}^{m,r} + \text{trace}\left\{\left[\sum_{r=1}^{n}\left(\mathbf{y}^r - \mu^m\right)\left(\mathbf{y}^r - \mu^m\right)^{\mathsf{T}}\tilde{w}^{m,r}\right]\left[\Sigma^m\right]^{-1}\right\}. \tag{3.26}$$

Two matrix derivative identities are useful here [10]

$$\frac{\text{d}\,\log\left\{|c\mathsf{A}|\right\}}{\text{d}\mathsf{A}} = \mathsf{A}^{-1}, \tag{3.27}$$

$$\frac{\text{d trace}\left(\mathsf{B}\mathsf{A}^{-1}\right)}{\text{d}\mathsf{A}} = -\mathsf{A}^{-1}\mathsf{B}\mathsf{A}^{-1}, \tag{3.28}$$

where c is a scalar constant. Note that the sum term inside the trace above is a constant matrix with respect to Σ^m, just like B in the derivative identity. Applying these, the derivative with respect to Σ^m is

$$\frac{\text{d}\mathcal{Q}^m_{\mu,\Sigma}}{\text{d}\Sigma^m} = \left[\Sigma^m\right]^{-1}\sum_{r=1}^{n}\tilde{w}^{m,r} - \left[\Sigma^m\right]^{-1}\left[\sum_{r=1}^{n}\left(\mathbf{y}^r - \mu^m\right)\left(\mathbf{y}^r - \mu^m\right)^{\mathsf{T}}\tilde{w}^{m,r}\right]\left[\Sigma^m\right]^{-1}. \tag{3.29}$$

This has a stationary point at

$$\hat{\Sigma}^m = \frac{\sum_{r=1}^{n}\left(\mathbf{y}^r - \hat{\mu}^m\right)\left(\mathbf{y}^r - \hat{\mu}^m\right)^{\mathsf{T}}\tilde{w}^{m,r}}{\sum_{r=1}^{n}\tilde{w}^{m,r}}, \tag{3.30}$$

which is the weighted sample variance.

The method for estimating the means and covariances of the components and the mixing proportions under EM can be summarised as follows:

1. Choose initial values for $\hat{\Pi}$, $\hat{\mu}$ and $\hat{\Sigma}$;
2. Use the current estimates to evaluate data association weights (3.8);
3. Calculate new estimates based on the weights using (3.23), (3.25) and (3.30);
4. Repeat steps 2 and 3 until $\mathcal{Q}$ converges.

The convergence test in step 4 above means that the value of the auxiliary function at its maximum changes by less than a threshold amount, mathematically we can write this as

$$\max\left\{\mathcal{Q}(\mu, \Sigma, \Pi|\hat{\mu}, \hat{\Sigma}, \hat{\Pi})\right\} - \mathcal{Q}(\hat{\mu}, \hat{\Sigma}, \hat{\Pi}|\hat{\mu}, \hat{\Sigma}, \hat{\Pi}) < \varepsilon \tag{3.31}$$

where ε is an arbitrary positive number. A smaller value of ε will require more iterations to achieve but will result in an estimate closer to a local maximum of the joint likelihood.

Figure 3.2 shows an example of applying EM mixture modelling to the measurements shown in Fig. 3.1. The initialization is poor, but nevertheless the converged

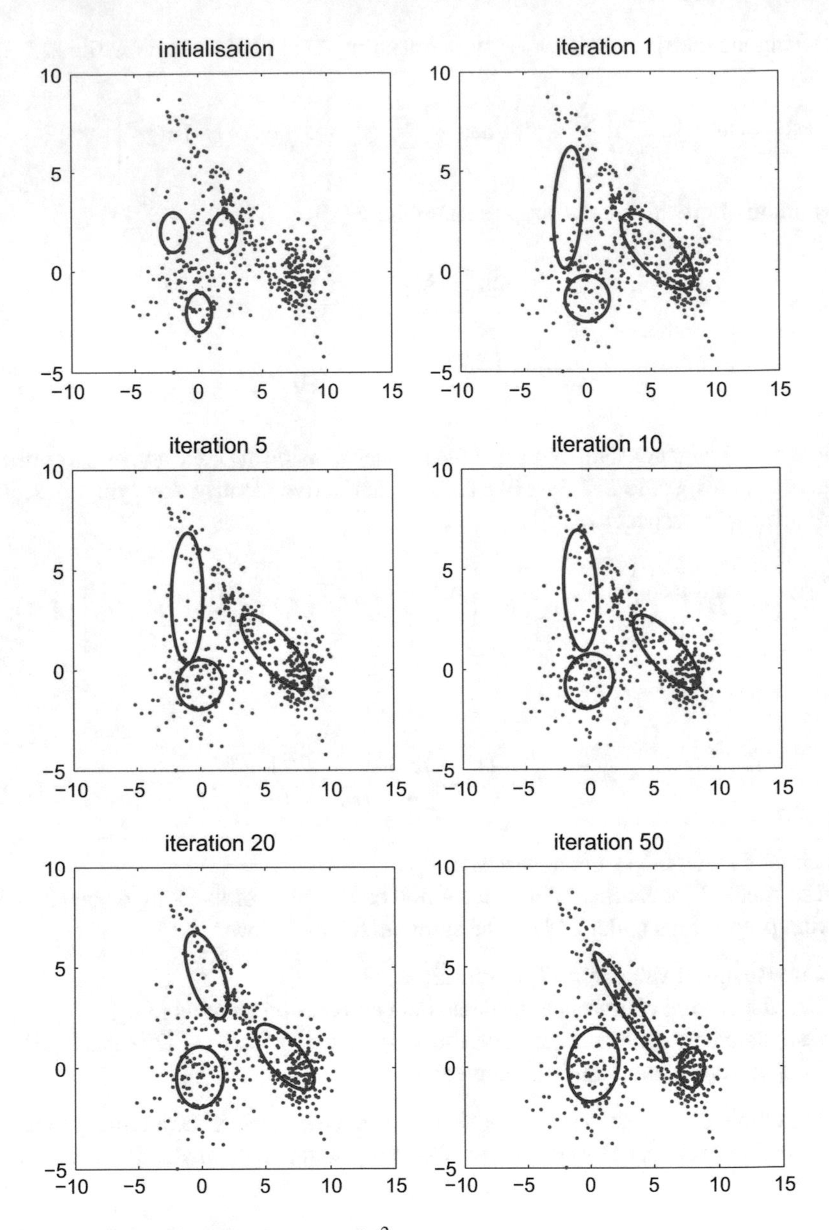

Fig. 3.2 EM fitting gaussian mixture model[2]

[2]*GaussMixDemo*
The GaussMixDemo function also contains an implementation of the maths we just derived to apply EM to fitting mixture components. This was used to generate Fig. 3.2.

estimates are quite close to the true mixture parameters. In the example there are 100 measurements and three components, so there is a relatively high amount of data. It is important to note that the method assumes that the number of components is known. This shortcoming will be addressed a little later.

3.2 Dynamic Mixture Model

The mixture model described in the previous section was a static one: the means and variances of the components were fixed as was the mixing proportion vector and a large number of measurements was collected. A dynamic mixture is one where the component parameters (the means and variances) and potentially the mixing parameters vary with time. Such a model is of much more practical interest and is the basis for PMHT. For the purposes of this chapter, assume that the covariance of measurements from each component is known. To simplify notation, further assume that these covariances are constant and the same for all components, so $\Sigma^m = \Sigma, \forall m$. Since the covariance is a known parameter it will mostly be hidden from the subsequent mathematics. Since the mixture is dynamic, the mean of each component is a random variable that evolves according to a known stochastic process and with a prior that is also known. The aim now is to track the changes in the component means.

Suppose that the sensor collects measurements at discrete times; the collection of measurements at one time instant is referred to as a frame. The tth frame contains measurements $\mathbf{y}_t^r$ for $r \in 1 \dots n_t$, where $n_t \geq 0$ is simply the measurement count at the tth frame. Revising our convention, the set of all of the measurements in the tth frame is $\mathbf{Y}_t$ and the collection across all frames is $\mathbb{Y}$. Similarly, k_t^r denotes the true source of $\mathbf{y}_t^r$ and $\mathbf{K}_t$ and $\mathbb{K}$ are the collections of measurement to component assignments. The state of component m at frame t is $\mathbf{x}_t^m$ (with collections $\mathbf{X}_t$ and $\mathbb{X}$) and the mean of measurements due to this component is $h\left(\mathbf{x}_t^m\right)$. The state dimensionality of the state vector depends on the application: in this monograph, we will assume that the state occupies a subspace of $\mathscr{R}^4$ (position and velocity in the plane) unless defined otherwise. The mixing proportion vector is now also time varying π_t^m and will be assumed independent from one time t to another t'. The abbreviated mixing proportion sets are denoted Π_t and Π. A batch of T frames is available and will be processed jointly.

Under the dynamic mixture model just defined, the complete data is $\{\mathbb{X}, \Pi, \mathbb{K}, \mathbb{Y}\}$ and the missing data remains the true measurement source assignments $\{\mathbb{K}\}$. Compared with the static case, the missing data is the same, but the complete data has changed. The covariance matrix Σ is now constant and the state $\mathbf{x}_t^m$ has indirectly replaced the mean μ^m.

3.2.1 Expectation Step

The expectation step is analytic and consists of the derivation of the EM auxiliary function, the expectation of the logarithm of the complete data likelihood over the conditional assignment probabilities,

$$\mathcal{Q}\left(\mathbb{X}, \Pi | \hat{\mathbb{X}}, \hat{\Pi}\right) = \sum_{\mathbb{K}} \log\left\{p_{\text{comp}}\left(\mathbb{X}, \mathbb{K}, \mathbb{Y}; \Pi\right)\right\} p_{\text{miss}}\left(\mathbb{K} | \mathbb{Y}, \hat{\mathbb{X}}, \hat{\Pi}\right). \quad (3.32)$$

Assume that the targets have independent priors, so $p\left(\mathbf{X}_0\right) = \prod_m p\left(\mathbf{x}_0^m\right)$. Also assume that the target dynamics follow independent Markov chains

$$p\left(\mathbb{X}\right) = \prod_{m=1}^{M}\left[p\left(\mathbf{x}_0^m\right)\prod_{t=1}^{T} p\left(\mathbf{x}_t^m | \mathbf{x}_{t-1}^m\right)\right]. \quad (3.33)$$

As before, the measurement source variables are conditionally independent from all other variables given the mixing proportions,

$$p\left(\mathbb{K}; \Pi\right) = \prod_{t=1}^{T}\prod_{r=1}^{n_t} \pi_t^{k_t^r}. \quad (3.34)$$

The measurements are also conditionally independent from all other random variables given the state of their source component

$$p\left(\mathbb{Y} | \mathbb{X}, \mathbb{K}\right) = \prod_{t=1}^{T}\prod_{r=1}^{n_t} \mathcal{N}\left(\mathbf{y}_t^r; h\left(\mathbf{x}_t^{k_t^r}\right), \Sigma\right). \quad (3.35)$$

These three terms combine to make the complete data likelihood

$$\begin{aligned} p_{\text{comp}}\left(\mathbb{X}, \mathbb{K}, \mathbb{Y}; \Pi\right) &= p\left(\mathbb{X}\right) p\left(\mathbb{K}; \Pi\right) p\left(\mathbb{Y} | \mathbb{X}, \mathbb{K}\right) \\ &= \prod_{m=1}^{M}\left[p\left(\mathbf{x}_0^m\right)\prod_{t=1}^{T} p\left(\mathbf{x}_t^m | \mathbf{x}_{t-1}^m\right)\right]\prod_{t=1}^{T}\left[\prod_{r=1}^{n_t} \pi_t^{k_t^r} \mathcal{N}\left(\mathbf{y}_t^r; h\left(\mathbf{x}_t^{k_t^r}\right), \Sigma\right)\right], \end{aligned} \quad (3.36)$$

and the logarithm of the complete data likelihood is

$$\begin{aligned} &\log\left\{p_{\text{comp}}\left(\mathbb{X}, \mathbb{K}, \mathbb{Y}; \Pi\right)\right\} \\ &= \sum_{m=1}^{M}\left[\log\left\{p\left(\mathbf{x}_0^m\right)\right\} + \sum_{t=1}^{T}\log\left\{p\left(\mathbf{x}_t^m | \mathbf{x}_{t-1}^m\right)\right\}\right] + \sum_{t=1}^{T}\sum_{r=1}^{n_t}\log\left\{\pi_t^{k_t^r}\right\} \end{aligned}$$

$$-\frac{1}{2}\log\left\{|2\pi\Sigma|\right\}\sum_{t=1}^{T}n_t-\frac{1}{2}\sum_{t=1}^{T}\sum_{r=1}^{n_t}\left(\mathbf{y}_t^r-h\left(\mathbf{x}_t^{k_t^r}\right)\right)^{\mathsf{T}}\Sigma^{-1}\left(\mathbf{y}_t^r-h\left(\mathbf{x}_t^{k_t^r}\right)\right). \tag{3.37}$$

The term involving the determinant $|2\pi\Sigma|$ is constant and is therefore discarded. Compared with the static case in (3.3), the complete data contains an extra term for the state evolution since this is now stochastic. This state term is independent of the assignment so it is its own expectation in the $\mathcal{Q}$ function.

The conditional probability of the missing data simplifies in a similar way to the static mixture case

$$\begin{aligned}
p_{\text{miss}}\left(\mathbb{K}|\mathbb{Y},\hat{\mathbb{X}},\hat{\Pi}\right) &= \frac{p\left(\mathbb{K},\mathbb{Y},\hat{\mathbb{X}},\hat{\Pi}\right)}{\sum_{\acute{\mathbb{K}}}p\left(\acute{\mathbb{K}},\mathbb{Y},\hat{\mathbb{X}},\hat{\Pi}\right)},\\
&= \frac{p\left(\hat{\mathbb{X}}\right)\prod_{t=1}^{T}\left[\prod_{r=1}^{n_t}\hat{\pi}_t^{k_t^r}\mathcal{N}\left(\mathbf{y}_t^r;h\left(\mathbf{x}_t^{k_t^r}\right),\Sigma\right)\right]}{p\left(\hat{\mathbb{X}}\right)\sum_{\acute{k}^{1:n}}\prod_{t=1}^{T}\left[\prod_{r=1}^{n_t}\hat{\pi}_t^{k_t^r}\mathcal{N}\left(\mathbf{y}_t^r;h\left(\mathbf{x}_t^{\acute{k}_t^r}\right),\Sigma\right)\right]},\\
&= \prod_{t=1}^{T}\prod_{r=1}^{n_t}\frac{\hat{\pi}_t^{k_t^r}\mathcal{N}\left(\mathbf{y}_t^r;h\left(\mathbf{x}_t^{k_t^r}\right),\Sigma\right)}{\sum_{m=1}^{M}\hat{\pi}_t^m\mathcal{N}\left(\mathbf{y}_t^r;h\left(\mathbf{x}_t^m\right),\Sigma\right)}\\
&= \prod_{t=1}^{T}\prod_{r=1}^{n_t}p\left(k_t^r|\mathbf{y}_t^r,\hat{\mathbf{X}},\hat{\Pi}_t\right).
\end{aligned} \tag{3.38}$$

The missing data probability is again given by the product of the marginal conditional probabilities of each measurement. Define the data association *weight* $\tilde{w}^{m,r}$

$$\tilde{w}_t^{m,r}=p\left(k_t^r|\mathbf{y}_t^r,\hat{\mathbf{X}}_t,\hat{\Pi}_t\right). \tag{3.39}$$

The data association weight is a weighted likelihood ratio, just as in the static case.

The logarithm of the complete data likelihood (3.37) contains three terms: a state evolution term, a mixing proportion term, and a measurement term. The mixing π_t^m term is essentially T copies of the static π^m term and can be simplified in the same manner as shown in (3.9)–(3.13) since again each term $\log\left\{\pi_t^{k_t^r}\right\}$ only depends on one assignment variable. The state evolution is constant with respect to the expectation and the measurement term is unchanged from the static case except that the mean is time varying. The dynamic mixture model auxiliary function is therefore

$$\begin{aligned}
&\mathcal{Q}\left(\mathbb{X},\Pi|\hat{\mathbb{X}},\hat{\Pi}\right)\\
&=\sum_{m=1}^{M}\left[\log\left\{p\left(\mathbf{x}_0^m\right)\right\}+\sum_{t=1}^{T}\log\left\{p\left(\mathbf{x}_t^m|\mathbf{x}_{t-1}^m\right)\right\}\right]+\sum_{t=1}^{T}\sum_{r=1}^{n_t}\sum_{m=1}^{M}\log\left\{\pi_t^m\right\}\tilde{w}_t^{m,r}
\end{aligned}$$

$$-\frac{1}{2}\sum_{t=1}^{T}\sum_{r=1}^{n_t}\sum_{m=1}^{M}\left(\mathbf{y}_t^r - h\left(\mathbf{x}_t^m\right)\right)^{\mathsf{T}} \Sigma^{-1}\left(\mathbf{y}_t^r - h\left(\mathbf{x}_t^m\right)\right)\tilde{w}_t^{m,r}, \tag{3.40}$$

$$= \sum_{t=1}^{T}\mathscr{Q}_{t,\pi} + \sum_{m=1}^{M}\mathscr{Q}_x^m. \tag{3.41}$$

3.2.2 Linear Gaussian Maximisation Step

As in the static case, the two components of the $\mathscr{Q}$ function decouple the unknowns and can be maximised separately. The mixing proportion function $\mathscr{Q}_{t,\pi}$ is given by

$$\mathscr{Q}_{t,\pi} = \sum_{r=1}^{n_t}\sum_{m=1}^{M}\log\left\{\pi_t^m\right\}\tilde{w}_t^{m,r}, \tag{3.42}$$

which is the same as the static mixture expression in (3.17). It therefore has the same solution, namely

$$\hat{\pi}_t^m = \frac{1}{n_t}\sum_{r=1}^{n_t}\tilde{w}_t^{m,r}. \tag{3.43}$$

The state component of the $\mathscr{Q}$ function is given by

$$\mathscr{Q}_x^m = \log\left\{p\left(\mathbf{x}_0^m\right)\right\} + \sum_{t=1}^{T}\log\left\{p\left(\mathbf{x}_t^m|\mathbf{x}_{t-1}^m\right)\right\}$$
$$-\frac{1}{2}\sum_{t=1}^{T}\sum_{r=1}^{n_t}\underbrace{\left(\mathbf{y}_t^r - h\left(\mathbf{x}_t^m\right)\right)^{\mathsf{T}}\Sigma^{-1}\left(\mathbf{y}_t^r - h\left(\mathbf{x}_t^m\right)\right)\tilde{w}_t^{m,r}}_{\text{quadratic in } \mathbf{x}_t^m}. \tag{3.44}$$

As marked, the measurement term above is a weighted sum of quadratics in $\mathbf{x}_t^m$ and can be simplified by expanding the quadratics and collecting together like terms as a function of $\mathbf{x}_t^m$, namely

$$\sum_{t=1}^{T}\sum_{r=1}^{n_t}\left(\mathbf{y}_t^r - h\left(\mathbf{x}_t^m\right)\right)^{\mathsf{T}}\Sigma^{-1}\left(\mathbf{y}_t^r - h\left(\mathbf{x}_t^m\right)\right)\tilde{w}_t^{m,r} = \sum_{t=1}^{T}\left\{\sum_{r=1}^{n_t}\left(\mathbf{y}_t^r\right)^{\mathsf{T}}\Sigma^{-1}\mathbf{y}_t^r\tilde{w}_t^{m,r}\right.$$
$$\left. - 2h\left(\mathbf{x}_t^m\right)^{\mathsf{T}}\Sigma^{-1}\left[\sum_{r=1}^{n_t}\left(\mathbf{y}_t^r\right)\tilde{w}_t^{m,r}\right] + h\left(\mathbf{x}_t^m\right)^{\mathsf{T}}\Sigma^{-1}h\left(\mathbf{x}_t^m\right)\left[\sum_{r=1}^{n_t}\tilde{w}_t^{m,r}\right]\right\}. \tag{3.45}$$

The first term in (3.45) is constant with respect to the state, so simply denote it as C: its details are irrelevant. The other two terms can be simplified by defining

$$\tilde{\Sigma}_t^m = \left[\sum_{r=1}^{n_t} \tilde{w}_t^{m,r}\right]^{-1} \Sigma, \tag{3.46}$$

and

$$\tilde{\mathbf{y}}_t^m = \left[\sum_{r=1}^{n_t} \tilde{w}_t^{m,r}\right]^{-1} \sum_{r=1}^{n_t} \left(\mathbf{y}_t^r\right) \tilde{w}_t^{m,r}. \tag{3.47}$$

Substituting (3.46) and (3.47) into (3.45) simplifies to a single quadratic

$$\begin{aligned}
&\sum_{t=1}^{T}\sum_{r=1}^{n_t} \left(\mathbf{y}_t^r - h\left(\mathbf{x}_t^m\right)\right)^{\mathsf{T}} \Sigma^{-1} \left(\mathbf{y}_t^r - h\left(\mathbf{x}_t^m\right)\right) \tilde{w}_t^{m,r} \\
&= C + \sum_{t=1}^{T} \left\{ -2h\left(\mathbf{x}_t^m\right)^{\mathsf{T}} \left(\tilde{\Sigma}_t^m\right)^{-1} \tilde{\mathbf{y}}_t^m + h\left(\mathbf{x}_t^m\right)^{\mathsf{T}} \left(\tilde{\Sigma}_t^m\right)^{-1} h\left(\mathbf{x}_t^m\right) \right\}, \\
&= C + \sum_{t=1}^{T} \left\{ \left(\tilde{\mathbf{y}}_t^m - h\left(\mathbf{x}_t^m\right)\right)^{\mathsf{T}} \left(\tilde{\Sigma}_t^m\right)^{-1} \left(\tilde{\mathbf{y}}_t^m - h\left(\mathbf{x}_t^m\right)\right) \right\},
\end{aligned} \tag{3.48}$$

where the contents of C have changed but they remain independent of $\mathbf{x}_t^m$. This little piece of chicanery is really nothing more than asserting that the weighted sum of a collection of quadratics is itself quadratic. However, it will prove to be a staple manoeuvre in later chapters. Putting this back into (3.44) gives

$$\begin{aligned}
\mathcal{Q}_x^m &= C + \log\left\{p\left(\mathbf{x}_0^m\right)\right\} \\
&+ \sum_{t=1}^{T} \left[\log\left\{p\left(\mathbf{x}_t^m | \mathbf{x}_{t-1}^m\right)\right\} - \frac{1}{2} \left(\tilde{\mathbf{y}}_t^m - h\left(\mathbf{x}_t^m\right)\right)^{\mathsf{T}} \left(\tilde{\Sigma}_t^m\right)^{-1} \left(\tilde{\mathbf{y}}_t^m - h\left(\mathbf{x}_t^m\right)\right) \right].
\end{aligned} \tag{3.49}$$

Assume that the prior probability $p\left(\mathbf{x}_0^m\right)$ is Gaussian with a known mean and covariance, $p\left(\mathbf{x}_0^m\right) = \mathcal{N}\left(\mathbf{x}_0^m; \bar{\mathbf{x}}_0^m, \Sigma_0^m\right)$ and assume that the transition probability $p\left(\mathbf{x}_t^m | \mathbf{x}_{t-1}^m\right)$ is also Gaussian with mean $\mathsf{F}_t \mathbf{x}_{t-1}^m$ and covariance Q_t^m. Then (3.49) is quadratic in $\mathbf{x}_t^m$

$$\begin{aligned}
\mathcal{Q}_x^m = C &- \frac{1}{2} \left(\mathbf{x}_0^m - \bar{\mathbf{x}}_0^m\right)^{\mathsf{T}} \left(\Sigma_0^m\right)^{-1} \left(\mathbf{x}_0^m - \bar{\mathbf{x}}_0^m\right) \\
&- \frac{1}{2} \sum_{t=1}^{T} \Big[\left(\mathbf{x}_t^m - \mathsf{F}_t \mathbf{x}_{t-1}^m\right)^{\mathsf{T}} \left(\mathsf{Q}_t^m\right)^{-1} \left(\mathbf{x}_t^m - \mathsf{F}_t \mathbf{x}_{t-1}^m\right) \\
&\qquad + \left(\tilde{\mathbf{y}}_t^m - h\left(\mathbf{x}_t^m\right)\right)^{\mathsf{T}} \left(\tilde{\Sigma}_t^m\right)^{-1} \left(\tilde{\mathbf{y}}_t^m - h\left(\mathbf{x}_t^m\right)\right) \Big].
\end{aligned} \tag{3.50}$$

If the measurement function is also linear, $h\left(\mathbf{x}_t^m\right) = \mathsf{H}_t^m \mathbf{x}_t^m$, then all of the derivatives of $\mathcal{Q}_x^m$ with respect to $\mathbf{x}_t^m$ at various t values are linear. These derivatives generate a set of $T+1$ linear equations in $T+1$ unknowns that can be stacked together to form a tri-diagonal block matrix which can be inverted to find the kinematic state sequence,

$$\begin{bmatrix} \mathsf{A}_{0,0} & \mathsf{A}_{0,1} & 0 & \cdots & 0 \\ \mathsf{A}_{1,0} & \mathsf{A}_{1,1} & \mathsf{A}_{1,2} & \ddots & \vdots \\ 0 & \ddots & \ddots & \ddots & 0 \\ \vdots & \ddots & \mathsf{A}_{T-1,T-2} & \mathsf{A}_{T-1,T-1} & \mathsf{A}_{T-1,T} \\ 0 & \cdots & 0 & \mathsf{A}_{T,T-1} & \mathsf{A}_{T,T} \end{bmatrix} \begin{bmatrix} \mathbf{x}_0^m \\ \mathbf{x}_1^m \\ \vdots \\ \mathbf{x}_T^m \end{bmatrix} = \begin{bmatrix} \left(\Sigma_0^m\right)^{-1} \bar{\mathbf{x}}_0^m \\ \left(\tilde{\Sigma}_1^m\right)^{-1} \tilde{\mathbf{y}}_1^m \\ \vdots \\ \left(\tilde{\Sigma}_T^m\right)^{-1} \tilde{\mathbf{y}}_T^m \end{bmatrix}. \tag{3.51}$$

This could complete the M-step, but Luginbuhl and Striet [12] demonstrated another solution path that is more intuitive and can lead to implementations for the non-linear $h\left(\mathbf{x}_t^m\right)$ case. Reconsider the cost function in (3.49): the measurement contribution here has been cast as a quadratic inner self-product of the difference between a predicted measurement and an equivalent measurement. This is only a constant away from the logarithm of a Gaussian distribution, so we could write (3.49) as

$$\mathcal{Q}_x^m = C + \log\left\{p\left(\mathbf{x}_0^m\right)\right\} + \sum_{t=1}^{T}\left[\log\left\{p\left(\mathbf{x}_t^m|\mathbf{x}_{t-1}^m\right)\right\} + \log\left\{\mathcal{N}\left(\tilde{\mathbf{y}}_t^m; h\left(\mathbf{x}_t^m\right), \tilde{\Sigma}_t^m\right)\right\}\right], \tag{3.52}$$

which is the logarithm of a joint likelihood $p\left(\mathbf{x}_{0:T}^m, \tilde{\mathbf{y}}_{1:T}^m\right)$. By definition, this is optimised by the Maximum A Posteriori (MAP) state sequence. If the MAP estimator is available then it can be applied to (3.52) even for non-linear non-Gaussian target dynamics or non-linear $h\left(\mathbf{x}_t^m\right)$. Effectively the PMHT E-step is a data association stage that for Gaussian measurement noise transforms the ambiguous estimation problem with unknown measurement sources into an equivalent known-source point measurement estimation problem. In particular, under linear Gaussian models, the MAP estimator is the Kalman Smoother.

3.3 Non-Gaussian Mixtures

So far we have discussed Gaussian mixtures, first with static and then dynamic parameters. Using a specific model makes the development more tangible and simplifies the maximisation step but the PMHT association approach doesn't require Gaussian assumptions. We now revisit the development using a generic measurement function. Let the point measurement density for target m be denoted $g^m\left(\mathbf{y}|\mathbf{x}_t^m\right)$. Substituting these back into the complete data probability gives

$$p_{\text{comp}}(\mathbb{X}, \mathbb{K}, \mathbb{Y}; \Pi) = \prod_{m=1}^{M}\left[p\left(\mathbf{x}_0^m\right)\prod_{t=1}^{T} p\left(\mathbf{x}_t^m|\mathbf{x}_{t-1}^m\right)\right]\prod_{t=1}^{T}\left[\prod_{r=1}^{n_t}\pi_t^{k_t^r} g^{k_t^r}\left(\mathbf{y}_t^r|\mathbf{x}_t^{k_t^r}\right)\right], \tag{3.53}$$

and the logarithm of the complete data likelihood is

$$\log\left\{p_{\text{comp}}(\mathbb{X}, \mathbb{K}, \mathbb{Y}; \Pi, \Sigma)\right\} = \sum_{m=1}^{M}\left[\log\left\{p\left(\mathbf{x}_0^m\right)\right\} + \sum_{t=1}^{T}\log\left\{p\left(\mathbf{x}_t^m|\mathbf{x}_{t-1}^m\right)\right\}\right] + \sum_{t=1}^{T}\sum_{r=1}^{n_t}\log\left\{\pi_t^{k_t^r}\right\} + \sum_{t=1}^{T}\sum_{r=1}^{n_t}\log\left\{g^{k_t^r}\left(\mathbf{y}_t^r|\mathbf{x}_t^{k_t^r}\right)\right\}. \tag{3.54}$$

The conditional probability of the missing data develops in the same way as the Gaussian case since its simplification is a result of the conditional independence assumptions, not the Gaussian model. This probability remains a product of marginal conditional probabilities for each measurement,

$$p_{\text{mico}}\left(\mathbb{K}|\mathbb{Y}, \hat{\mathbb{X}}, \hat{\Pi}\right) = \prod_{t=1}^{T}\prod_{r=1}^{n_t}\tilde{w}_t^{m,r}, \tag{3.55}$$

where the data association weight is now given by

$$\tilde{w}_t^{m,r} = \frac{\hat{\pi}_t^{k_t^r} g^{k_t^r}\left(\mathbf{y}_t^r|\hat{\mathbf{x}}_t^{k_t^r}\right)}{\sum_{m=1}^{M}\hat{\pi}_t^m g^m\left(\mathbf{y}_t^r|\hat{\mathbf{x}}_t^m\right)}. \tag{3.56}$$

Although these weights are a combination of non-linear non-Gaussian functions, they are evaluated using estimates of the states from the previous EM iteration, so in the context of the auxiliary function, the weights are constants.

The independence assumptions are the same so the auxiliary function simplifies into two components $\mathcal{Q}_{t,\pi}$ and $\mathcal{Q}_x^m$. The first of these has the same form as the linear Gaussian expression given in (3.42), which is itself the same as the static mixture expression in (3.17). The difference is the weight definition above (3.56).

The state component of the auxiliary function is

$$\mathcal{Q}_x^m = \log\left\{p\left(\mathbf{x}_0^m\right)\right\} + \sum_{t=1}^{T}\log\left\{p\left(\mathbf{x}_t^m|\mathbf{x}_{t-1}^m\right)\right\} + \sum_{t=1}^{T}\sum_{r=1}^{n_t}\log\left\{g^m\left(\mathbf{y}_t^r|\mathbf{x}_t^m\right)\right\}\tilde{w}_t^{m,r}. \tag{3.57}$$

Comparing this with the Gaussian case shown in (3.44), the quadratic measurement term has been replaced with the generic $\log\left\{g^m\left(\mathbf{y}_t^r|\mathbf{x}_t^m\right)\right\}$. In the Gaussian noise case we were able to simplify the measurement term because it was a linear combination of quadratics. This sleight of hand boils down to equating coefficients of an equivalent quadratic, so it can be replicated if $\log\left\{g^m\left(\mathbf{y}_t^r|\mathbf{x}_t^m\right)\right\}$ is a polynomial, that is the measurement noise is part of the exponential family [5, 9]. However, in general this will not be the case and no simplification is possible. That doesn't mean that (3.57) can't be solved, instead it means that there is no closed form estimator that will provide the MAP estimate. In such a situation the state estimates can be obtained by numerically optimising (3.57). Several options are available depending on the form of $g^m(\cdot)$ in a specific application. If its derivatives are known then gradient based methods could be used. Alternatively a discrete grid in the state space could be used to brute-force an approximate solution, or a particle filter based algorithm could be developed. These will be explored further in Chap. 7 for the H-PMHT algorithm, which we will soon see is a very close relative of point measurement PMHT.

3.4 Incorporating Clutter

The discussion so far has assumed that each component of the mixture is a target with a corresponding state to be estimated. Often there are also false alarm measurements that are caused by noise and interference or objects simply not of interest to the sensor user. The term clutter can be used to specifically refer to false measurements caused by ground reflections in a radar system. Within this monograph we use the label more broadly to refer to non-target generated data.

The previous section described methods for non-Gaussian mixtures: a subset of this is the case of Gaussian targets and a more widespread clutter distribution. Under the framework already developed one could use a component of the mixture to account for clutter. For the time being, we will assume the simplest model for clutter where false alarms follow a uniform distribution. Clutter measurements are identified by $k^r = 0$, which has a corresponding prior of π_t^0. The clutter density is written as $g^0\left(\mathbf{y}|\emptyset\right) = g^0$, which is constant because the clutter is uniform. There is no clutter state because we have assumed that the clutter distribution is known: there is nothing to estimate.

3.5 Examples of PMHT Point Measurement Tracking

We now illustrate the PMHT algorithm on the canonical multi-target scenario. For the point measurement case, assume that the sensor collects position based measurements with additive Gaussian noise. Since the target position is contained within the state, this measurement function is linear,

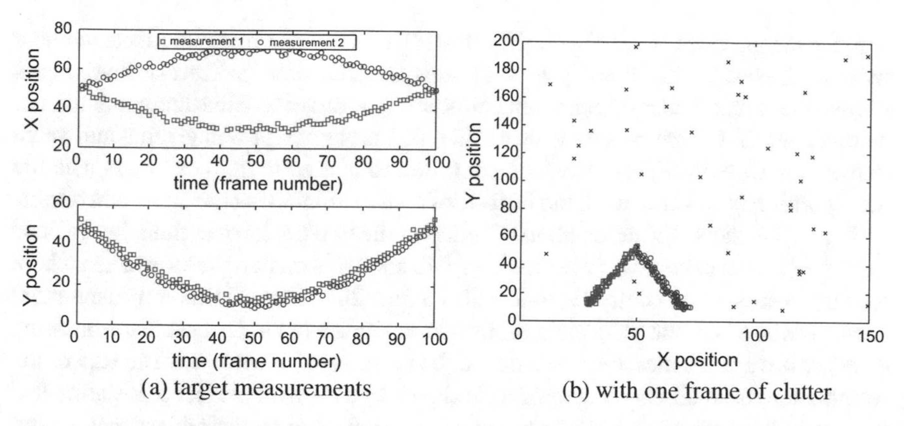

(a) target measurements (b) with one frame of clutter

Fig. 3.3 Example frame for the two target scenario

$$\mathbf{y}_t^r = \mathsf{H}\mathbf{x}_t^m + \zeta_t^r, \quad k_t^r = m > 0, \tag{3.58}$$

where ζ_t^r is a zero mean white Gaussian noise process with covariance R. However, the sensor detection process is imperfect and these measurements are not always present. The target measurement is available with probability of detection P_D. In practice, the detection probability will be state dependent for some applications but for simplicity assume it is known and independent of the state. The detector also produces false alarms that are distributed uniformly across the measurement space. The number of false alarms is Poisson distributed with mean Λ^0. These measurement assumptions are typical in point measurement problems. For the canonical multi-target scenario, the measurement space is a rectangular region in the plane spanning $X \in [0, 150]$ and $Y \in [0, 200]$.

3.5.1 Two Targets

Begin with a simplification of the canonical multi-target scenario where only the first two targets are present. Figure 3.3 shows the positions of these targets as a function of time and shows an example realisation of the measurement process for $\mathsf{R} = \begin{bmatrix} 1 & 0 \\ 0 & 1 \end{bmatrix}$ and $\Lambda^0 = 50$.

The PMHT[3] was run on this example by initializing the state estimates at time 0 with the true states and the mixing proportions with $\hat{\pi}_0^0 = \hat{\pi}_0^1 = \hat{\pi}_0^2 = 1/3$. At each frame, the initial state estimate was obtained by predicting forward the estimate from the previous frame $\hat{\mathbf{x}}_t^m = \mathsf{F}\hat{\mathbf{x}}_{t-1}^m$ and the initial mixing proportion was obtained by

[3] *PMHTTracker*
The H-PMHT toolbox function PMHTTracker contains the implementation of point measurement PMHT used to generate the figures for this chapter.

copying the previous frame $\hat{\pi}_t^m = \hat{\pi}_{t-1}^m$. PMHT was run one frame at a time to refine these estimates. To assist analysis, the measurements were ordered so that $\mathbf{y}_t^1$ was always the target 1 measurement and similarly for target 2. Measurements $3 \ldots n_t$ were clutter. If a target was undetected then its measurement was given a null value to maintain ordering. This measurement ordering is used to help us assess whether the algorithm is working well, the PMHT, of course, thinks that the order is arbitrary.

Figure 3.4 shows the association weights for the two tracks over time. In Fig. 3.4a the whole scenario is shown whereas Fig. 3.4b zooms in on two time segments where the targets cross over, centred at time 100 and time 200. The weight for measurement 1 is shown as a box and the weight for measurement 2 as a circle. Due to the ordering these are always the measurements due to the two targets, if detected. The sum of the weights across all other measurements is shown as a cross. If perfect association had been achieved then track 1 would have boxes at unity and circles and crosses at zero at all times. Similarly track 2 should have circles at unity with boxes and crosses at zero. The two tracks consistently give high association probability to the true measurement for their corresponding target and the clutter weights associate most of the false alarms. The sum of the clutter weights fluctuates because this reflects the posterior number of measurements associated with clutter and the true number of false alarms is a Poisson random variable. Both tracks incorrectly associate false alarm measurements if they fall close enough to the track position: for example track 1 in Fig. 3.4a has a cross a little before frame 50 indicating a clutter measurement with association weight approximately 0.8 and two clutter measurements at frame 350 with weights close to unity leading to a cross at approximately 2. Each 100 frames the targets cross over and the weights shown in Fig. 3.4b show time segments centred on these crossings. When the targets are very close, the measurements are sometimes associated to the wrong track.

3.5.2 Numerous Targets

The simulation is now repeated using the full twenty target canonical multi-target scenario. Figures 3.5 and 3.6 show a representation of the measurement to track association weights. As before the tracks were initialized using the true target states and the average number of clutter measurements was $\Lambda^0 = 50$. In these plots the weight for the true measurement from the target used to initialize each track is shown as a square. The sum of the weights over the other 19 targets is shown as a dot and the sum over the clutter measurements is shown as a cross. Under perfect association, each track should have squares at unity and dots and crosses at zero. A track that has incorrectly associated clutter measurements will have crosses higher than zero. If a track swaps from one target to another then the squares move from unity to zero and the dots swap from zero to unity, for example track 1. If a track is seduced by clutter and diverges from the target then all of the weights will tend to zero after it diverges, for example track 8. With the full 20 targets the number of target crossings is much higher and some of these crossings occur as the targets manoeuvre whereas in the

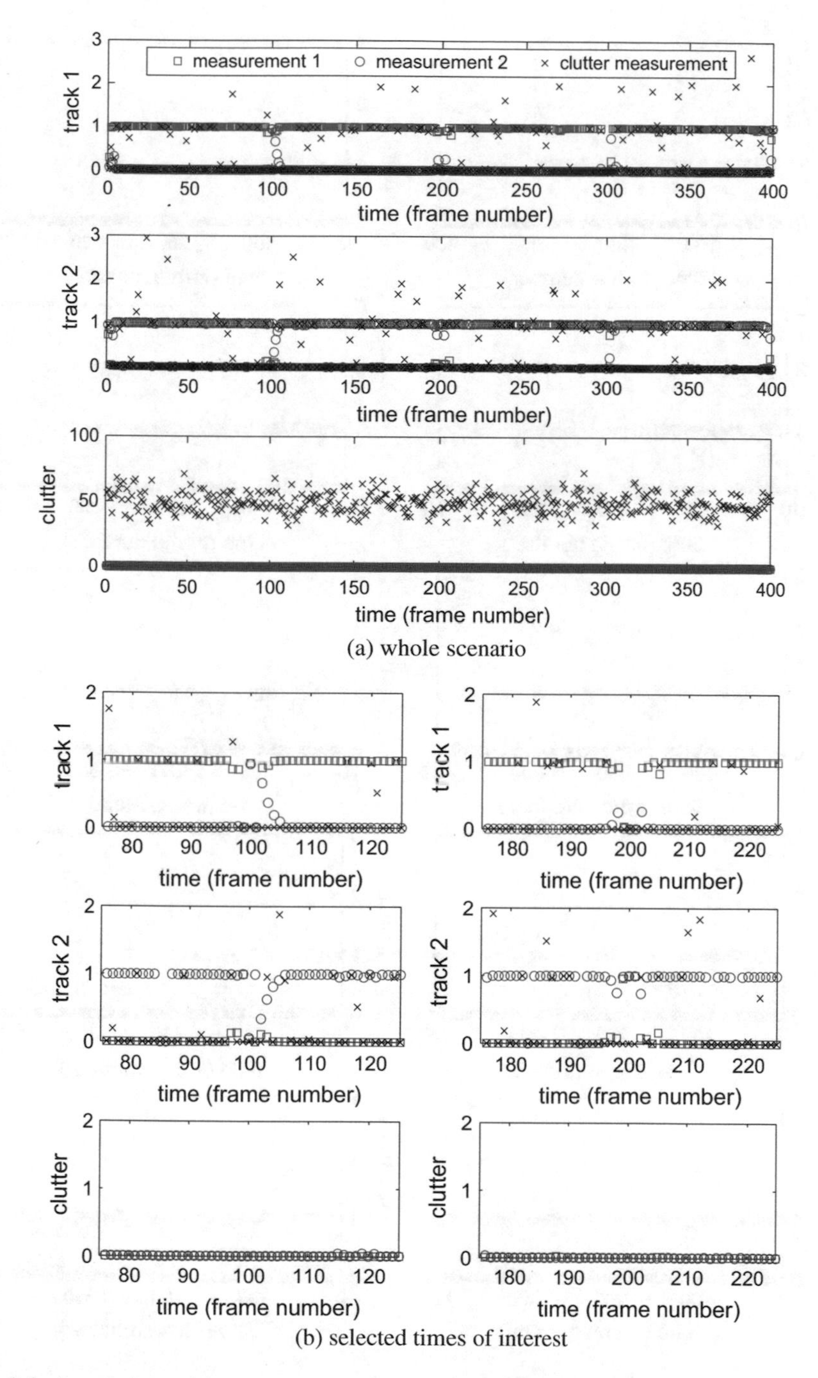

Fig. 3.4 Association weights

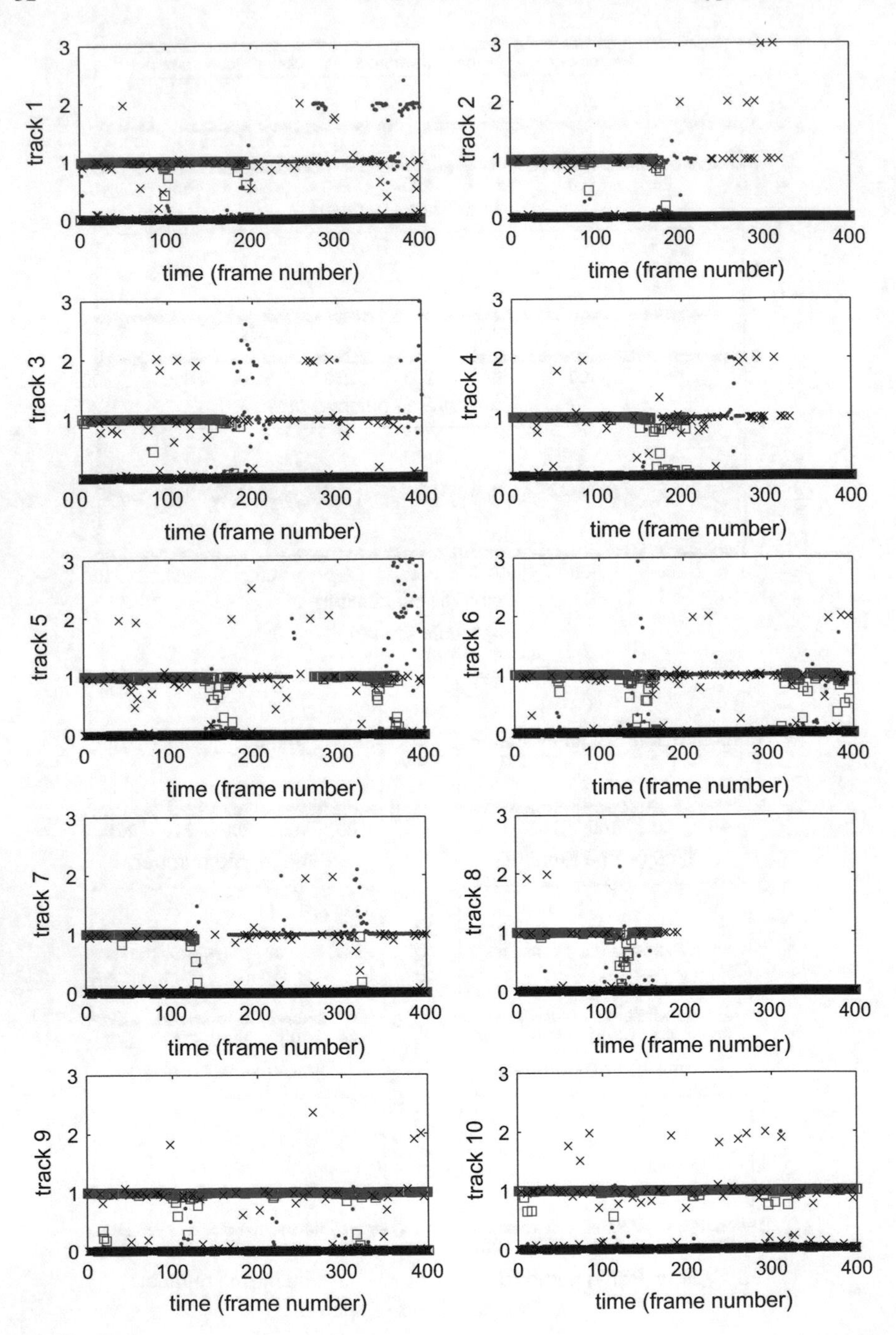

Fig. 3.5 Association weights, targets 1 . . . 10

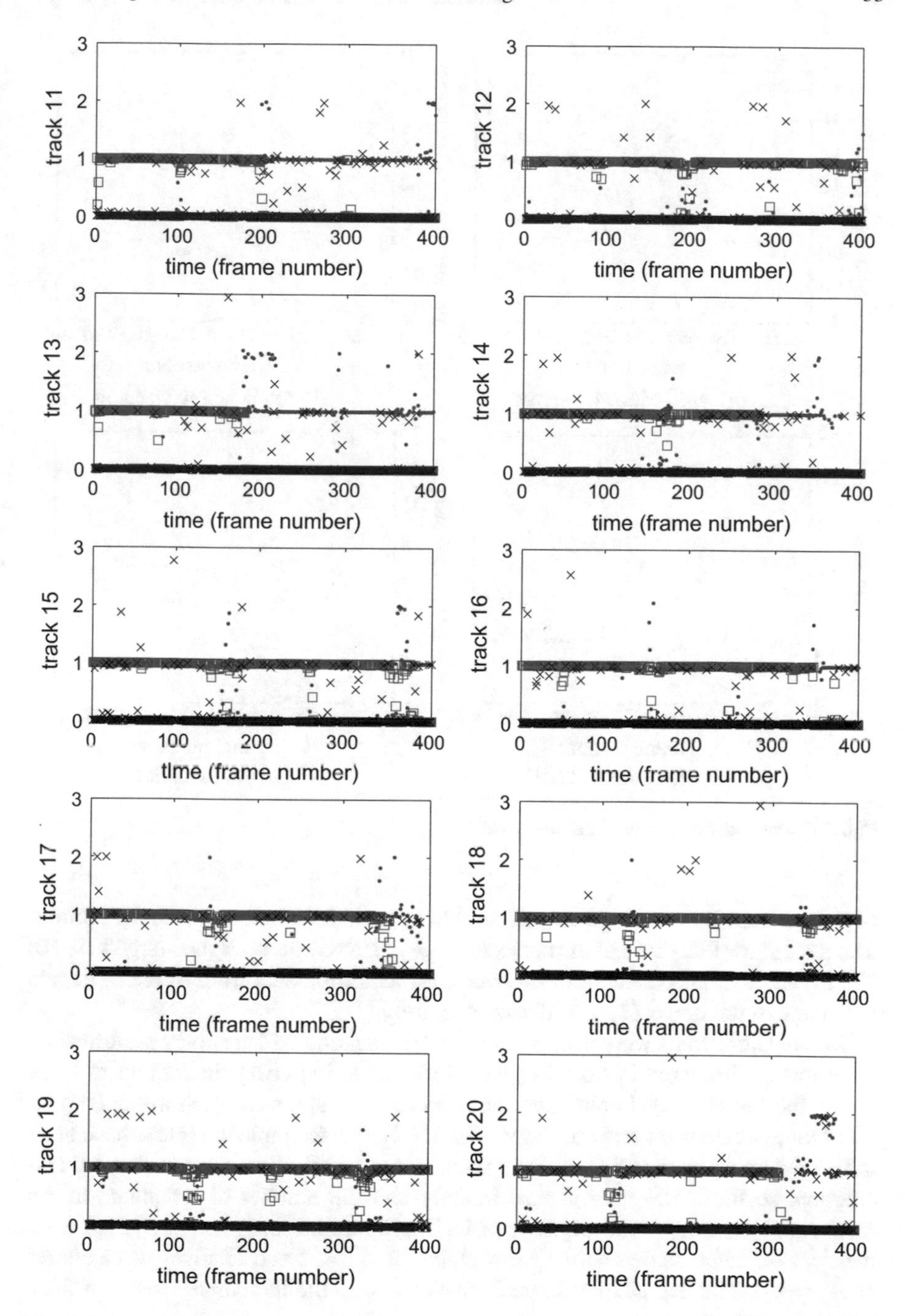

Fig. 3.6 Association weights, targets 11 . . . 20

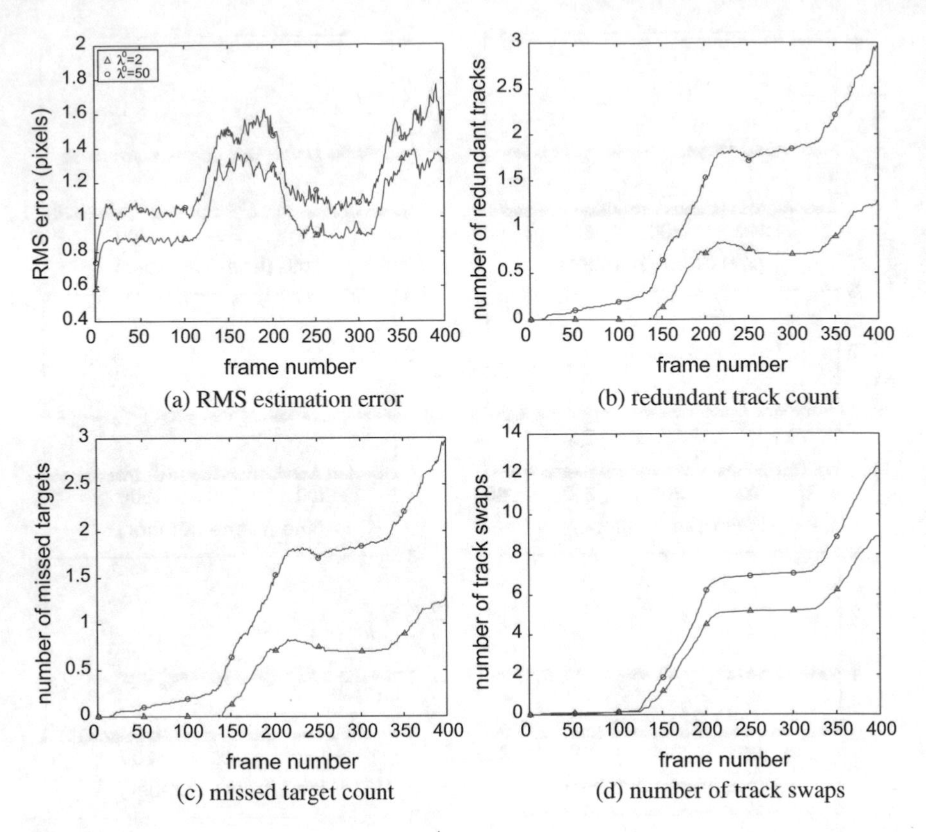

(a) RMS estimation error

(b) redundant track count

(c) missed target count

(d) number of track swaps

Fig. 3.7 Statistical measures of performance[4]

two target case the crossings only occur during periods of constant velocity motion. The plots show that some of the tracks manage to stay on their original targets (9, 10, 12, 15, 18, 19) but others swapped over onto measurements from different targets and a few of the tracks (2, 4, 8) diverged entirely.

Monte Carlo trials were used to calculate association and accuracy performance measures as discussed in Sect. 1.6; the RMS estimation error, the redundant track count, the missed target count and the number of swaps were each averaged over 100 Monte Carlo trials and are shown Fig. 3.7. In this example the tracks have been initialized using truth and there is no track management, that is, track removal and creation, so the RMS error is zero initially and the number of redundant tracks must equal the number of missed targets. The figure presents two curves for each metric, calculated using relatively low clutter at $\Lambda^0 = 2$ and relatively high clutter at $\Lambda^0 = 50$. The degradation in performance due to the increased density of false alarms is apparent.

[4]*GenerateMOPs*
The statistics in Fig. 3.7 can be reproduced using the H-PMHT toolbox function Generate-

The statistical results are consistent with the single realisation plots in Figs. 3.5 and 3.6. There are two periods that are consistently well tracked from around $t = 0 \ldots 100$ and $t = 200 \ldots 300$; due to the periodic nature of the target paths the second of these is where the scenario retraces the first 100 frames. The other two time periods, corresponding to the second half of the scenario pattern around $t = 100 \ldots 200$ and $t = 300 \ldots 400$ are not tracked well and there is a steady increase in both target swaps and divergent tracks that lead to redundant tracks and missed targets.

3.6 Problems with PMHT

The beneficial feature of the PMHT approach is that the cost scales modestly with the number of targets. For the case where every target is detected and the targets are sufficiently close together, the number of weights is $Mn_t = M^2$ ignoring clutter. This is considerably better than the $M!$ permutations that need to be enumerated in an explicit multi-target association approach, such as Joint Probabilistic Data Association [1] or hypothesis orientated Multi-Hypothesis Tracking [2]. However, the method is not without its problems. The most important of these are now discussed and the approaches that address them are described.

3.6.1 Model Order Estimation

The core PMHT as derived in [12] and described in this chapter assumes a fixed and known number of targets M. In practice the number of targets varies with time and is not known. This is essentially the same situation as many other methods, such as Joint Probabilistic Data Association. The approaches used for existing tracking methods can be applied to PMHT. The basic principle is to over model the data by introducing a large collection of candidate tracks and then to discard poor candidates based on a track quality metric. References [4, 7] describes several different metrics that are suitable for PMHT. In the context of tracking, model order estimation is often referred to as track maintenance, which is the task of starting tracks on new targets and terminating stale tracks. Methods for track maintenance with Histogram PMHT will be discussed in Chap. 5 and a superior approach based on a Poisson prior [6] will be derived in Chap. 6.

MOPs. It requires a set of tracks and truth information as inputs. Figure 3.7 used PMHTTracker to make the tracks and GenerateCanon to provide the truth.

3.6.2 Adaptivity

The association weights in PMHT are calculated conditioned on an assumed target state. This means that they are a likelihood ratio using the measurement covariance matrix, Σ. Other tracking algorithms usually make use of the innovation covariance matrix for association probabilities, which combines the measurement variance with the error variance of the existing state estimate. This difference allows the algorithms to adaptively increase the distance at which measurements are associated if the track has been struggling to find measurements, such as when the target manoeuvres. A method to account for the state covariance was introduced by Ruan and Willet [11] based on the turbo coding used in communications.

3.6.3 Optimism

If a particular track associates more than one measurement then the PMHT treats this as an increase in the available information and filters using the centroid of the two measurements with a tighter measurement covariance matrix. If one believes that the target can only produce one measurement, then this situation leads to ambiguity: only one of the two measurements can be the correct one, so the filter should distrust the data, not increase its confidence. In contrast, Probabilistic Data Association (as an example) increases the covariance by a term dependent on the scatter of multiple associated measurements. This is a philosophical difference: the PMHT has assumed that the target can produce more than one measurement and it proceeds accordingly. The appropriateness of that assumption is application dependent.

3.7 Summary

This chapter has reviewed the application of Expectation Maximisation to point measurement data association. Under the assumption that the source of each measurement is an independent realisation of a random process, the PMHT algorithm associates the measurements to components of a mixture model and is able to estimate the mixture parameters independently for each component. An important consequence of the measurement model is that the association probabilities are independent between measurements so the computation complexity of the algorithm is linear in the number of measurements. Algorithms built on the more standard one measurement per target assumption have to deal with a joint assignment space that grows as the number of permutations of measurements and targets. There are numerous papers in the literature that extend this core PMHT approach to wider classes of problems, such as [3, 4, 8, 13].

References

1. Bar-Shalom, Y., Willett, P.K., Tian, X.: Tracking and Data Fusion: A Handbook of Algorithms. YBS, Storrs (2011)
2. Blackman, S.S., Popoli, R.: Design and Analysis of Modern Tracking Systems. Artech House, Norwood (1999)
3. Crouse, D.F., Guerriero, M., Willett, P.: A critical look at the PMHT. ISIF J. Adv. Inf. Fusion **4**, 93–116 (2009)
4. Davey, S.J.: Extensions to the probabilistic multi-hypothesis tracker for improved data association. School of Electrical and Electronic Engineering, the University of Adelaide (2003)
5. Davey, S.J.: Histogram PMHT with particles. In: Proceedings of the 14th International Conference on Information Fusion, Chicago, USA (2011)
6. Davey, S.J.: Probabilistic multihypothesis tracker with an evolving poisson prior. IEEE Trans. Aerosp. Electron. Syst. **51**, 747–759 (2015)
7. Davey, S.J., Gray, D.A.: Integrated track maintenance for the PMHT via the hysteresis model. IEEE Trans. Aerosp. Electron. Syst. **43**(1), 93–111 (2007)
8. Davey, S.J., Gray, D.A., Streit, R.L.: Tracking, association and classification - a combined PMHT approach. Digit. Signal Process. **12**, 372–382 (2002)
9. Luginbuhl, T.E.: Estimation of general discrete-time FM processes. Ph.D. thesis, University of Connecticut (1999)
10. Peterson, K.B., Pederson, M.S.: The Matrix Cookbook (2012). http://matrixcookbook.com
11. Ruan, Y., Willett, P.: An improved PMHT using an idea from coding. In: Proceedings of the 2001 Aerospace Conference, Big Sky, Montana, USA (2001)
12. Streit, R.L., Luginbuhl, T.E.: Probabilistic multi-hypothesis tracking. Technical report 10428, NUWC, Newport, Rhode Island, USA (1995)
13. Willett, P., Ruan, Y., Streit, R.L.: PMHT: problems and some solutions. IEEE Trans. Aerosp. Electron. Syst. **38**, 738–754 (2002)

Chapter 4
Histogram Probabilistic Multi-hypothesis Tracking

The previous chapter revised the PMHT data association method for point measurement tracking. The starting premise is that the sensor provides a collection of points in a vector space. For many sensors, this is not the intrinsic measurement created by the sensor. Instead, the sensor provides an image, which is a collection of pixels each containing a scalar value. As discussed in the introduction, the use of the terms image and pixels does not necessarily imply a two-dimensional data product. For the moment, we will consider images with scalar values in each pixel. A key stage in the development of Histogram PMHT is to treat this integer as a count of a number of point measurements that have arrived inside the spatial extent of that pixel. In other words, the image is interpreted as a histogram. As the name suggests, Histogram-PMHT results from applying PMHT data association to a histogram measurement model. A histogram is created by sorting measurements into bins and then counting the number inside each bin. The histogram itself consists of the measurement count for each bin, it does not retain information about the precise locations of the measurements. Figure 4.1 shows the Gaussian mixture example from Sect. 3.1 and a two-dimensional histogram formed from these measurements. As the figure shows, this histogram is really an image. This is the key to the H-PMHT measurement model, it treats imaging sensors as though they collect histograms.

In the point measurement problem, the measurement model assumes that the location of every measurement $\mathbf{y}_t^r$ is precisely known. In the histogram case, the measurement data is a count of the number of measurements in each pixel. For pixel, i this is denoted by n_t^i, with $i = 1 \dots I$ where I is the number of pixels in the frame. Following the usual convention, the collection of counts across all of the pixels (i.e. the image) is written as $\mathbf{N}_t \equiv n_t^{1:I}$ and the collection of images across time is written as $\mathbb{N} \equiv \mathbf{N}_{1:T}$. Recall from the previous chapter that for the point measurement case n_t was the total number of point measurements in frame t. For histogram (image) measurements, it will also be useful to refer to the total count of point measurements

S. J. Davey and H. Gaetjens, *Track-Before-Detect Using Expectation Maximisation*, Signals and Communication Technology,
https://doi.org/10.1007/978-981-10-7593-3_4

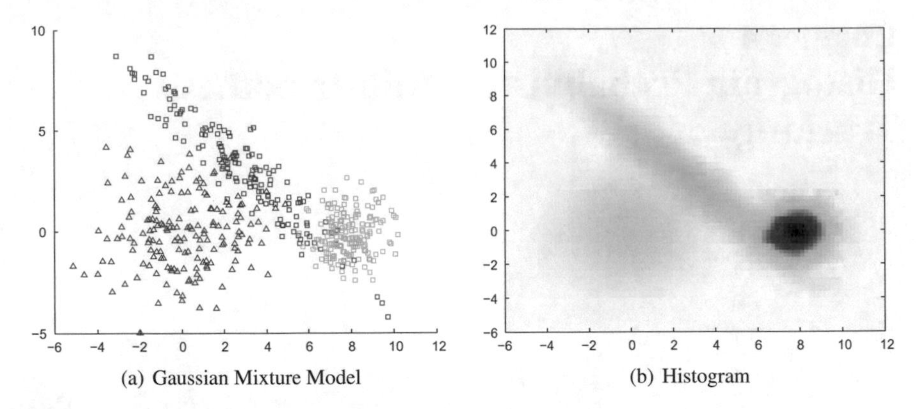

(a) Gaussian Mixture Model (b) Histogram

Fig. 4.1 Histogram measurement model

across the whole image so the notation n_t retains its meaning. For a histogram, the obvious relationship is that $n_t = \sum_{i=1}^{I} n_t^i$. Note the difference between n_t, which is the total measurement count at frame t, and $\mathbf{N}_t$, which is a vector of the measurement counts in each pixel. An important feature of point measurement PMHT and core H-PMHT is that the total number of measurements is assumed to be known: n_t is not considered a random variable. This will change later in Chap. 6.

4.1 Histogram Data Association

In the PMHT point measurement case the missing data is the source of each measurement k_t^r. The histogram measurement model adds a second layer of missing data: the precise location of each point measurement within the image pixel. Let W^i denote the physical extent of pixel i. In some applications, the pixels can change from frame to frame but this complication is largely a matter of bookkeeping: the reader can insert the additional subscripts if she pleases. We continue using the notation from Chap. 3 and key terms are provided in Table 4.1; a full list of notation is available at the front of the book.

The EM auxiliary function is again the expectation of the logarithm of the complete data likelihood over the conditional assignment probabilities. The complete data is the same as the point measurement problem augmented with the pixel counts, which are actually the measurements now, $\{\mathbb{X}, \Pi, \mathbb{K}, \mathbb{Y}, \mathbb{N}\}$. The missing data is the true measurement source assignments and the precise point measurement locations $\{\mathbb{K}, \mathbb{Y}\}$.

There are two potential ways to formulate the complete data likelihood depending on how one chooses to condition the random variables. An intuitive perspective could be to treat the counts as a consequence of the point measurements. That is, the complete data likelihood could be written as

Table 4.1 Notation refresher

n_t^i, n_t	Measurement counts for pixel i in frame t, and the total across pixels
$\mathbf{N}_t, \mathbb{N}$	Sets of pixel counts in frame t, and across frames
$\mathbf{y}_t^r$	The rth point measurement in frame t
$\mathbf{Y}_t, \mathbb{Y}$	Sets of point measurements in frame t, and across frames
k_t^i	The true source of measurement $\mathbf{y}_t^i$
$\mathbf{K}_t, \mathbb{K}$	Sets of measurement sources at frame t, and across frames
$\mathbf{x}_t^m$	The kinematic state of target m at frame t
$\mathbf{X}_t, \mathbb{X}$	Sets of target states at frame t, and across frames
π_t^m	The mixing proportion for target m
Π_t, Π	Sets of mixing proportions at frame t, and across frames

$$
\begin{aligned}
p_{\mathsf{comp}}\left(\mathbb{X}, \mathbb{K}, \mathbb{Y}, \mathbb{N}; \Pi\right) &= p\left(\mathbb{X}\right) p\left(\mathbb{K}; \Pi\right) p\left(\mathbb{Y}|\mathbb{X}, \mathbb{K}\right) p\left(\mathbb{N}|\mathbb{X}, \mathbb{K}, \mathbb{Y}\right), \\
&= p\left(\mathbb{X}\right) p\left(\mathbb{K}; \Pi\right) p\left(\mathbb{Y}|\mathbb{X}, \mathbb{K}\right) p\left(\mathbb{N}|\mathbb{Y}\right),
\end{aligned} \tag{4.1}
$$

where the counts are conditionally independent of the states and the assignments given the measurement locations since the counts are a deterministic function of these locations. The problem with this arrangement is that the counts are redundant given the point measurements: the point measurements themselves give complete information about measurement locations and the counts give imprecise information about the same measurement locations. The probability $p\left(\mathbb{N}|\mathbb{Y}\right)$ is essentially an indicator function that enforces the constraint that the number of point measurements inside each pixel is known.

The alternative formulation is to condition the point measurements on the counts. From a generative point of view, the sensor samples the number of point measurements in each cell and then the location of each point measurement is sampled from the conditional distribution $p(\mathbf{y}, k|\mathbf{X}, \mathbf{y} \in W^i)$ where the conditioning $\mathbf{y} \in W^i$ represents the limitation that the measurement must fall inside the spatial region W^i.

$$
p_{\mathsf{comp}}\left(\mathbb{X}, \mathbb{K}, \mathbb{Y}, \mathbb{N}; \Pi\right) = p\left(\mathbb{X}\right) p\left(\mathbb{N}|\mathbb{X}\right) p\left(\mathbb{K}|\mathbb{N}, \mathbb{X}; \Pi\right) p\left(\mathbb{Y}|\mathbb{K}, \mathbb{N}, \mathbb{X}\right), \tag{4.2}
$$

This form does not suffer from the degeneracy issues described for the first version of conditioning. Note that the count information is ambiguous in two senses: first, it is imprecise about the location of point measurements; and second, it does not prescribe exactly which point measurements fall in each cell. As a result of this second ambiguity, the density $p\left(\mathbb{Y}|\mathbb{K}, \mathbb{N}, \mathbb{X}\right)$ has a permutation symmetry, that is if one were to swap the locations of $\mathbf{y}_t^r$ and $\mathbf{y}_t^s$ then the probability would be the same. For the H-PMHT derivation, we will use this second form of conditioning.

The labelling of measurements and assignments is arbitrary, so we are free to adopt a different set of arbitrary labels. The measurement density that will be required moving forward is one conditioned on knowing the number of measurements in each pixel so we now adopt a set of indices that use this information. The measurements that fall in pixel i in frame t are denoted by $\mathbf{y}_t^{i,r}$ and the assignments of these are similarly double indexed $k_t^{i,r}$ with $r = 1 \ldots n_t^i$. Again, we emphasise that this indexing implicitly defines how many measurements are inside the pixel but this does not introduce new information because the double-indexed measurements will always be used conditioned on the counts $\mathbb{N}$. As before, $\mathbf{Y}_t$ and $\mathbf{K}_t$ denote single frame collections of measurements and assignments and $\mathbb{Y}$ and $\mathbb{K}$ denote batch collections.

4.1.1 Expectation Step

As for the point measurement case, the aim is to estimate the target states $\mathbb{X}$ and the mixing proportions Π using EM. The assignment of measurements to targets $\mathbb{K}$ was the missing data for PMHT. For histogram measurements, the precise measurement locations $\mathbb{Y}$ is an additional piece of missing data. The EM auxiliary function is therefore

$$\begin{aligned}\mathcal{Q}\left(\mathbb{X}, \Pi | \hat{\mathbb{X}}, \hat{\Pi}\right) &= E_{\mathbb{K},\mathbb{Y}|\hat{\mathbb{X}},\hat{\Pi}}\left[\log\left\{p_{\text{comp}}\left(\mathbb{X}, \mathbb{K}, \mathbb{Y}, \mathbb{N}; \Pi\right)\right\}\right] \\ &= \sum_{\mathbb{K}} \int \log\left\{p_{\text{comp}}\left(\mathbb{X}, \mathbb{K}, \mathbb{Y}, \mathbb{N}; \Pi\right)\right\} p_{\text{miss}}\left(\mathbb{K}, \mathbb{Y} | \mathbb{N}, \hat{\mathbb{X}}, \hat{\Pi}\right) d\mathbb{Y}.\end{aligned} \tag{4.3}$$

4.1.1.1 Complete Data Likelihood

The complete data likelihood is defined by (4.2)

$$p_{\text{comp}}\left(\mathbb{X}, \mathbb{K}, \mathbb{Y}, \mathbb{N}; \Pi\right) = p\left(\mathbb{X}\right) p\left(\mathbb{N}|\mathbb{X}; \Pi\right) p\left(\mathbb{K}|\mathbb{N}, \mathbb{X}; \Pi\right) p\left(\mathbb{Y}|\mathbb{K}, \mathbb{N}, \mathbb{X}\right).$$

The first term is the state likelihood, which is (3.33), exactly the same as in the point measurement case

$$p\left(\mathbb{X}\right) = \prod_{m=1}^{M}\left[p\left(\mathbf{x}_0^m\right)\prod_{t=1}^{T} p\left(\mathbf{x}_t^m | \mathbf{x}_{t-1}^m\right)\right].$$

The second term in the complete data likelihood is the probability of the counts given the states. As mentioned in the introduction, a characteristic of core H-PMHT is that it implicitly assumes that the total number of measurements is known, so really the probability required is $p(\mathbb{N}|\mathbb{X}; \Pi, n_{1:T})$. The measurements are conditionally independent over time, so

$$p\left(\mathbb{N}|\mathbb{X}, n_{1:T}\right) = \prod_{t=1}^{T} p\left(\mathbf{N}_t | \mathbf{X}_t, n_t\right). \tag{4.4}$$

The point measurements are independent identically distributed, so the probability of the count vector has a multinomial distribution where the parameters are the probabilities that a particular point measurement falls in each pixel. That is

$$p\left(\mathbf{N}_t|\mathbf{X}_t; \Pi_t, n_t\right) = \frac{n_t!}{\prod_{i=1}^{I} n_t^i!} \prod_{i=1}^{I} \left(G_t^i\right)^{n_t^i} = C(\mathbf{N}_t) \prod_{i=1}^{I} \left(G_t^i\right)^{n_t^i}, \tag{4.5}$$

where G_t^i is defined as the probability that a particular measurement falls in pixel i at frame t,

$$G_t^i = p\left(\mathbf{y} \in W^i|\mathbf{X}_t; \Pi_t\right), \tag{4.6}$$

and is referred to as the cell probability. The term $C(\mathbf{N}_t)$ is sometimes referred to as the multinomial coefficient. The multinomial distribution naturally arises when one performs a number of independent realisations of the same random trial with a discrete set of outcomes and then counts the number of each outcome observed. A simple example is rolling a die. If the die is rolled n_t times then the distribution of the number of times each face of the die is rolled is the multinomial distribution. The parameters are the six single-roll probabilities of rolling each particular number. Note that the cell probability G_t^i does not specify which component has caused the measurement or the precise measurement location but the existing measurement model only defines the probability density at a point for each component separately. Using the law of total probability, the required cell probability can be expressed by marginalising over the measurement location and source, that is

$$\begin{aligned}
G_t^i &= \sum_{m=0}^{M} \int_{W^i} p\left(\mathbf{y}, k = m|\mathbf{X}_t; \Pi_t\right) \mathrm{d}\mathbf{y}, \\
&= \sum_{m=0}^{M} \int_{W^i} p\left(k = m; \Pi_t\right) p\left(\mathbf{y}|\mathbf{X}_t, k = m\right) \mathrm{d}\mathbf{y}, \\
&= \pi_t^0 \int_{W^i} g^0\left(\mathbf{y}\right) \mathrm{d}\mathbf{y} + \sum_{m=1}^{M} \pi_t^m \int_{W^i} g^m\left(\mathbf{y}|\mathbf{x}_t^m\right) \mathrm{d}\mathbf{y}, \qquad (4.7) \\
&\equiv \sum_{m=0}^{M} \pi_t^m G_t^{m,i}, \qquad (4.8)
\end{aligned}$$

where $G_t^{m,i}$ is the probability that a measurement known to come from component m falls in pixel i

$$G_t^{m,i} = p\left(\mathbf{y} \in W^i|\mathbf{X}_t, k = m\right), \tag{4.9}$$

and is referred to as the cell probability for component m. The measurement function $g^0\left(\mathbf{y}\right)$ is the probability density of clutter measurements.

The third term in the complete data likelihood (4.2) is the probability of the measurement sources given knowledge of which pixel each has landed in, $p\left(\mathbb{K}|\mathbb{N}, \mathbb{X}; \Pi\right)$. Once again the measurements are conditionally independent over time, so

$$p\left(\mathbb{K}|\mathbb{N}, \mathbb{X}; \Pi\right) = \prod_{t=1}^{T} p\left(\mathbf{K}_t|\mathbf{N}_t, \mathbf{X}_t; \Pi_t\right). \tag{4.10}$$

The assignments are independent identically distributed so their probability is a product over individual assignments

$$p\left(\mathbf{K}_t|\mathbf{N}_t, \mathbf{X}_t; \Pi_t\right) = \prod_{i=1}^{I} \prod_{r=1}^{n_t^i} p\left(k_t^{i,r}|\mathbf{y} \in W^i, \mathbf{X}_t; \Pi_t\right). \tag{4.11}$$

Using Bayes rule, the individual assignment conditional probability can be written in terms of the assignment prior Π_t and the cell probabilities

$$\begin{aligned} p\left(k_t^{i,r}|\mathbf{y} \in W^i, \mathbf{X}_t; \Pi_t\right) &= \frac{p\left(k_t^{i,r}, \mathbf{y} \in W^i|\mathbf{X}_t; \Pi_t\right)}{p\left(\mathbf{y} \in W^i|\mathbf{X}_t; \Pi_t\right)} \\ &= \frac{\pi_t^{k_t^{i,r}} G_t^{k_t^{i,r},i}}{G_t^i}. \end{aligned} \tag{4.12}$$

The final term in the complete data likelihood (4.2) is the probability of drawing the precise measurement locations $\mathbb{Y}$ conditioned on knowledge of which pixel each measurement belongs to and which component generated each measurement, $p\left(\mathbb{Y}|\mathbb{K}, \mathbb{N}, \mathbb{X}\right)$. Using the same independence assumptions as for the assignments this can be written as a product over individual measurements

$$p\left(\mathbb{Y}|\mathbb{K}, \mathbb{N}, \mathbb{X}\right) = \prod_{t=1}^{T} \prod_{i=1}^{I} \prod_{r=1}^{n_t^i} p\left(\mathbf{y}_t^{i,r}|\mathbf{y}_t^{i,r} \in W^i, k_t^{i,r}, \mathbf{X}_t\right). \tag{4.13}$$

Bayes rule again allows us to express this in terms of the individual measurement functions and the cell probabilities

$$\begin{aligned} p\left(\mathbf{y}_t^{i,r}|\mathbf{y}_t^{i,r} \in W^i, k_t^{i,r}, \mathbf{X}_t\right) &= \frac{p\left(\mathbf{y}_t^{i,r} \in W^i, \mathbf{y}_t^{i,r}|k_t^{i,r}, \mathbf{X}_t\right)}{p\left(\mathbf{y}_t^{i,r} \in W^i|k_t^{i,r}, \mathbf{X}_t\right)} \\ &= \frac{g^{k_t^{i,r}}\left(\mathbf{y}_t^{i,r}|\mathbf{x}_t^{k_t^{i,r}}\right)}{G_t^{k_t^{i,r},i}}, \end{aligned} \tag{4.14}$$

where for notational convenience we have defined $g^0\left(\mathbf{y}|\mathbf{x}_t^0\right) = g^0\left(\mathbf{y}\right)$.

Putting all the pieces back into the complete data likelihood we have

$$p_{\text{comp}}(\mathbb{X},\mathbb{K},\mathbb{Y},\mathbb{N};\Pi)=\prod_{m=1}^{M}\left[p\left(\mathbf{x}_0^m\right)\prod_{t=1}^{T}p\left(\mathbf{x}_t^m|\mathbf{x}_{t-1}^m\right)\right]$$
$$\times\prod_{t=1}^{T}\left[\frac{n_t!}{\prod_{i=1}^{I}n_t^i!}\prod_{i=1}^{I}\left(\left(G_t^i\right)^{n_t^i}\prod_{r=1}^{n_t^i}\frac{\pi_t^{k_t^{i,r}}G_t^{k_t^{i,r},i}}{G_t^i}\frac{g^{k_t^{i,r}}\left(\mathbf{y}_t^{i,r}|\mathbf{x}_t^{k_t^{i,r}}\right)}{G_t^{k_t^{i,r},i}}\right)\right]$$
$$=C(\mathbb{N})\prod_{m=1}^{M}\left[p\left(\mathbf{x}_0^m\right)\prod_{t=1}^{T}p\left(\mathbf{x}_t^m|\mathbf{x}_{t-1}^m\right)\right]\prod_{t=1}^{T}\prod_{i=1}^{I}\prod_{r=1}^{n_t^i}\pi_t^{k_t^{i,r}}g^{k_t^{i,r}}\left(\mathbf{y}_t^{i,r}|\mathbf{x}_t^{k_t^{i,r}}\right),\quad(4.15)$$

where $C(\mathbb{N})$ is a product of multinomial coefficients and is independent of the unknowns. It is a scaling factor that can be ignored. This means that the logarithm of the complete data likelihood is

$$\log\left\{p_{\text{comp}}(\mathbb{X},\mathbb{K},\mathbb{Y},\mathbb{N};\Pi)\right\}$$
$$=\sum_{m=1}^{M}\left[\log\left\{p\left(\mathbf{x}_0^m\right)\right\}+\sum_{t=1}^{T}\log\left\{p\left(\mathbf{x}_t^m|\mathbf{x}_{t-1}^m\right)\right\}\right]+\sum_{t=1}^{T}\sum_{i=1}^{I}\sum_{r=1}^{n_t^i}\log\left\{\pi_t^{k_t^{i,r}}\right\}$$
$$+\sum_{t=1}^{T}\sum_{i=1}^{I}\sum_{r=1}^{n_t^i}\log\left\{g^{k_t^{i,r}}\left(\mathbf{y}_t^{i,r}|\mathbf{x}_t^{k_t^{i,r}}\right)\right\},\quad(4.16)$$

plus a constant $\log\{C(\mathbb{N})\}$.

That completes the first half of the auxiliary function. Go have a cup of tea and then we will move on to the second half: the missing data likelihood.

4.1.1.2 Missing Data Likelihood

The conditional probability of the missing data is again simplified using Bayes rule

$$p_{\text{miss}}\left(\mathbb{K},\mathbb{Y}|\mathbb{N},\hat{\mathbb{X}};\hat{\Pi}\right)=\frac{p\left(\hat{\mathbb{X}},\mathbb{K},\mathbb{Y},\mathbb{N};\hat{\Pi}\right)}{p\left(\hat{\mathbb{X}},\mathbb{N};\hat{\Pi}\right)}.\quad(4.17)$$

The numerator is the complete data likelihood evaluated at the estimate $\hat{\mathbb{X}}$ and the denominator is the joint probability of the observed counts and the state estimates. This probability is made up of two terms we already know

$$p\left(\hat{\mathbb{X}},\mathbb{N};\hat{\Pi}\right)=p\left(\hat{\mathbb{X}}\right)p\left(\mathbb{N}|\hat{\mathbb{X}};\hat{\Pi}\right)\quad(4.18)$$

where $p\left(\hat{\mathbb{X}}\right)$ is given by (3.33) and $p\left(\mathbb{N}|\hat{\mathbb{X}};\hat{\Pi}\right)$ is given by (4.4) and (4.5). These combine to give

$$p\left(\hat{\mathbb{X}},\mathbb{N};\hat{\Pi}\right)=C(\mathbb{N})p(\hat{\mathbb{X}})\prod_{t=1}^{T}\prod_{i=1}^{I}\left(G_t^i\right)^{n_t^i}. \tag{4.19}$$

Substituting (4.19) and (4.15) into (4.17) gives

$$\begin{aligned}p_{\mathsf{miss}}\left(\mathbb{K},\mathbb{Y}|\mathbb{N},\hat{\mathbb{X}};\hat{\Pi}\right)&=\frac{C(\mathbb{N})p(\hat{\mathbb{X}})\prod_{t=1}^{T}\prod_{i=1}^{I}\prod_{r=1}^{n_t^i}\pi_t^{k_t^{i,r}}g^{k_t^{i,r}}\left(\mathbf{y}_t^{i,r}|\mathbf{x}_t^{k_t^{i,r}}\right)}{C(\mathbb{N})p(\hat{\mathbb{X}})\prod_{t=1}^{T}\prod_{i=1}^{I}\left(G_t^i\right)^{n_t^i}},\\&=\prod_{t=1}^{T}\prod_{i=1}^{I}\prod_{r=1}^{n_t^i}\left[\frac{\hat{\pi}_t^{k_t^{i,r}}g^{k_t^{i,r}}\left(\mathbf{y}_t^{i,r}|\hat{\mathbf{x}}_t^{k_t^{i,r}}\right)}{G_t^i}\right].\end{aligned} \tag{4.20}$$

This combines with (4.16) to give the EM auxiliary function.

4.1.1.3 Auxiliary Function

The auxiliary function is the expectation of (4.16) over the conditional likelihood of the missing data. Based on the form of (4.16), this can be expressed as

$$\begin{aligned}\mathcal{Q}\left(\mathbb{X},\Pi|\hat{\mathbb{X}},\hat{\Pi}\right)&=E_{\mathbb{K},\mathbb{Y}|\hat{\mathbb{X}},\hat{\Pi}}\Big[\log\left\{p_{\mathsf{comp}}\left(\mathbb{X},\mathbb{K},\mathbb{Y},\mathbb{N};\Pi\right)\right\}\Big]\\&=E_{\mathbb{K},\mathbb{Y}|\hat{\mathbb{X}},\hat{\Pi}}\Big[\log\{p(\mathbb{X})\}\Big]+E_{\mathbb{K},\mathbb{Y}|\hat{\mathbb{X}},\hat{\Pi}}\left[\sum_{t=1}^{T}\sum_{i=1}^{I}\sum_{r=1}^{n_t^i}\log\left\{\pi_t^{k_t^{i,r}}\right\}\right]\\&\quad+E_{\mathbb{K},\mathbb{Y}|\hat{\mathbb{X}},\hat{\Pi}}\left[\sum_{t=1}^{T}\sum_{i=1}^{I}\sum_{r=1}^{n_t^i}\log\left\{g^{k_t^{i,r}}\left(\mathbf{y}_t^{i,r}|\mathbf{x}_t^{k_t^{i,r}}\right)\right\}\right].\end{aligned} \tag{4.21}$$

The first term in (4.21) above depends only on the target states, not the missing data, so it is its own expectation, as in the point measurement case. The second term in the auxiliary function (4.21) is a function of the priors Π and the assignments $\mathbb{K}$. As in the point measurement case, it is the only term dependent on Π and leads to the auxiliary sub-function $\mathcal{Q}_\pi$, which is given by

$$\mathcal{Q}_\pi=\sum_{\mathbb{K}}\int\left(\sum_{t=1}^{T}\sum_{i=1}^{I}\sum_{r=1}^{n_t^i}\log\left\{\pi_t^{k_t^{i,r}}\right\}\right)p_{\mathsf{miss}}\left(\mathbb{K},\mathbb{Y}|\mathbb{N},\hat{\mathbb{X}};\hat{\Pi}\right)\mathrm{d}\mathbb{Y}$$

$$= \sum_{\mathbb{K}} \left(\sum_{t=1}^{T} \sum_{i=1}^{I} \sum_{r=1}^{n_t^i} \log \left\{ \pi_t^{k_t^{i,r}} \right\} \right) \int p_{\text{miss}} \left(\mathbb{K}, \mathbb{Y} | \mathbb{N}, \hat{\mathbb{X}}, \hat{\Pi} \right) \mathrm{d}\mathbb{Y}. \tag{4.22}$$

The integral above can be simplified as

$$\int p_{\text{miss}} \left(\mathbb{K}, \mathbb{Y} | \mathbb{N}, \hat{\mathbb{X}}, \hat{\Pi} \right) \mathrm{d}\mathbb{Y} = \prod_{t=1}^{T} \prod_{i=1}^{I} \prod_{r=1}^{n_t^i} \left\{ \int_{W^i} \left[\frac{\hat{\pi}_t^{k_t^{i,r}} g^{k_t^{i,r}} \left(\mathbf{y}_t^{i,r} | \hat{\mathbf{x}}_t^{k_t^{i,r}} \right)}{G_t^i} \right] \mathrm{d}\mathbf{y}_t^{i,r} \right\}$$
$$= \prod_{t=1}^{T} \prod_{i=1}^{I} \prod_{r=1}^{n_t^i} \left[\frac{\hat{\pi}_t^{k_t^{i,r}} G_t^{k_t^{i,r},i}}{G_t^i} \right] \equiv \prod_{t=1}^{T} \prod_{i=1}^{I} \prod_{r=1}^{n_t^i} w_t^{i,k_t^{i,r}}. \tag{4.23}$$

In the point measurement case, the data association lead to target-measurement weights given by (3.56), repeated below

$$\tilde{w}_t^{m,r} = \frac{\hat{\pi}_t^{k_t^r} g^{k_t^r} \left(\mathbf{y}_t^r | \hat{\mathbf{x}}_t^{k_t^r} \right)}{\sum_{s=1}^{M} \hat{\pi}_t^s g^s \left(\mathbf{y}_t^r | \hat{\mathbf{x}}_t^s \right)}.$$

The point measurement weight is the relative likelihood of the measurement scaled by the target mixing proportion. In the histogram case, the point location is not precisely known, rather a spatial region corresponding to the pixel is known. The corresponding weight is analogous to the point measurement case with the point density replaced with its integral over the pixel

$$w_t^{i,m} = \frac{\hat{\pi}_t^m G_t^{m,i}}{\sum_{s=0}^{M} \hat{\pi}_t^s G_t^{s,i}}. \tag{4.24}$$

Using the weights, the assignment prior auxiliary sub-function becomes

$$\mathcal{Q}_\pi = \sum_{k_{1:T}} \left(\sum_{t=1}^{T} \sum_{i=1}^{I} \sum_{r=1}^{n_t^i} \log \left\{ \pi_t^{k_t^{i,r}} \right\} \right) \prod_{t=1}^{T} \prod_{i=1}^{I} \prod_{r=1}^{n_t^i} w_t^{i,k_t^{i,r}},$$
$$= \sum_{t=1}^{T} \sum_{i=1}^{I} \sum_{r=1}^{n_t^i} \sum_{m=0}^{M} \log \left\{ \pi_t^m \right\} w_t^{i,m} \sum_{k_{1:T} \setminus k_t^{i,r}} \prod_{t=1}^{T} \prod_{i=1}^{I} \prod_{r=1}^{n_t^i} w_t^{i,k_t^{i,r}}$$
$$= \sum_{t=1}^{T} \sum_{i=1}^{I} \sum_{r=1}^{n_t^i} \sum_{m=0}^{M} \log \left\{ \pi_t^m \right\} w_t^{i,m},$$

$$= \sum_{t=1}^{T} \sum_{i=1}^{I} \sum_{m=0}^{M} \log \left\{ \pi_t^m \right\} n_t^i w_t^{i,m}. \tag{4.25}$$

The second line recognises that each term in the sum depends on only one particular measurement and the others are marginalised out in the third line. The final line recognises that the weight for target m and measurement i, r is the same for every measurement inside pixel r since it is marginalised over the exact measurement location: the sum over r is n_t^i repetitions of the same term. The resulting function is exactly the same as the point measurement case and again is maximised using a Lagrangian as demonstrated in Sect. 3.1. The refined estimate is then

$$\hat{\pi}_t^m := \frac{\sum_{i=1}^{I} n_t^i w_t^{i,m}}{\sum_{s=0}^{M} \sum_{i=1}^{I} n_t^i w_t^{i,s}}. \tag{4.26}$$

The final term in (4.21) is a measurement term and can be simplified in a similar way but is a little more troublesome since the integrals are not so easily disposed of

$$E_{\mathbb{K},\mathbb{Y}|\hat{\mathbb{X}},\hat{\Pi}} \left[\sum_{t=1}^{T} \sum_{i=1}^{I} \sum_{r=1}^{n_t^i} \log \left\{ g^{k_t^{i,r}} \left(\mathbf{y}_t^{i,r} | \mathbf{x}_t^{k_t^{i,r}} \right) \right\} \right]$$

$$= \sum_{k_{1:T}} \int \left[\sum_{t=1}^{T} \sum_{i=1}^{I} \sum_{r=1}^{n_t^i} \log \left\{ g^m \left(\mathbf{y}_t^{i,r} | \mathbf{x}_t^m \right) \right\} \right] \prod_{t=1}^{T} \prod_{i=1}^{I} \prod_{r=1}^{n_t^i} \left[\frac{\hat{\pi}_t^{k_t^{i,r}} g^{k_t^{i,r}} \left(\mathbf{y}_t^{i,r} | \hat{\mathbf{x}}_t^{k_t^{i,r}} \right)}{G_t^i} \right] d\mathbb{Y}$$

$$= \left[\sum_{t=1}^{T} \sum_{i=1}^{I} \sum_{r=1}^{n_t^i} \sum_{m=0}^{M} \int_{W^i} \frac{\hat{\pi}_t^m}{G_t^i} \log \left\{ g^m \left(\mathbf{y}_t^{i,r} | \mathbf{x}_t^m \right) \right\} g^m \left(\mathbf{y}_t^{i,r} | \hat{\mathbf{x}}_t^m \right) d\mathbf{y}_t^{i,r} \right]$$

$$\times \sum_{k_{1:T} \backslash k_t^{i,r}} \int \prod_{t=1}^{T} \prod_{i=1}^{I} \prod_{r=1}^{n_t^i} \left[\frac{\hat{\pi}_t^{k_t^{i,r}} g^{k_t^{i,r}} \left(\mathbf{y}_t^{i,r} | \hat{\mathbf{x}}_t^{k_t^{i,r}} \right)}{G_t^i} \right] d\mathbb{Y} \backslash \mathbf{y}_t^{i,r} \tag{4.27}$$

$$= \sum_{t=1}^{T} \sum_{i=1}^{I} \sum_{m=0}^{M} \frac{n_t^i \hat{\pi}_t^m}{G_t^i} \int_{W^i} \log \left\{ g^m \left(\mathbf{y} | \mathbf{x}_t^m \right) \right\} g^m \left(\mathbf{y} | \hat{\mathbf{x}}_t^m \right) d\mathbf{y}, \tag{4.28}$$

$$= \sum_{t=1}^{T} \sum_{i=1}^{I} \sum_{m=0}^{M} \frac{n_t^i w_t^{i,m}}{G_t^{m,i}} \int_{W^i} \log \left\{ g^m \left(\mathbf{y} | \mathbf{x}_t^m \right) \right\} g^m \left(\mathbf{y} | \hat{\mathbf{x}}_t^m \right) d\mathbf{y}. \tag{4.29}$$

This remains a sum over independent terms for each target, so the state-dependent part of the auxiliary function can still be expressed as a sum over $\mathscr{Q}_x^m$ defined as

$$\mathscr{Q}_x^m = \log\left\{p\left(\mathbf{x}_0^m\right)\right\} + \sum_{t=1}^{T} \log\left\{p\left(\mathbf{x}_t^m|\mathbf{x}_{t-1}^m\right)\right\}$$
$$+ \sum_{t=1}^{T}\sum_{i=1}^{I} n_t^i w_t^{i,m} \frac{\int_{W^i} \log\left\{g^m\left(\mathbf{y}|\mathbf{x}_t^m\right)\right\} g^m\left(\mathbf{y}|\hat{\mathbf{x}}_t^m\right) \mathrm{d}\mathbf{y}}{G_t^{m,i}}. \tag{4.30}$$

Note that there are two instances of the appearance function $g^m(\cdot)$ in (4.30). The first is inside a logarithm and is conditioned on the (unknown) target state. The second appearance function is conditioned on the previous state estimate $\hat{\mathbf{x}}_t^m$ and so is constant with respect to $\mathbf{x}_t^m$. The arrangement of (4.30) has been made to provide the intuition that the integral can be viewed as a weighted mean since

$$\frac{\int_{W^i} \log\left\{g^m\left(\mathbf{y}|\mathbf{x}_t^m\right)\right\} g^m\left(\mathbf{y}|\hat{\mathbf{x}}_t^m\right) \mathrm{d}\mathbf{y}}{G_t^{m,i}} = \frac{\int_{W^i} \log\left\{g^m\left(\mathbf{y}|\mathbf{x}_t^m\right)\right\} g^m\left(\mathbf{y}|\hat{\mathbf{x}}_t^m\right) \mathrm{d}\mathbf{y}}{\int_{W^i} g^m\left(\mathbf{y}|\hat{\mathbf{x}}_t^m\right) \mathrm{d}\mathbf{y}}.$$

So this fraction term can be interpreted as the weighted average of the log-appearance function over the pixel.

Having now defined the auxiliary function, the E-step is complete.

4.1.2 Maximisation Step

The maximisation step consists of finding the parameter estimates that maximise the auxiliary function that has just been developed. As for point measurements, this auxiliary function can be expressed as a mixing prior term $\mathscr{Q}_\pi$ and a sum of individual target terms $\mathscr{Q}_x^m$:

$$\mathscr{Q}\left(\mathbb{X}, \Pi|\hat{\mathbb{X}}, \hat{\Pi}\right) = E_{\mathbb{K},\mathbb{Y}|\hat{\mathbb{X}},\hat{\Pi}} = \mathscr{Q}_\pi + \sum_{m=1}^{M} \mathscr{Q}_x^m. \tag{4.31}$$

The mixing prior term is essentially the same as in the point measurement case and its maximisation has already been discussed in the previous section: the solution is given in (4.26). The individual target term $\mathscr{Q}_x^m$ is defined in (4.30). The difficulty with maximising this expression is that in general, the integral in (4.30) is not tractable and numerical maximisation is required. This will be discussed in Chap. 7. For the special case that $g^m(\cdot)$ is part of the exponential family, then $\log\{g^m(\cdot)\}$ is a polynomial and the integral can be analytically solved. As we saw in Chap. 3, the sum term can be expressed as a single polynomial by equating coefficients. We now demonstrate this for the Gaussian case.

Consider a single time slice of the measurement term in (4.30) for the case where $g^m(\mathbf{y}|\mathbf{x}) = \mathscr{N}(\mathbf{y}; h(\mathbf{x}), \Sigma)$. The time slice can be simplified as

$$\sum_{i=1}^{I} \frac{n_t^i w_t^{i,m}}{G_t^{m,i}} \int_{W^i} \log \left\{ \mathcal{N}\left(\mathbf{y}; h\left(\mathbf{x}_t^m\right), \Sigma\right) \right\} \mathcal{N}\left(\mathbf{y}; h\left(\hat{\mathbf{x}}_t^m\right), \Sigma\right) d\mathbf{y}$$

$$= \sum_{i=1}^{I} \frac{n_t^i w_t^{i,m}}{G_t^{m,i}} \int_{W^i} \left(\mathbf{y} - h\left(\mathbf{x}_t^m\right)\right)^{\mathsf{T}} \Sigma^{-1} \left(\mathbf{y} - h\left(\mathbf{x}_t^m\right)\right) \mathcal{N}\left(\mathbf{y}; h\left(\hat{\mathbf{x}}_t^m\right), \Sigma\right) d\mathbf{y}. \tag{4.32}$$

The quadratic in $\mathbf{x}_t^m$ expands into three terms: a $h\left(\mathbf{x}_t^m\right)^{\mathsf{T}} h\left(\mathbf{x}_t^m\right)$ term that is constant with respect to the integral over $\mathbf{y}$; a $h\left(\mathbf{x}_t^m\right)^{\mathsf{T}} \mathbf{y}$ cross-term and a $\mathbf{y}^{\mathsf{T}}\mathbf{y}$ term that is independent of $\mathbf{x}_t^m$. As in the point measurement analysis, let C denote an arbitrary term that is independent of $\mathbf{x}_t^m$. As the development unfolds, the exact contents of C change, but since it has no bearing on the optimisation problem, we do not need to worry about these details. The time slice is then

$$h\left(\mathbf{x}_t^m\right)^{\mathsf{T}} \Sigma^{-1} h\left(\mathbf{x}_t^m\right) \sum_{i=1}^{I} \frac{n_t^i w_t^{i,m}}{G_t^{m,i}} \int_{W^i} \mathcal{N}\left(\mathbf{y}; h\left(\hat{\mathbf{x}}_t^m\right), \Sigma\right) d\mathbf{y}$$

$$- 2h\left(\mathbf{x}_t^m\right)^{\mathsf{T}} \Sigma^{-1} \sum_{i=1}^{I} \frac{n_t^i w_t^{i,m}}{G_t^{m,i}} \int_{W^i} \mathbf{y} \mathcal{N}\left(\mathbf{y}; h\left(\hat{\mathbf{x}}_t^m\right), \Sigma\right) d\mathbf{y} + C. \tag{4.33}$$

Define

$$\bar{\mathbf{y}}_t^{m,i} = \frac{\int_{W^i} \mathbf{y} \mathcal{N}\left(\mathbf{y}; h\left(\hat{\mathbf{x}}_t^m\right), \Sigma\right) d\mathbf{y}}{G_t^{m,i}} = \frac{\int_{W^i} \mathbf{y} \mathcal{N}\left(\mathbf{y}; h\left(\hat{\mathbf{x}}_t^m\right), \Sigma\right) d\mathbf{y}}{\int_{W^i} \mathcal{N}\left(\mathbf{y}; h\left(\hat{\mathbf{x}}_t^m\right), \Sigma\right) d\mathbf{y}}. \tag{4.34}$$

In his development in [5] Streit referred to this as the cell-level centroid, it is the mean point measurement location given that the measurement falls in pixel i. Also define

$$\bar{n}_t^m = \sum_{i=1}^{I} n_t^i w_t^{i,m}, \tag{4.35}$$

which can be interpreted as the number of measurements probabilistically assigned to target m. Notice that the integral in the first term of (4.33) is the definition of the cell probability for target m, so this cancels the denominator of that term. Using these, the time slice can be expressed as

$$h\left(\mathbf{x}_t^m\right)^{\mathsf{T}} \Sigma^{-1} h\left(\mathbf{x}_t^m\right) \bar{n}_t^m - 2h\left(\mathbf{x}_t^m\right)^{\mathsf{T}} \Sigma^{-1} \bar{n}_t^m \sum_{i=1}^{I} \frac{n_t^i w_t^{i,m} \bar{\mathbf{y}}_t^{m,i}}{\bar{n}_t^m} + C, \tag{4.36}$$

which can be factorised into

$$\left(h\left(\mathbf{x}_t^m\right)-\tilde{\mathbf{y}}_t^m\right)^{\mathsf{T}}\left(\tilde{\Sigma}_t^m\right)^{-1}\left(h\left(\mathbf{x}_t^m\right)-\tilde{\mathbf{y}}_t^m\right)+C, \tag{4.37}$$

where the equivalent covariance is

$$\tilde{\Sigma}_t^m=\frac{1}{\bar{n}_t^m}\Sigma, \tag{4.38}$$

and the equivalent measurement is

$$\tilde{\mathbf{y}}_t^m=\frac{\sum_{i=1}^I n_t^i w_t^{i,m}\bar{\mathbf{y}}_t^{m,i}}{\bar{n}_t^m}. \tag{4.39}$$

Using this result, the state-dependent sub-function can be written as

$$\mathscr{Q}_x^m=\log\left\{p\left(\mathbf{x}_0^m\right)\right\}+\sum_{t=1}^T\log\left\{p\left(\mathbf{x}_t^m|\mathbf{x}_{t-1}^m\right)\right\}+\sum_{t=1}^T\log\left\{\mathscr{N}\left(\tilde{\mathbf{y}}_t^m;h\left(\mathbf{x}_t^m\right),\tilde{\Sigma}_t^m\right)\right\}. \tag{4.40}$$

This is the same piece of mathematical manoeuvring as the point measurement case in Chap. 3. However, the consequence of it is surprising: the cost function we started with in (4.30) is an apparently complicated weighted expectation over an image measurement but the resulting cost function in (4.40) is the logarithm of a Gaussian point measurement problem. It can be maximised using a point measurement MAP estimator. For example, if $h(\mathbf{x})$ is linear then a Kalman Smoother can be used. The expectation step in the Gaussian histogram measurement problem is a procedure for deriving a single unambiguously assigned point measurement for each target component that can then be used in a point measurement estimator for the maximisation step.

4.2 Unobserved Pixels

A seemingly innocuous assumption in the preceding discussion is that the I pixels are disjoint and cover the whole observation space W, so that

$$\sum_{i=1}^I G_t^{m,i}=1. \tag{4.41}$$

The problem with this is that it implicitly requires the pixel coverage, and hence the sensor field of view, to be infinite. This may not be the case for some applications. When a target is away from the edges of the pixel coverage, then the amount of target energy spilling out of the field of view is negligible and there is no problem. However, when a target is close to the edge then the sum above can be significantly

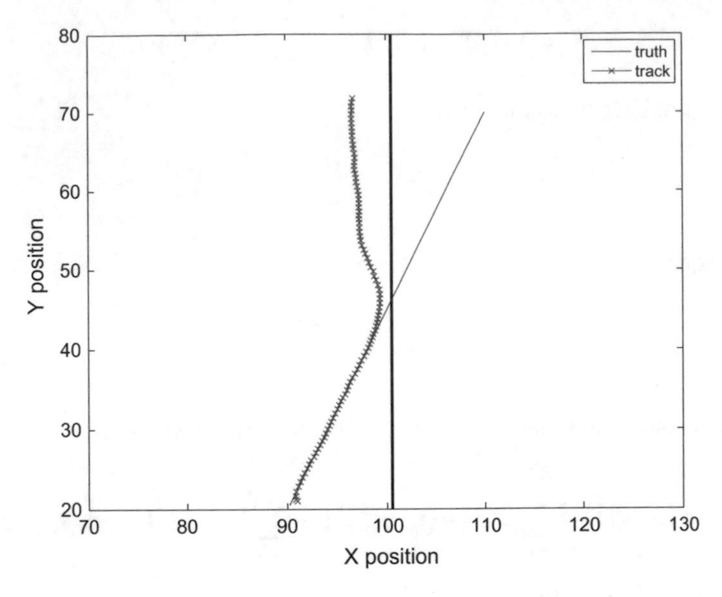

Fig. 4.2 Bias in a track leaving the sensor field of view

less than one and the target estimate becomes biased. As should be intuitive, this bias pushes the estimate away from the boundary of coverage. In the worst case, a target can transition from inside the coverage to outside and the corresponding track can become stuck on the edge, drifting off onto noise once the target influence is completely beyond coverage. Figure 4.2 shows an example of exactly this behaviour. One way to compensate for this would be to re-normalise the appearance function $g^m(\cdot)$ to guarantee the sum (4.41). Unfortunately, this can make the maximisation step much more difficult if, for example, a Gaussian is replaced with a truncated Gaussian. Worse, it does not even work very well. It can be demonstrated that the estimates are still biased and the problems persist.

An alternative way of compensating for edge effects is to introduce extra pixels along the edges. There are no measurements for these fake pixels, but the value we would have observed if there really was a pixel can be treated as further missing data in the EM process. These unobserved pixels were added to the derivation in [5] and [6, 7] showed examples where estimation performance near the boundary was greatly improved. Let the number of unobserved pixels be I^U and assume that they are indexed from $i = I+1 \ldots I+I^U$. The missing values are then $\mathbf{N}_t^U = n_t^{(I+1):(I+I^U)}$ and $\mathbb{N}^U = \mathbf{N}_{1:T}^U$ is the total collection of unobserved counts. In the same way as for the observed pixels, define $n_t^U = \sum_{i=I+1}^{I+I^U} n_t^i$ to be the total number of unobserved measurements. The extra piece of missing data means that the auxiliary function becomes a triple expectation

$$\mathscr{Q}\left(\mathbb{X},\Pi|\hat{\mathbb{X}},\hat{\Pi}\right) = E_{\mathbb{K},\mathbb{Y},\mathbb{N}^U|\hat{\mathbb{X}},\hat{\Pi}}\left[\log\left\{p_{\text{comp}}\left(\mathbb{X},\mathbb{K},\mathbb{Y},\mathbb{N},\mathbb{N}^U;\Pi\right)\right\}\right]$$
$$= \sum_{\mathbb{N}^U}\sum_{\mathbb{K}}\int \log\left\{p_{\text{comp}}\left(\mathbb{X},\mathbb{K},\mathbb{Y},\mathbb{N},\mathbb{N}^U;\Pi\right)\right\} p_{\text{miss}}\left(\mathbb{K},\mathbb{Y},\mathbb{N}^U|\mathbb{N},\hat{\mathbb{X}};\hat{\Pi}\right)\mathbf{dy}_{1:T}, \tag{4.42}$$

where the sum over $\mathbb{N}^U$ is really $T \times I^U$ nested sums where each n_t^i $(i > I)$ ranges from 0 to infinity,

$$\sum_{\mathbb{N}^U}(\cdot) \equiv \sum_{n_1^{I+1}=0}^{\infty}\cdots\sum_{n_1^{I+I^U}=0}^{\infty}\sum_{n_2^{I+1}=0}^{\infty}\cdots\sum_{n_T^{I+I^U}=0}^{\infty}(\cdot).$$

In this book, we treat the unobserved pixels exactly like the observed pixels, so the joint probability of all of the counts, observed and unobserved, remains a multinomial distribution. With the inclusion of unobserved pixels, the missing data can be expressed as

$$p_{\text{miss}}\left(\mathbb{K},\mathbb{Y},\mathbb{N}^U|\mathbb{N},\hat{\mathbb{X}};\hat{\Pi}\right) = p\left(\mathbb{K},\mathbb{Y}|\mathbb{N}^U,\mathbb{N},\hat{\mathbb{X}};\hat{\Pi}\right)p\left(\mathbb{N}^U|\mathbb{N},\hat{\mathbb{X}},\hat{\Pi}\right), \tag{4.43}$$

where the first term on the RHS is the missing data from Sect. 4.1 but now defined over an enlarged domain of pixels, and the second term is only a function of unobserved counts since the states and mixing proportions are represented by previous estimates. Victimising notation, we can then write

$$\mathscr{Q}^{\text{unobs}}\left(\mathbb{X},\Pi|\hat{\mathbb{X}},\hat{\Pi}\right) = \sum_{\mathbb{N}^U}\mathscr{Q}^{3.1}(\cdot)\,p\left(\mathbb{N}^U|\mathbb{N},\hat{\mathbb{X}},\hat{\Pi}\right),$$
$$= \sum_{\mathbb{N}^U}\mathscr{Q}_\pi\,p\left(\mathbb{N}^U|\mathbb{N},\hat{\mathbb{X}},\hat{\Pi}\right) + \sum_{m=1}^{M}\sum_{\mathbb{N}^U}\mathscr{Q}_x^m\,p\left(\mathbb{N}^U|\mathbb{N},\hat{\mathbb{X}},\hat{\Pi}\right). \tag{4.44}$$

Consider first the assignment prior term, which is itself a triple nested sum

$$\sum_{\mathbb{N}^U}\mathscr{Q}_\pi\,p\left(\mathbb{N}^U|\mathbb{N},\hat{\mathbb{X}},\hat{\Pi}\right) = \sum_{\mathbb{N}^U}\sum_{t=1}^{T}\sum_{i=1}^{I+I^U}\sum_{m=0}^{M}\log\left\{\pi_t^m\right\}n_t^i w_t^{i,m}\,p\left(\mathbb{N}^U|\mathbb{N},\hat{\mathbb{X}},\hat{\Pi}\right). \tag{4.45}$$

Each term in the summation depends only on a single pixel i. For the observed pixels, where $1 \leq i \leq I$, the summand is independent of the unobserved counts and for unobserved pixels, where $I < i \leq I + I^U$, the summand only depends on that particular unobserved pixel, so the prior term can be written as

$$\sum_{\mathbb{N}^U} \mathcal{Q}_\pi p\left(\mathbb{N}^U|\mathbb{N}, \hat{\mathbb{X}}, \hat{\Pi}\right) = \sum_{t=1}^{T}\sum_{i=1}^{I}\sum_{m=0}^{M} \log\left\{\pi_t^m\right\} n_t^i w_t^{i,m}$$
$$+ \sum_{t=1}^{T}\sum_{i=I+1}^{I+I^U}\sum_{m=0}^{M} \log\left\{\pi_t^m\right\} w_t^{i,m} \sum_{n_t^i=0}^{\infty} n_t^i p\left(n_t^i|\mathbb{N}, \hat{\mathbb{X}}, \hat{\Pi}\right). \tag{4.46}$$

The second term contains the expected value of the measurement count for pixel i at time t conditioned on the observed counts and estimates of the target states and mixing proportions. Using the multinomial model, this is given by

$$\begin{aligned}
E\left(n_t^i|n_t\right) &= \sum_{n_t^U} n_t^i p\left(n_t^U|n_t, \mathbf{x}_t^{1:M}, \pi_t^{0:M}\right), \\
&= \sum_{n_t^U} n_t^i \left(n_t + ||n_t^U||\right)! \frac{(G_t)^{n_t}}{n_t!} \prod_{j=I+1}^{I+I^U} \frac{\left(G_t^j\right)^{n_t^i}}{n_t^j!}, \\
&= \frac{(G_t)^{n_t}}{n_t!} \sum_{n_t^U \backslash n_t^i} \left[\prod_{j=I+1\backslash i}^{I+I^U} \frac{\left(G_t^j\right)^{n_t^i}}{n_t^j!}\right]\left[\sum_{n_t^i=1}^{\infty}\left(n_t + ||n_t^U||\right)! n_t^i \frac{\left(G_t^j\right)^{n_t^i}}{n_t^j!}\right], \\
&= \frac{G_t^i}{G_t}(n_t+1)\frac{(G_t)^{n_t+1}}{(n_t+1)!} \\
&\quad \times \sum_{n_t^U \backslash n_t^i} \left[\prod_{j=I+1\backslash i}^{I+I^U} \frac{\left(G_t^j\right)^{n_t^i}}{n_t^j!}\right]\left[\sum_{n_t^i=1}^{\infty}\left(n_t + ||n_t^U||\right)! \frac{\left(G_t^j\right)^{(n_t^i-1)}}{(n_t^j-1)!}\right], \\
&= \frac{G_t^i}{G_t}(n_t+1) \sum_{n_t^U} p\left(n_t^U|n_t+1, \mathbf{x}_t^{1:M}, \pi_t^{0:M}\right), \\
&= \frac{G_t^i}{\sum_{j=1}^{I} G_t^j}(n_t+1).
\end{aligned} \tag{4.47}$$

This result is intuitive: the expected count for an unobserved pixel is the total observed count scaled by the ratio of the unobserved pixel prior to the total observed pixels prior. Note that Streit assumed a negative multinomial over n_t^U, which is almost the same as above except that it implicitly assumes that the last point measurement landed in one of the observed pixels and the resulting expectation in (4.47) is scaled by n_t instead of $(n_t + 1)$. We prefer to avoid the confusion of introducing another distribution. For most applications, $n_t \gg 1$ and there is no practical difference between the two.

Define the expected measurement count as

$$\bar{n}_t^i = \begin{cases} n_t^i & 1 \leq i \leq I, \\ \frac{G_t^i}{\sum_{j=1}^I G_t^j}(n_t + 1) & I < i \leq I + I^U. \end{cases} \tag{4.48}$$

This can then be used to write a simpler expression for the prior auxiliary sub-function

$$\mathcal{Q}_\pi = \sum_{t=1}^{T} \sum_{i=1}^{I+I^U} \sum_{m=0}^{M} \log\{\pi_t^m\} \bar{n}_t^i w_t^{i,m}. \tag{4.49}$$

Once again, the familiar nested sum prior expression has returned and by inspection the maximising solution is

$$\hat{\pi}_t^m := \frac{\sum_{i=1}^{I+I^U} \bar{n}_t^i w_t^{i,m}}{\sum_{s=0}^{M} \sum_{i=1}^{I+I^U} \bar{n}_t^i w_t^{i,s}}. \tag{4.50}$$

The inclusion of unobserved pixels has resulted in an extended summation where the extra terms are the mean counts in the unobserved pixels. These mean counts are derived from estimates of the kinematic states and priors from the previous iteration, so they will tend to bias the estimate towards the previous estimated prior. However, since the EM process repeatedly updates the estimates, the only effect this is likely to have on the prior is to slow convergence.

The second term in the unobserved pixel auxiliary function (4.44) is the state-dependent part

$$\begin{aligned} \sum_{\mathbb{N}^U} \mathcal{Q}_x^m p\left(\mathbb{N}^U | \mathbb{N}, \hat{\mathbb{X}}, \hat{\Pi}\right) &= \log\{p(\mathbf{x}_0^m)\} + \sum_{t=1}^{T} \log\{p(\mathbf{x}_t^m | \mathbf{x}_{t-1}^m)\} \\ &+ \sum_{\mathbb{N}^U} \sum_{t=1}^{T} \sum_{i=1}^{I+I^U} n_t^i w_t^{i,m} \frac{\int_{W^i} \log\{g^m(\mathbf{y}|\mathbf{x}_t^m)\} g^m(\mathbf{y}|\hat{\mathbf{x}}_t^m)\, d\mathbf{y}}{G_t^{m,i}} p\left(\mathbb{N}^U | \mathbb{N}, \hat{\mathbb{X}}, \hat{\Pi}\right), \\ &= \log\{p(\mathbf{x}_0^m)\} + \sum_{t=1}^{T} \log\{p(\mathbf{x}_t^m | \mathbf{x}_{t-1}^m)\} \\ &+ \sum_{t=1}^{T} \sum_{i=1}^{I+I^U} \bar{n}_t^i w_t^{i,m} \frac{\int_{W^i} \log\{g^m(\mathbf{y}|\mathbf{x}_t^m)\} g^m(\mathbf{y}|\hat{\mathbf{x}}_t^m)\, d\mathbf{y}}{G_t^{m,i}}, \end{aligned} \tag{4.51}$$

where similar simplifications have been applied as for the prior auxiliary sub-function. The form of this expression is the same as the corresponding function without unobserved pixels (4.30) and so the solution method is the same. For the case of a Gaussian appearance function, the image estimation problem can be expressed as a point measurement smoother with equivalent measurements and measurement covariances given by

$$\tilde{\Sigma}_t^m = \frac{1}{\sum_{i=1}^{I+I^U} \bar{n}_t^i w_t^{i,m}} \Sigma, \tag{4.52}$$

$$\tilde{\mathbf{y}}_t^m = \frac{\sum_{i=1}^{I+I^U} \bar{n}_t^i w_t^{i,m} \bar{\mathbf{y}}_t^{m,i}}{\sum_{i=1}^{I+I^U} \bar{n}_t^i w_t^{i,m}}, \tag{4.53}$$

where the cell centroids are defined as before in (4.34) but over more pixels

$$\bar{\mathbf{y}}_t^{m,i} = \frac{\int_{W^i} \mathbf{y} \mathcal{N}\left(\mathbf{y}; h\left(\hat{\mathbf{x}}_t^m\right), \Sigma\right) d\mathbf{y}}{G_t^{m,i}}.$$

Figure 4.3 shows the result of applying unobserved pixels along the edge of the image in the boundary crossing example from Fig. 4.2. The blue crosses show the original track that uses the sensor images directly and fails to properly exit the image. The green circles show a track formed using H-PMHT with a band of unobserved pixels along the image edge. The track bias is significantly improved. The unobserved-pixel track diverges from the truth after the target has left the image completely since there is no measurement information any longer. In practice, a track would be terminated by the track manager once it has left the image.

Figure 4.4 shows a sequence of image frames as the target crosses the region boundary. These are cropped to only show the area close to the crossing. The vertical

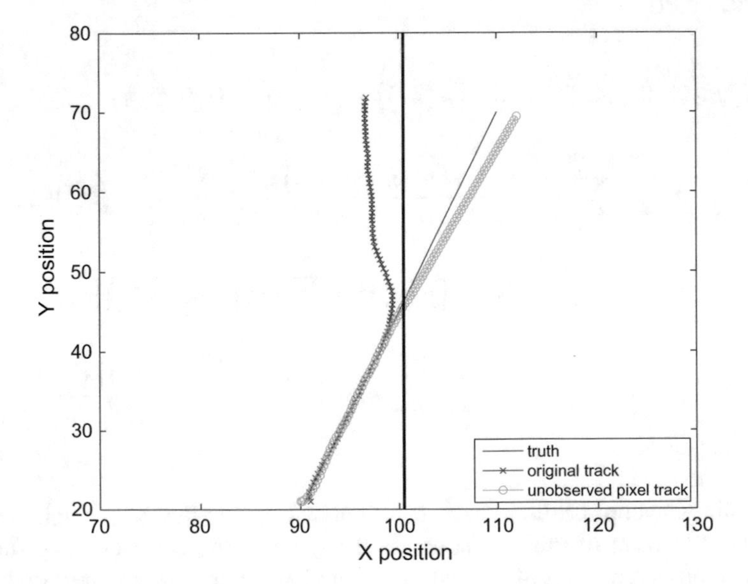

Fig. 4.3 Bias in a track leaving the sensor field of view

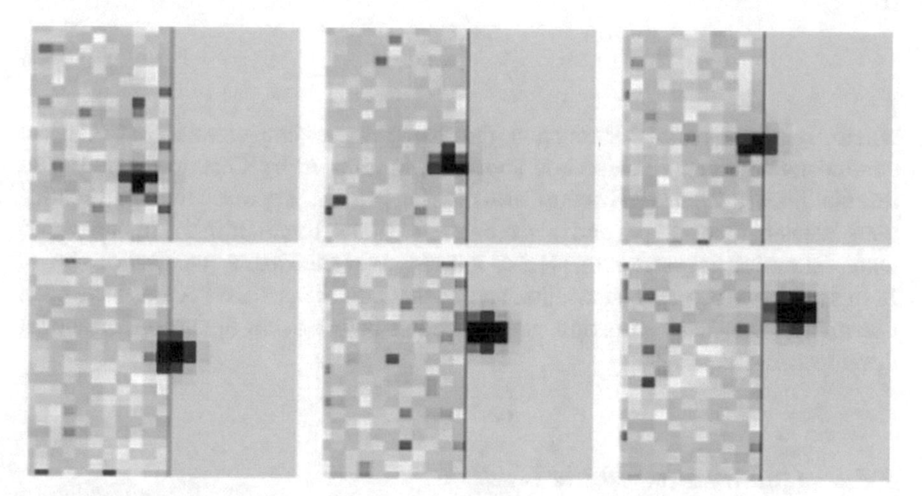

Fig. 4.4 Selected frames with unobserved pixels as the target crosses the edge of the image

line marks the edge of the sensor image: to the left of the line is the original image and to the right is the expected measurement count in the unobserved pixels, $\bar{n}_t^i$. Far from the target, the expected value of the unobserved pixels is simply the mean noise level but there is a high-energy peak close to the predicted target position, which is what allows the track to cross the boundary gracefully.

The unobserved pixels can be used for other purposes besides fixing edge effects. For example, in Doppler radar the sensor will be unable to collect data at near zero Doppler because of the overwhelming ground clutter: the low Doppler pixels can be treated as unobserved. Unobserved pixels can also be used as a mechanism for compression by discarding data in very low power pixels and labelling these as unobserved.

4.3 Image Quantising

The preceding discussion has assumed that the measurements consist of a collection of integers n_t^i. For some applications, such as optical video, this is exactly what the sensor provides. For others, such as radar or sonar, the sensor image is more likely to consist of floating point values, which we can treat as continuous valued. In order to apply the histogram measurement model to these sensors, the continuous valued pixel values need to be converted into integers. There are at least two ways to do this. The first method, introduced by Luginbuhl and Willett [3] is to quantise the image values at a prescribed discrete level $\hbar^2$. Denoting the continuous valued pixel intensity as $\mathbf{z}_t^i$ the derived pixel count is then

$$n_t^i = \left\lfloor \frac{\mathbf{z}_t^i}{\hbar^2} \right\rfloor, \tag{4.54}$$

where $\lfloor \cdot \rfloor$ denotes the floor function, that is the largest integer value less than the continuous argument. The second approach, introduced by Ceravic [1], is to use resampling, a la sequential Monte Carlo filtering [4]. This approach treats the continuous intensity map as a pdf and then draws n_t samples from it. If this is implemented using deterministic resampling [4] then it is closely related to the quantisation above. In this book, we use the quantisation approach. Once the per pixel counts have been determined, the state estimation can proceed as before using the histogram model developed in this chapter.

4.3.1 Quantisation in the Limit

The quantisation step above requires the choice of the discretisation step $\hbar^2$. It is unclear whether or how the particular value chosen here influences performance. Streit removed this ambiguity by considering the limiting case as $\hbar^2 \rightarrow 0$. It is not possible to take the limit on the auxiliary function directly because as $\hbar^2$ is reduced the resulting n_t increases and because each point measurement has a finite negative cost, the auxiliary function would become infinite for all sequences of $\mathbb{X}$ and Π. The solution is to define the continuous pixel measurement auxiliary function as

$$\mathcal{Q}\left(\mathbb{X}, \Pi | \hat{\mathbb{X}}, \hat{\Pi}\right) = \lim_{\hbar^2 \rightarrow 0} \hbar^2 \mathcal{Q}^n \left(\mathbb{X}, \Pi | \hat{\mathbb{X}}, \hat{\Pi}; \hbar^2\right), \tag{4.55}$$

where $\mathcal{Q}^n \left(\mathbb{X}, \Pi | \hat{\mathbb{X}}, \hat{\Pi}; \hbar^2\right)$ is the discrete count auxiliary function derived using quantisation level $\hbar^2$. We will see shortly that this limit is finite, so can be sensibly optimised with respect to the parameters. The discrete auxiliary function is composed of a prior sub-function and a kinematic state sub-function:

$$\mathcal{Q}\left(\mathbb{X}, \Pi | \hat{\mathbb{X}}, \hat{\Pi}\right) = \lim_{\hbar^2 \rightarrow 0} \hbar^2 \mathcal{Q}_\pi + \sum_{m=1}^{M} \lim_{\hbar^2 \rightarrow 0} \hbar^2 \mathcal{Q}_x^m. \tag{4.56}$$

Before addressing these sub-functions directly, we derive some limits of terms that will prove to be useful. The derived count can be expressed as

$$n_t^i = \left\lfloor \frac{\mathbf{z}_t^i}{\hbar^2} \right\rfloor = \frac{\mathbf{z}_t^i}{\hbar^2} - \varepsilon, \tag{4.57}$$

where $0 \leq \varepsilon < \hbar^2$. Using this it is straightforward to observe that

$$\lim_{\hbar^2 \rightarrow 0} \hbar^2 n_t^i = \lim_{\hbar^2 \rightarrow 0} \left(\mathbf{z}_t^i - \varepsilon \hbar^2\right) = \mathbf{z}_t^i, \tag{4.58}$$

$$\lim_{\hbar^2\to 0} \hbar^2 (n_t + 1) = \sum_{i=1}^{I+I^U} \lim_{\hbar^2\to 0} \hbar^2 n_t^i + \lim_{\hbar^2\to 0} \hbar^2 = \sum_{i=1}^{I+I^U} \mathbf{z}_t^i \equiv ||\mathbf{z}_t||, \tag{4.59}$$

where $||\mathbf{z}_t||$ is the L1 norm of the observed image, or in less erudite language, the sum of all the observed pixel intensity values $||\mathbf{z}_t|| = \sum_{i=1}^{I} \mathbf{z}_t^i$.

Consider the prior auxiliary sub-function

$$\lim_{\hbar^2\to 0} \hbar^2 \mathscr{Q}_\pi = \sum_{t=1}^{T} \sum_{i=1}^{I+I^U} \sum_{m=0}^{M} \log\left\{\pi_t^m\right\} w_t^{i,m} \lim_{\hbar^2\to 0} \hbar^2 \bar{n}_t^i. \tag{4.60}$$

The expected count limit is

$$\lim_{\hbar^2\to 0} \hbar^2 \bar{n}_t^i = \begin{cases} \lim_{\hbar^2\to 0} \hbar^2 n_t^i = \mathbf{z}_t^i & 1 \le i \le I, \\ \lim_{\hbar^2\to 0} \hbar^2 \frac{G_t^i}{\sum_{j=1}^{I} G_t^j} (n_t + 1) = \frac{G_t^i}{\sum_{j=1}^{I} G_t^j} ||\mathbf{z}_t|| & I < i \le I + I^U. \end{cases} \tag{4.61}$$

Denoting the expected count limit as $\bar{\mathbf{z}}_t^i = \lim_{\hbar^2\to 0} \hbar^2 \bar{n}_t^i$, the updated prior estimate is then

$$\hat{\pi}_t^m := \frac{\sum_{i=1}^{I+I^U} \bar{\mathbf{z}}_t^i w_t^{i,m}}{\sum_{s=0}^{M} \sum_{i=1}^{I+I^U} \bar{\mathbf{z}}_t^i w_t^{i,s}}. \tag{4.62}$$

Thus the prior estimate takes exactly the same form as in the discrete count case but using the continuous intensity values instead of counts.

The kinematic state sub-function can be simplified as

$$\begin{aligned} \lim_{\hbar^2\to 0} \hbar^2 \mathscr{Q}_x^m &= \log\left\{p\left(\mathbf{x}_0^m\right)\right\} \lim_{\hbar^2\to 0} \hbar^2 + \sum_{t=1}^{T} \log\left\{p\left(\mathbf{x}_t^m | \mathbf{x}_{t-1}^m\right)\right\} \lim_{\hbar^2\to 0} \hbar^2 \\ &\quad + \sum_{t=1}^{T} \sum_{i=1}^{I+I^U} \lim_{\hbar^2\to 0} \hbar^2 \bar{n}_t^i w_t^{i,m} \frac{\int_{W^i} \log\left\{g^m\left(\mathbf{y}|\mathbf{x}_t^m\right)\right\} g^m\left(\mathbf{y}|\hat{\mathbf{x}}_t^m\right) d\mathbf{y}}{G_t^{m,i}}, \\ &= \sum_{t=1}^{T} \sum_{i=1}^{I+I^U} \bar{\mathbf{z}}_t^i w_t^{i,m} \frac{\int_{W^i} \log\left\{g^m\left(\mathbf{y}|\mathbf{x}_t^m\right)\right\} g^m\left(\mathbf{y}|\hat{\mathbf{x}}_t^m\right) d\mathbf{y}}{G_t^{m,i}}. \end{aligned} \tag{4.63}$$

Just like with the prior expression, the measurement term is the same as the discrete count case but using the continuous intensity values instead of counts. However, we have a problem: the state transition density has vanished from the auxiliary function. Because the discrete point measurements were treated as independent, the limit has essentially resulted in an assumed infinite amount of measurement data which easily swamps the finite transition density.

4.3.2 Resampled Target Prior

Streit solved the case of the vanishing target prior by making the prior data dependent [5]. The idea is that every shot draws its own realisation of the kinematic state distribution, so there are $n_t + n_t^U$ IID target kinematic states for each target m. These realisations are constrained to all take the same value $\mathbf{x}_t^m$, which means that the transition density $p\left(\mathbf{x}_t^m|\mathbf{x}_{t-1}^m\right)$ is replaced with the density $p\left(\mathbf{x}_t^m|\mathbf{x}_{t-1}^m\right)^{n_t+||n_t^U||}$. Applying this to the kinematic state auxiliary sub-function results in

$$\begin{aligned}\lim_{\hbar^2\to 0} \hbar^2 \mathcal{Q}_x^m &= \log\left\{p\left(\mathbf{x}_0^m\right)\right\} \lim_{\hbar^2\to 0} \hbar^2 + \sum_{t=1}^{T} \log\left\{p\left(\mathbf{x}_t^m|\mathbf{x}_{t-1}^m\right)\right\} \lim_{\hbar^2\to 0} \hbar^2 (n_t + ||n_t^U||)\\ &\quad + \sum_{t=1}^{T}\sum_{i=1}^{I+I^U} \lim_{\hbar^2\to 0} \hbar^2 \bar{n}_t^i w_t^{i,m} \frac{\int_{W^i} \log\left\{g^m\left(\mathbf{y}|\mathbf{x}_t^m\right)\right\} g^m\left(\mathbf{y}|\hat{\mathbf{x}}_t^m\right) \mathrm{d}\mathbf{y}}{G_t^{m,i}},\\ &= \sum_{t=1}^{T} \frac{||\mathbf{z}_t||}{\sum_{i=1}^{I} G_t^i} \log\left\{p\left(\mathbf{x}_t^m|\mathbf{x}_{t-1}^m\right)\right\} + \sum_{t=1}^{T}\sum_{i=1}^{I+I^U} \bar{\mathbf{z}}_t^i w_t^{i,m} \frac{\int_{W^i} \log\left\{g^m\left(\mathbf{y}|\mathbf{x}_t^m\right)\right\} g^m\left(\mathbf{y}|\hat{\mathbf{x}}_t^m\right) \mathrm{d}\mathbf{y}}{G_t^{m,i}}.\end{aligned} \tag{4.64}$$

In the Gaussian case $p\left(\mathbf{x}_t^m|\mathbf{x}_{t-1}^m\right) \sim \mathcal{N}\left(\mathbf{x}_t^m; \mathsf{F}_t\mathbf{x}_{t-1}^m; \mathsf{Q}_t^m\right)$, the logarithm of the transition density is quadratic in the target state and the data dependent resampled prior can be expressed as

$$\begin{aligned}\frac{||\mathbf{z}_t||}{\sum_{i=1}^{I} G_t^i} \log\left\{p\left(\mathbf{x}_t^m|\mathbf{x}_{t-1}^m\right)\right\} &= \frac{||\mathbf{z}_t||}{\sum_{i=1}^{I} G_t^i} \left(\mathbf{x}_t^m - \mathsf{F}_t\mathbf{x}_{t-1}^m\right)^{\mathsf{T}} \left(\mathsf{Q}_t^m\right)^{-1} \left(\mathbf{x}_t^m - \mathsf{F}_t\mathbf{x}_{t-1}^m\right)\\ &= \log\left\{\mathcal{N}\left(\mathbf{x}_t^m; \mathsf{F}_t\mathbf{x}_{t-1}^m; \frac{\sum_{i=1}^{I} G_t^i}{||\mathbf{z}_t||}\mathsf{Q}_t^m\right)\right\},\end{aligned} \tag{4.65}$$

where we have ignored terms independent of the target state. This shows that the data dependent prior influences implementation by scaling the process noise covariance matrix

$$\tilde{\mathsf{Q}}_t^m = \frac{\sum_{i=1}^{I} G_t^i}{||\mathbf{z}_t||}\mathsf{Q}_t^m. \tag{4.66}$$

A consequence is that the process noise used is inversely proportional to the total image intensity. The equivalent measurement covariance for the Gaussian is defined in (4.52), which is

$$\tilde{\Sigma}_t^m = \frac{1}{\sum_{i=1}^{I+I^U} \bar{n}_t^i w_t^{i,m}} \Sigma.$$

This can be interpreted as scaling the measurement noise by the power associated with that target. For a single target, if the SNR is high, then the total image power is approximately equal to the power associated with the target and the ratio of the process noise to the measurement noise is approximately constant irrespective of these scaling factors. In such a case, the tracking performance is not influenced by the SNR. For a lower SNR target or a case where there are multiple targets, the total image power is more influenced by the background noise or other targets. If the noise changes or the SNR of a different target changes, even if these changes are remote, then tracking performance will change. A simple example is to double the image size without changing resolution. This doubles the noise power and hence scales Q by a factor of two: adding more pixels at a location unrelated to the target will influence tracking. An approach for removing this dependency is discussed in Chap. 5.

4.4 Associated Images

We now present a slightly different perspective on the H-PMHT which gives an intuitive interpretation of the mathematics [2]. For each model in the mixture and each pixel, the H-PMHT E-step produces a pixel association weight defined in (4.24):

$$w_t^{i,m} = \frac{\hat{\pi}_t^m G_t^{m,i}}{\sum_{s=0}^{M} \hat{\pi}_t^s G_t^{s,i}}.$$

These weights can be combined with the image at frame t to define the quantity

$$\mathfrak{z}_t^{i,m} = w_t^{i,m} \bar{\mathbf{z}}_t^i, \tag{4.67}$$

which will be referred to as the associated image frame for model m at frame t. As the name suggests, this quantity is an image, it has the same shape as $\mathbf{z}_t$, and it contains the measurement power associated with model m. Summing the associated images over the models gives the original sensor image

$$\sum_{m=0}^{M} \mathfrak{z}_t^{i,m} = \bar{\mathbf{z}}_t^i \sum_{m=0}^{M} w_t^{i,m} = \bar{\mathbf{z}}_t^i. \tag{4.68}$$

So we can think of this breaking the sensor image into component parts. Figure 4.5 shows an example of the associated images obtained through applying H-PMHT to the mixture in Fig. 4.1.

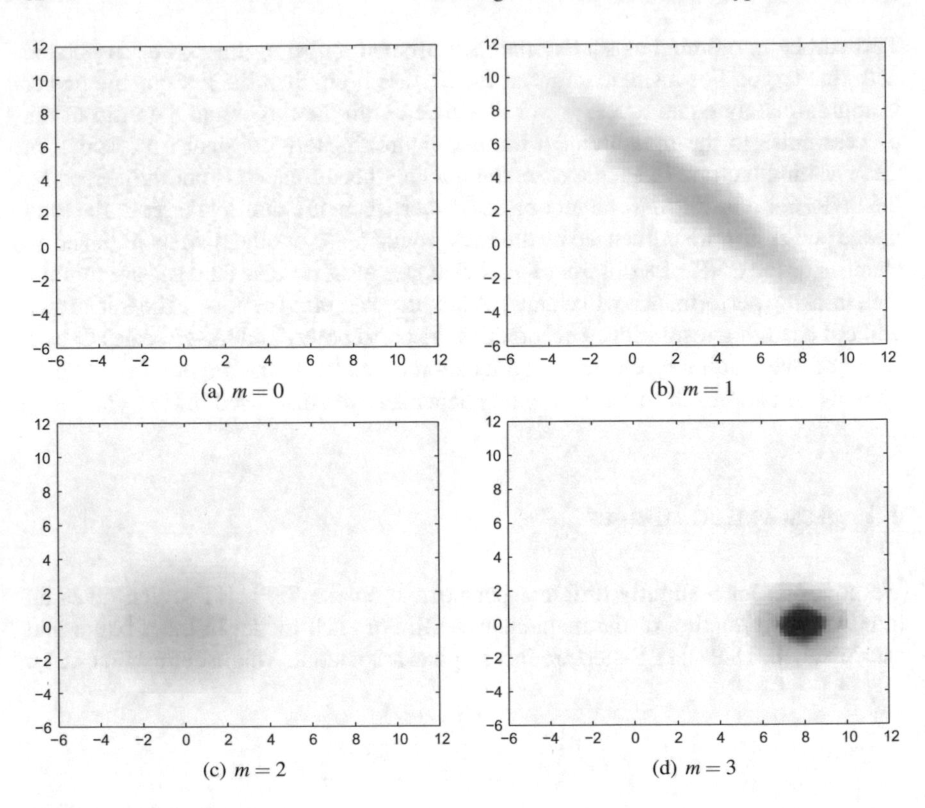

(a) $m = 0$ (b) $m = 1$

(c) $m = 2$ (d) $m = 3$

Fig. 4.5 Associated images

The total image intensity $||\mathbf{z}_t||$ was previously defined as simply the sum of the pixel intensities over the whole observed sensor image $||\mathbf{z}_t|| = \sum_{i=1}^{I} \mathbf{z}_t^i$. Similarly define $||\bar{\mathbf{z}}_t||$ as the total sum of the intensity values across both the observed pixels and the unobserved pixels, $||\bar{\mathbf{z}}_t|| = \sum_{i=1}^{I+I^U} \bar{\mathbf{z}}_t^i$ and define $||\mathfrak{z}_t^m||$ to be the total sum of the associated image intensity values for model m across both the observed pixels and the unobserved pixels, $||\mathfrak{z}_t^m|| = \sum_{i=1}^{I+I^U} \mathfrak{z}_t^{i,m}$. Using these, it is easy to see that the updated prior estimate can be expressed as

$$\hat{\pi}_t^m := \frac{||\mathfrak{z}_t^m||}{||\bar{\mathbf{z}}_t||}. \tag{4.69}$$

In words, this means that the estimated prior is the ratio of the estimated total intensity due to model m to the total image intensity. This result is simple and elegant.

Algorithm 1 Core H-PMHT for Gaussian appearance

1: Initialise the state estimates $\hat{\mathbb{X}}$ and the mixing proportion estimates $\hat{\Pi}$
2: **while** not converged **do**
3: Calculate the cell probabilities G_t^m and G_t for all t and m using (4.8)
4: Calculate the pixel weights $w_t^{i,m}$ given by (4.24)
5: Update the mixing proportion estimates $\hat{\Pi}$ using (4.62)
6: **for** each target m **do**
7: Determine the synthetic covariance matrix $\tilde{\Sigma}_t^m$ using (4.52)
8: Determine the synthetic measurement vector $\tilde{\mathbf{y}}_t^m$ using (4.53)
Update the state estimates $\hat{\mathbf{x}}_{1:T}^m$ using a Kalman smoother over the synthetic measurements
9: **end for**
10: **end while**

4.5 Algorithm Summary for Gaussian Appearance

The core H-PMHT for Gaussian appearance is summarised in Algorithm 1.

4.6 Simulated Example

The core H-PMHT[1] derived in this chapter was applied to the canonical multi-target scenario as was done for the point measurement PMHT in Chap. 3. This time around, we skip straight to the 20 target case. The higher clutter example of $\Lambda^0 = 50$ in Chap. 3 corresponds to on average 1 pixel in 600 forming a false detection since the image size is 150×200 pixels. For unit variance complex Gaussian noise, this would be achieved with a threshold of 3.6 in amplitude. For the example here, we simulate a peak amplitude of 5 for every target, which would give a probability of detection of 98%. In the point measurement case, the tracks were pre-initialised with the truth: here we will use the track manager that is described in Chap. 5. This means that it takes a while for all the tracks to form and the number of missed tracks starts at 20 and decays down as targets are detected. Figures 4.6 and 4.7 show the tracks formed on the 20 targets for one example realisation. The plots show X-position as a function of time. Clearly, the tracks align very closely with the truth and the initiation delay is quite small.

Monte Carlo trials were used to calculate association and accuracy performance measures as discussed in Sect. 1.6; the RMS estimation error, the redundant track count, the missed target count and the number of swaps were each averaged over 100 Monte Carlo trials and are shown in Fig. 4.8. The figure also repeats the results for

[1] *MakeHPMHTParams*
The H-PMHT toolbox contains two main functions for running H-PMHT. *MakeHPMHTParams* defines tracking parameters, including which version of H-PMHT to execute, and *HPMHTTracker* actually does the tracking. The default tracking parameters for the core H-PMHT are created by *MakeHPMHTParams('core')* or *MakeHPMHTParams()* For details on how to use this function, refer to the H-PMHT toolbox documentation.

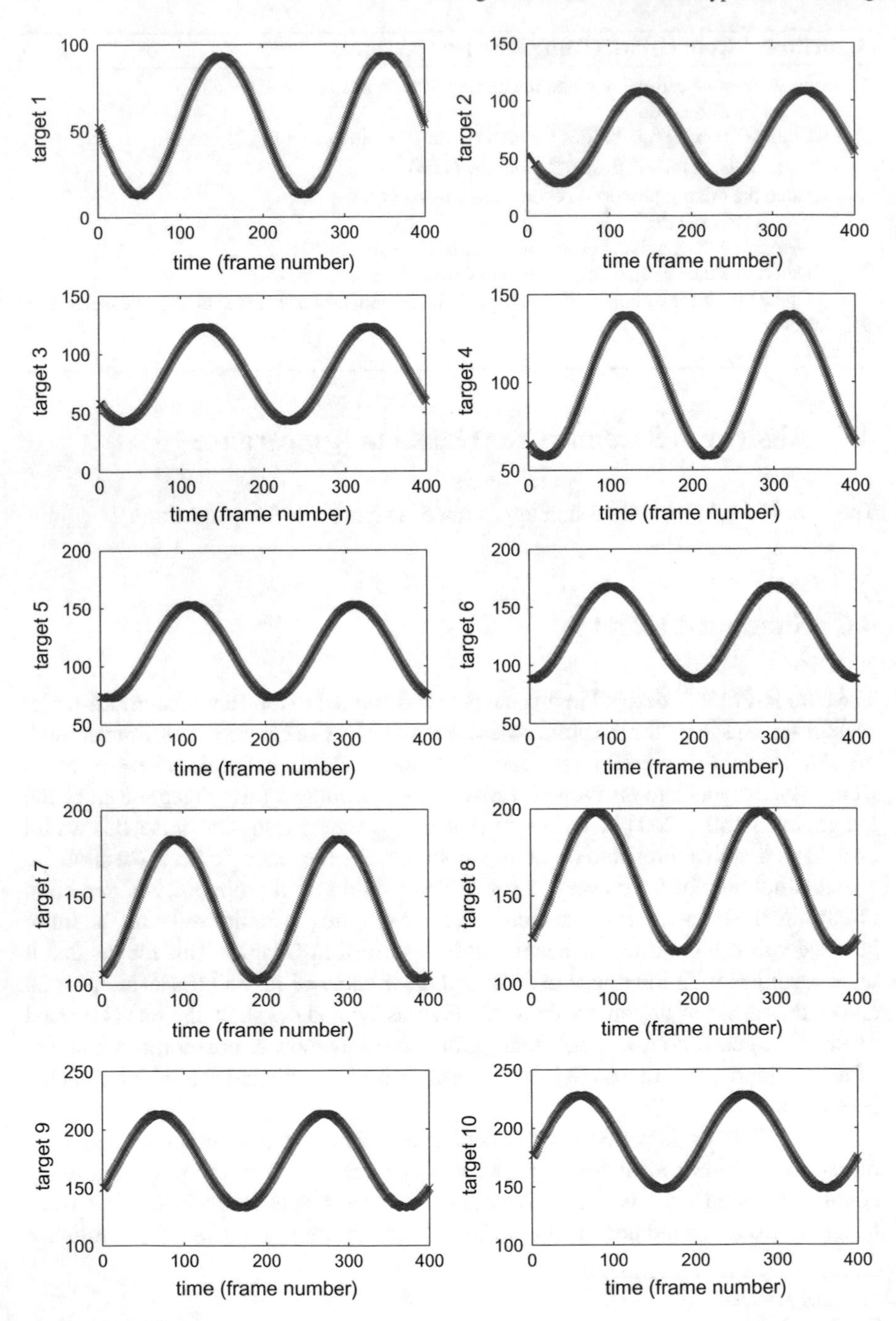

Fig. 4.6 Targets 1 . . . 10 and associated tracks

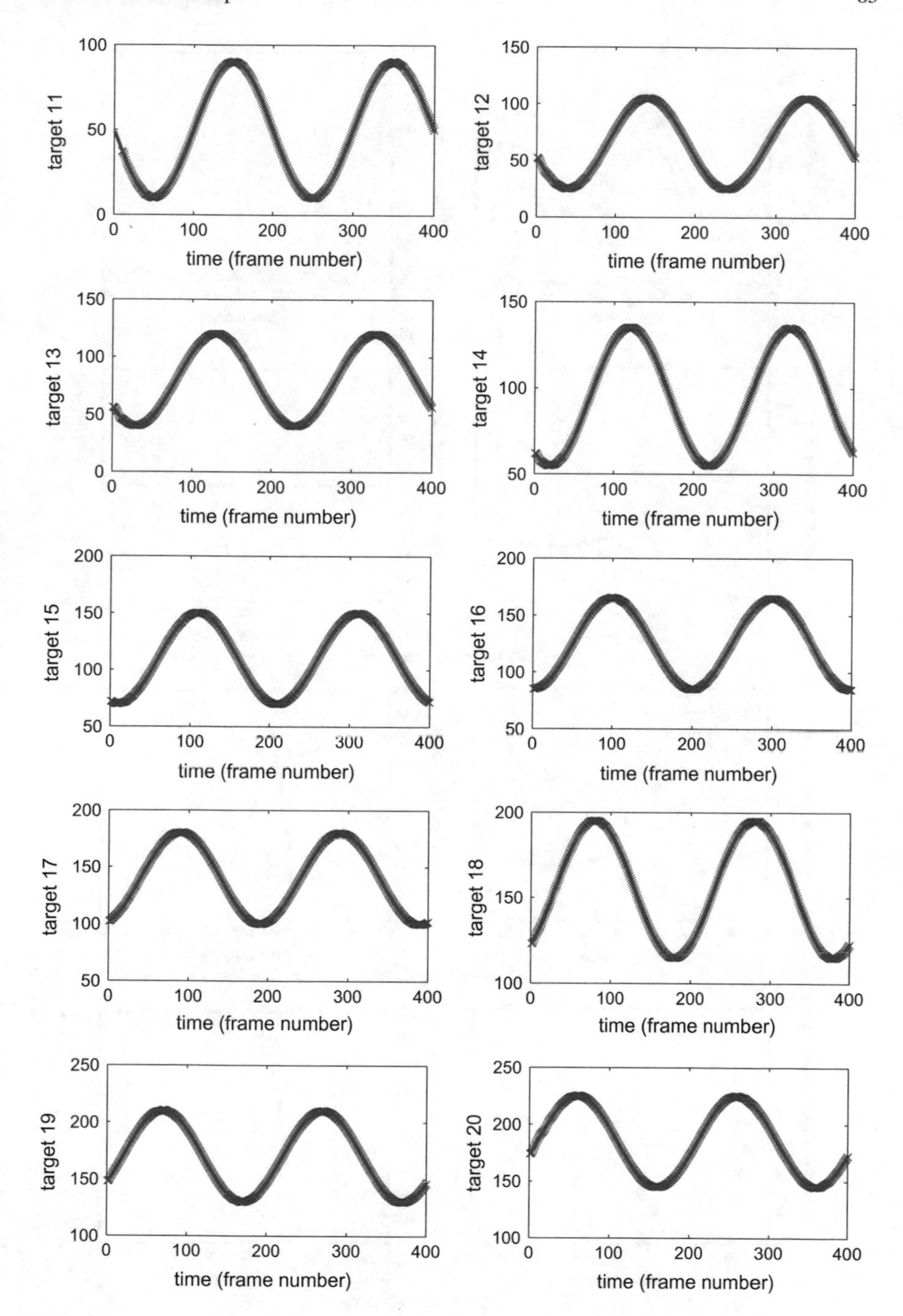

Fig. 4.7 Targets 11 . . . 20 and associated tracks

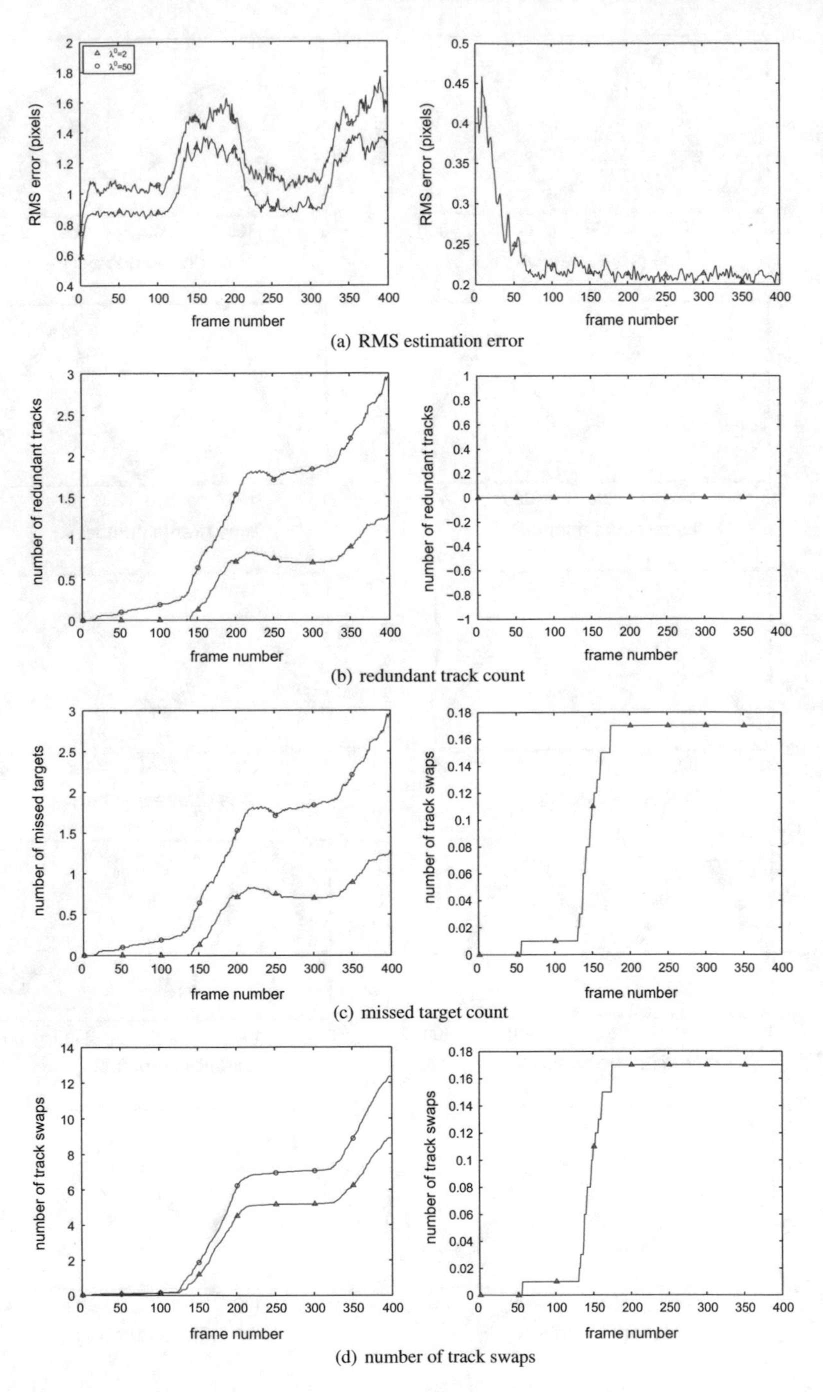

Fig. 4.8 Performance comparison, PMHT and H-PMHT

the point measurement PMHT from Fig. 3.7: the performance is immensely better. Estimation error is greatly reduced, there are very few track swaps, no redundant tracks and tracks are only missed during the initial acquisition period. H-PMHT gives much better performance than point measurement PMHT on this scenario because it explicitly models the way that the energy of closely spaced targets interacts.

4.7 Summary

This chapter has derived the core H-PMHT. We started with the point measurement PMHT and showed how it is extended to the situation of histogram measurement data where the location of each point is only partially available. This was then further extended to image data by taking the limit of quantised images. The resulting algorithm is core H-PMHT. We applied core H-PMHT to the canonical multi-target scenario and showed that it gives significant improvement over point measurement PMHT because its measurement model better describes closely interacting targets with finite sensor resolution.

References

1. Ceravic, D., Davey, S.J.: Two algorithms for modeling and tracking of dynamic time-frequency spectra. IEEE Trans. Signal Process. **64** (2016)
2. Davey, S.J., Wieneke, M., Vu, H.X.: Histogram-PMHT unfettered. IEEE J. Sel. Top. Signal Process. **7**(3), 435–447 (2013)
3. Luginbuhl, T.E., Willett, P.: Estimating the parameters of general frequency modulated signals. IEEE Trans. Signal Process. **52**(1), 117–131 (2004)
4. Ristic, B., Arulampalam, S., Gordon, N.J.: Beyond the Kalman Filter: Particle Filters for Tracking Applications. Artech House, Boston (2004)
5. Streit, R.L.: Tracking on intensity-modulated data streams. Technical report 11221, NUWC, Newport, Rhode Island, USA (2000)
6. Streit, R.L., Graham, M.L., Walsh, M.J.: Multiple target tracking of distributed targets using Histogram-PMHT. In: Proceedings of the 4th International Conference on Information Fusion. Montreal, Canada (2001)
7. Streit, R.L., Graham, M.L., Walsh, M.J.: Multitarget tracking of distributed targets using histogram-PMHT. Digit. Signal Process. **12**(2), 394–404 (2002)

Chapter 5
Implementation Considerations

This chapter considers some practical aspects of implementing the core H-PMHT in software. The mathematics in the previous chapters has been sufficient to describe the algorithm in detail, but there are a few other considerations required to produce a fully functional tracking system. There are also important decisions to be made during implementation that can result in significant impact of the execution cost of the algorithm and the memory footprint.

The first issue considered is the resampled target prior, which was introduced in Chap. 4 to compensate for the artificial abundance of measurement data introduced by taking the limit of the quantised measurements. The original resampled prior proposed by Streit [20] for the core H-PMHT results in data-dependent scaling terms for both the measurement covariance matrix and the target process noise covariance matrix. Here we describe an alternative proposed by Gaetjens [13, 22] that mitigates the data dependence by modifying the target prior. The next issues discussed are implementation strategies that are motivated by achieving computation efficiency. These are integration implementation, vectorization of the mathematics for two-dimensional measurements and a single-target factorisation that allows for parallel computation of the targets. The EM approach underlying PMHT and H-PMHT is a method for finding a point estimate, but in many applications it is important to understand the accuracy of this estimate and a covariance is also required. After implementation, we review approaches to evaluate the covariance of PMHT based estimates. The final issue considered in this chapter is that of model order estimation, that is, estimating how many components should be used. In the context of target tracking this task is often referred to as track management since it can be interpreted as starting new tracks on targets as they appear and terminating old tracks on targets that are no longer present.

S. J. Davey and H. Gaetjens, *Track-Before-Detect Using Expectation Maximisation*, Signals and Communication Technology,
https://doi.org/10.1007/978-981-10-7593-3_5

5.1 Alternative Resampled Prior

A key characteristic of H-PMHT is the use of a quantisation step to transform an image association problem into a point histogram association problem. The derivation of the core H-PMHT takes a limit over this quantisation step. There are two consequences of this limit: first the quantisation disappears from the final algorithm; but second there is an inflation in the apparent information content of the measurement data because the quantised shots are treated as independent. We saw in Chap. 4 that this inflation emasculates the kinematic prior and the solution was to introduce a special resampled kinematic prior, which assumes a different realisation of the process model for every shot in the image and then constrains these realisations to be identical. The resampled prior ties the kinematics to the quantisation and consequently the information content in the process model is inflated, compensating for the inflated measurement information. In the case of a Gaussian appearance function, the spread of target energy is defined by the measurement covariance matrix Σ. In the case of a Gaussian kinematic process model, the covariance of the process noise is denoted by Q_t^m. When both of these conditions are met, the core H-PMHT can be implemented using a Kalman filter with a data-dependent equivalent measurement covariance given by (4.52)

$$\tilde{\Sigma}_t^m = \frac{1}{\sum_{i=1}^{I+I^U} \bar{n}_t^i w_t^{i,m}} \Sigma = ||\mathfrak{z}_t^m||^{-1} \Sigma,$$

and a data-dependent equivalent process covariance given by (4.66)

$$\tilde{\mathsf{Q}}_t^m = ||\mathbf{z}_t||^{-1} \mathsf{Q}_t^m.$$

Note that the measurement covariance is scaled by the image energy associated with that target whereas the process noise covariance is scaled by the total image energy. In a typical point measurement tracker the measurement covariance matrix influences two parts of the estimation process: the measurement noise contributes to the association of measurements with tracks; and it is used to determine the filter gain and updated covariance. In H-PMHT the actual spreading matrix Σ is used to associate image energy but the equivalent covariance $\tilde{\Sigma}$ is used to update the estimate.

It is important to remember that the EM approach used to derive H-PMHT results in a point estimate not a pdf. We can use the Kalman filter to evaluate this estimate but that does not mean that the matrices involved in that filter reflect the covariance of the estimation error. Rather, the covariance matrix inside the equivalent Kalman filter is simply an intermediate term that defines the appropriate gain. A method to estimate the covariance will be discussed later in this chapter. With this intuition, what is critical is the ratio of $\tilde{\mathsf{Q}}$ and $\tilde{\mathsf{R}}$, or rather its matrix extension, through the Kalman gain. What we mean is that if both $\tilde{\mathsf{Q}}$ and $\tilde{\mathsf{R}}$ are scaled by the same amount then the steady state Kalman gain is not changed. The scaling will affect the strength of the initial prior and the rate of convergence to that steady state. From the definitions of these equivalent covariances, it is clear that the Kalman gain in the equivalent filter is a function of the proportion of the total image energy associated with the target, that is π_t^m.

$$\tilde{\Sigma}_t^m = ||\mathfrak{z}_t^m||^{-1}\Sigma = \left(\pi_t^m||\bar{\mathbf{z}}_t||\right)^{-1}\Sigma, \tag{5.1}$$

The different data dependence of the equivalent covariances means that unexpected performance changes can occur when characteristics of the sensor or scene change that are not directly related to the target. For example, if the sensor resolution is fixed and the target peak SNR is constant, then the energy associated with the target should be approximately constant. However, the total image energy is the target energy plus the noise energy. The noise energy is proportional to the total number of pixels in the image. So the synthetic process noise will be reduced if the size of the sensor image is increased. For fixed resolution this means that the number of pixels on the target is constant and that the coverage footprint of the image increases. Intuitively, this should not influence estimation performance but it will because of the change in the synthetic process noise. Figure 5.1 shows an example of this using a single target from the canonical multi-target scenario. Three track plots are shown: in Fig. 5.1a the SNR was 10 dB and the image size was 200×150; in Fig. 5.1b the SNR was 10 dB and the image size was 600×450; and in Fig. 5.1c the SNR was 20 dB and the image size was 600×450. Clearly the target can be tracked at 10 dB but increasing the image size caused track breaks because it reduces the H-PMHT $\tilde{\mathsf{Q}}$ even though nothing has changed for this target: we really just added more noise pixels in a different location. With a lower $\tilde{\mathsf{Q}}$ and the tracker can no longer follow the manoeuvres. Increasing the SNR restores the track performance because this reduces $\tilde{\tilde{}}$ and compensates for the lower $\tilde{\mathsf{Q}}$. This performance aspect is obviously undesirable.

An alternative way that the same effect could be achieved would be to use a fixed image size and to add a second high SNR target distant from the test target, for example we could use targets 1 and 20 from the canonical scenario. This again increases the total image energy without changing the target energy and has the same end result.

Performance changes also occur with target SNR because this influences the energy associated with the target and hence $\tilde{\Sigma}$. If the target energy is a small fraction of the total image energy, such as in the canonical scenario when all 20 targets are present, then the synthetic process noise will be approximately constant with target SNR but the synthetic measurement noise will be inversely proportional to it. It may seem intuitively appealing that the assumed accuracy of measurements decreases with SNR, but it is relatively simple to cook up a situation where a target can be tracked with 40 dB peak SNR and not 20 dB. With 20 dB SNR the target would be easily detected using a single-frame threshold and tracking should be straightforward.

The point of this discussion is that for the core H-PMHT the covariance of the measurement spread and the target process noise may have to be tuned to give good performance. Further, this tuning will be application dependent since different applications have differing sensor resolutions and target densities. This is less than

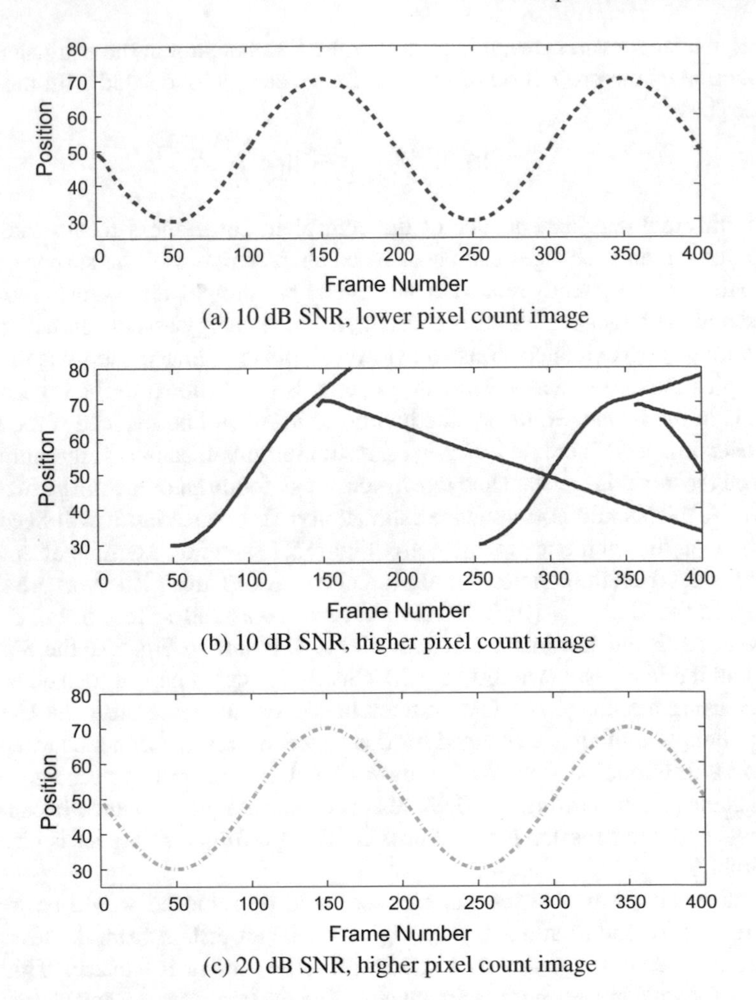

(a) 10 dB SNR, lower pixel count image

(b) 10 dB SNR, higher pixel count image

(c) 20 dB SNR, higher pixel count image

Fig. 5.1 Estimation performance changes with image size for the core H-PMHT

ideal. The process noise should be based on target dynamics and the measurement spread covariance based on sensor resolution. We can ameliorate this situation by changing the resampled target prior.

One possible modification to the data-dependent prior is to only draw realisations of the kinematic state for shot measurements caused by that target. This approach was describe in [13, 22]. Recall that the target prior is defined in the quantized measurement problem and then we take the limit of the quantization. Denote the total number of shot measurements caused by target m as $\tilde{n}_t^m$. This quantity can be expressed as a function of the assignments $\mathbf{K}_t$ and is conditionally independent of all other variables given the assignments. The kinematic state auxiliary sub-function is

$$\mathscr{Q}_x^m = E_{\mathbb{K},\mathbb{Y},\mathbb{N}^U|\hat{\mathbb{X}},\hat{\Pi}} \log\left\{p\left(\mathbb{X}^m\right)\right\} + \mathscr{Q}\left(\mathbb{X}^m, \mathbb{Y}, \mathbb{N}^U\right), \tag{5.2}$$

where the measurement dependent term does not change as a result of the target prior. We can ignore it for this discussion because it remains the same as in Sect. 4.3.1. The state evolution term is

$$E_{\mathbb{K},\mathbb{Y},\mathbb{N}^U|\hat{\mathbb{X}},\hat{\Pi}} \log\left\{p\left(\mathbb{X}^m\right)\right\} = \log\left\{p\left(\mathbf{x}_0^m\right)\right\} + \sum_{t=1}^{T} \log\left\{p\left(\mathbf{x}_t^m|\mathbf{x}_{t-1}^m\right)\right\} E_{\mathbb{K},\mathbb{Y},\mathbb{N}^U|\hat{\mathbb{X}},\hat{\Pi}} \tilde{n}_t^m, \tag{5.3}$$

so what remains is to find the expected value of the target measurement count over the missing data. This count can be expressed as the sum of indicator functions

$$\tilde{n}_t^m = \sum_{i=1}^{I+I^U} \sum_{r=1}^{n_t^i} \delta\left(k_t^{i,r}, m\right), \tag{5.4}$$

where

$$\delta(a,b) = \begin{cases} 1 & a = b, \\ 0 & \text{otherwise,} \end{cases} \tag{5.5}$$

is the Kronecker delta function. Since the expectation is a linear operator, it can interchange order with the sum

$$\begin{aligned} E_{\mathbb{K},\mathbb{Y},\mathbb{N}^U|\hat{\mathbb{X}},\hat{\Pi}} \tilde{n}_t^m &= \sum_{i=1}^{I+I^U} \sum_{r=1}^{n_t^i} E_{\mathbb{K},\mathbb{Y},\mathbb{N}^U|\hat{\mathbb{X}},\hat{\Pi}} \delta\left(k_t^{i,r}, m\right), \\ &= \sum_{i=1}^{I} \sum_{r=1}^{n_t^i} p\left(k_t^{i,r} = m\right) + \sum_{i=I+1}^{I+I^U} \sum_{n_t^i=0}^{\infty} \sum_{r=1}^{n_t^i} p\left(k_t^{i,r} = m\right) p\left(n_t^i\right), \\ &= \sum_{i=1}^{I+I^U} \bar{n}_t^i w_t^{i,m}. \end{aligned} \tag{5.6}$$

This is the expected number of measurements caused by target m given the measurement image and estimates of the kinematic states and mixing proportions. We have already shown in Sect. 4.3.1 that in the limit of quantization, the quantized measurement count is replaced with the image value itself and

$$\lim_{\hbar^2 \to 0} \hbar^2 E_{\mathbb{K},\mathbb{Y},\mathbb{N}^U|\hat{\mathbb{X}},\hat{\Pi}} \tilde{n}_t^m = \sum_{i=1}^{I+I^U} \bar{\mathbf{z}}_t^i w_t^{i,m}. \tag{5.7}$$

For the case of a Gaussian appearance function and a Gaussian target process we can now write

$$\mathcal{Q}_x^m = \log\left\{p\left(\mathbf{X}_0^m\right)\right\} + \sum_{t=1}^{T}\sum_{i=1}^{I+I^U} \bar{\mathbf{z}}_t^i w_t^{i,m}\left[-\frac{1}{2}\left(\mathbf{x}_t^m - \mathsf{F}_t\mathbf{x}_{t-1}^m\right)^{\mathsf{T}}\left(\mathsf{Q}_t^m\right)^{-1}\left(\mathbf{x}_t^m - \mathsf{F}_t\mathbf{x}_{t-1}^m\right)\right.$$
$$\left.-\frac{1}{2}\left(h\left(\mathbf{x}_t^m\right) - \tilde{\mathbf{y}}_t^m\right)^{\mathsf{T}}(\Sigma)^{-1}\left(h\left(\mathbf{x}_t^m\right) - \tilde{\mathbf{y}}_t^m\right)\right]. \quad (5.8)$$

Within a single frame the data-dependent scaling term no longer modifies the ratio of the measurement and process noise. For a time recursive implementation, the scaling term cancels completely and the optimisation can be performed using a Kalman filter with unmodified Q and Σ. For the batch case, the relative powers between frames influence the priority of each frame within the estimation.

The other part of the algorithm that is affected by the target prior is the pixel association weights that comprise the missing data likelihood. In the core H-PMHT, the target prior is constant and cancels out of the weights expression. This modified prior depends on the number of measurements due to each target and so is not constant with respect to $\mathbb{K}$. The modified prior is

$$p\left(\mathbf{X}_t|\mathbf{X}_{t-1}, \mathbf{K}_t\right) = \prod_{m=1}^{M} p\left(\mathbf{x}_t^m|\mathbf{x}_{t-1}^m\right)^{\tilde{n}_t^m}$$
$$= \prod_{i=1}^{I}\prod_{r=1}^{n_t} p\left(\mathbf{x}_t^{k_t^{i,r}}|\mathbf{x}_{t-1}^{k_t^{i,r}}\right). \quad (5.9)$$

The core H-PMHT complete data likelihood is given by (4.15):

$$p_{\mathsf{comp}}\left(\mathbb{X}, \mathbb{K}, \mathbb{Y}, \mathbb{N}; \Pi\right)$$
$$= C(\mathbb{N})\prod_{m=1}^{M}\left[p\left(\mathbf{x}_0^m\right)\prod_{t=1}^{T} p\left(\mathbf{x}_t^m|\mathbf{x}_{t-1}^m\right)\right]\prod_{t=1}^{T}\prod_{i=1}^{I}\prod_{r=1}^{n_t^i}\pi_t^{k_t^{i,r}} g^{k_t^{i,r}}\left(\mathbf{y}_t^{i,r}|\mathbf{x}_t^{k_t^{i,r}}\right).$$

We must now modify this to reflect the new prior. Noticing that the product in (5.9) is the same as the measurement product in (4.15), the modified complete data likelihood is

$$p_{\mathsf{comp}}\left(\mathbb{X}, \mathbb{K}, \mathbb{Y}, \mathbb{N}; \Pi\right)$$
$$= C(\mathbb{N})\prod_{m=1}^{M} p\left(\mathbf{x}_0^m\right)\prod_{t=1}^{T}\prod_{i=1}^{I}\prod_{r=1}^{n_t^i} p\left(\mathbf{x}_t^{k_t^{i,r}}|\mathbf{x}_{t-1}^{k_t^{i,r}}\right)\pi_t^{k_t^{i,r}} g^{k_t^{i,r}}\left(\mathbf{y}_t^{i,r}|\mathbf{x}_t^{k_t^{i,r}}\right). \quad (5.10)$$

The missing data likelihood is given by (4.17)

$$p_{\text{miss}}\left(\mathbb{K}, \mathbb{Y}|\mathbb{N}, \hat{\mathbb{X}}; \hat{\Pi}\right) = \frac{p\left(\hat{\mathbb{X}}, \mathbb{K}, \mathbb{Y}, \mathbb{N}; \hat{\Pi}\right)}{p\left(\hat{\mathbb{X}}, \mathbb{N}; \hat{\Pi}\right)}.$$

The numerator is the complete data; we need a modified expression for the denominator. This can be expressed as

$$\begin{aligned} p\left(\hat{\mathbb{X}}, \mathbb{N}; \hat{\Pi}\right) &= \sum_{\mathbb{K}} p\left(\hat{\mathbb{X}}, \mathbb{K}, \mathbb{N}; \hat{\Pi}\right) \\ &= \prod_{t=1}^{T} \sum_{\mathbf{K}_t} p\left(\mathbf{X}_t|\mathbf{X}_{t-1}, \mathbf{K}_t\right) p\left(\mathbf{K}_t|\mathbf{N}_t; \hat{\Pi}_t\right) p\left(\mathbf{N}_t; \hat{\Pi}_t\right) \end{aligned} \tag{5.11}$$

The first term is given by (5.9) above. The second term was defined in (4.11) and is

$$p\left(\mathbf{K}_t|\mathbf{N}_t, \mathbf{X}_t; \Pi_t\right) = \prod_{i=1}^{I} \prod_{r=1}^{n_t^i} \frac{\pi_t^{k_t^{i,r}} G_t^{k_t^{i,r},i}}{G_t^i}.$$

The final term is our familiar multinomial from (4.5)

$$p\left(\mathbf{N}_t; \hat{\Pi}_t\right) = C(\mathbf{N}_t) \prod_{i=1}^{I} \left(G_t^i\right)^{n_t^i},$$

Combining these, (5.11) becomes

$$p\left(\hat{\mathbb{X}}, \mathbb{N}; \hat{\Pi}\right) = C(\mathbf{N}_t) \prod_{m=1}^{M} p\left(\mathbf{x}_0^m\right) \prod_{t=1}^{T} \prod_{i=1}^{I} \prod_{r=1}^{n_t^i} \sum_{k=0}^{M} p\left(\mathbf{x}_t^k|\mathbf{x}_{t-1}^k\right) \pi_t^k G_t^{k,i} \tag{5.12}$$

We can now write the missing data likelihood as

$$p_{\text{miss}}\left(\mathbb{K}, \mathbb{Y}|\mathbb{N}, \hat{\mathbb{X}}; \hat{\Pi}\right) = \prod_{t=1}^{T} \prod_{i=1}^{I} \prod_{r=1}^{n_t^i} \frac{p\left(\mathbf{x}_t^{k_t^{i,r}}|\mathbf{x}_{t-1}^{k_t^{i,r}}\right) \pi_t^{k_t^{i,r}} g^{k_t^{i,r}}\left(\mathbf{y}_t^{i,r}|\mathbf{x}_t^{k_t^{i,r}}\right)}{\sum_{k=0}^{M} p\left(\mathbf{x}_t^k|\mathbf{x}_{t-1}^k\right) \pi_t^k G_t^{k,i}} \tag{5.13}$$

From this, it is a small leap to see that the weights become

$$w_t^{i,m} = \frac{p\left(\hat{\mathbf{x}}_t^m|\hat{\mathbf{x}}_{t-1}^m\right) \hat{\pi}_t^m G_t^{m,i}}{\sum_{s=0}^{M} p\left(\hat{\mathbf{x}}_t^s|\hat{\mathbf{x}}_{t-1}^s\right) \hat{\pi}_t^s G_t^{s,i}}. \tag{5.14}$$

The core H-PMHT weight in (4.24) is

$$w_t^{i,m} = \frac{\hat{\pi}_t^m G_t^{m,i}}{\sum_{s=0}^{M} \hat{\pi}_t^s G_t^{s,i}},$$

so the change is to scale each term by the corresponding state transition probability. For closely spaced targets this means that the modified prior will prefer associations that minimise the manoeuvres. For spaced targets it will make minimal difference.

5.2 Integrals

There are two types of integrals that need to be implemented in the H-PMHT algorithm. The first of these is the cell probability for each component $G_t^{m,i}$ defined in (4.9). This is the probability that a measurement known to come from component m falls in pixel i

$$\begin{aligned} G_t^{m,i} &= p\left(\mathbf{y} \in W^i | \mathbf{X}_t, k = m\right), \\ &= \int_{W^i} g^m\left(\mathbf{y}|\mathbf{x}_t^m\right) \mathrm{d}\mathbf{y}. \end{aligned}$$

The actual evaluation of this integral in software obviously depends on the functional form of $g^m(\cdot)$. If an analytic solution exists then no further discussion is required but often it does not. Numerical integration is a well studied field, for example [12]. We have no intention to expound the merits of one technique over another for H-PMHT. However, note that these integrals are required for each target over every pixel at every iteration. In practice we can reduce the number of calculations by approximating $G_t^{m,i} \approx 0$ for pixels a very long way away from $\mathbf{x}_t^m$ but the number of integrals is still very large. Experience shows this is the computation bottleneck of the algorithm.

One might expect things to be simpler for the Gaussian spread function case but unfortunately they are not simple without additional assumptions. In the one-dimensional image case, the integral above can be expressed in terms of the error function *erf*, which is efficiently implemented in readily available numerical libraries. However, this does not extend to multiple dimensions unless $g^m(\cdot)$ can be factorised, as discussed in the next section. In physical terms, this means that if the spread function is rotated from the sensor axes then numerical methods are required.

The second type of integral used by the H-PMHT algorithm is the expectation of the log-likelihood (4.30)

$$\int_{W^i} \log\left\{g^m\left(\mathbf{y}|\mathbf{x}_t^m\right)\right\} g^m\left(\mathbf{y}|\hat{\mathbf{x}}_t^m\right) \mathrm{d}\mathbf{y}.$$

In the general case, this will be treated in the same way as the cell probability, but in the Gaussian $g^m(\cdot)$ case, this term can be simplified. Recall that the Gaussian $g^m(\cdot)$ leads to a cell-level centroid defined in (4.34):

$$\bar{\mathbf{y}}_t^{m,i} = \frac{\int_{W^i} \mathbf{y}\mathcal{N}\left(\mathbf{y}; h\left(\hat{\mathbf{x}}_t^m\right), \Sigma\right) \mathrm{d}\mathbf{y}}{G_t^{m,i}} = \frac{\int_{W^i} \mathbf{y}\mathcal{N}\left(\mathbf{y}; h\left(\hat{\mathbf{x}}_t^m\right), \Sigma\right) \mathrm{d}\mathbf{y}}{\int_{W^i} \mathcal{N}\left(\mathbf{y}; h\left(\hat{\mathbf{x}}_t^m\right), \Sigma\right) \mathrm{d}\mathbf{y}}.$$

The vector derivative of a Gaussian pdf is given by

$$\begin{aligned}
\frac{\mathrm{d}\mathcal{N}(\mathbf{y}; \mu, \Sigma)}{\mathrm{d}\mathbf{y}} &= |2\pi\Sigma|^{-1} \frac{\mathrm{d}}{\mathrm{d}\mathbf{y}} \exp\left\{-\frac{1}{2}(\mathbf{y}-\mu)^{\mathsf{T}} \Sigma^{-1} (\mathbf{y}-\mu)\right\}, \\
&= \mathcal{N}(\mathbf{y}; \mu, \Sigma) \left[-\frac{1}{2}\frac{\mathrm{d}}{\mathrm{d}\mathbf{y}}(\mathbf{y}-\mu)^{\mathsf{T}} \Sigma^{-1} (\mathbf{y}-\mu)\right], \\
&= \mathcal{N}(\mathbf{y}; \mu, \Sigma) \left[-\Sigma^{-1} (\mathbf{y}-\mu)\right],
\end{aligned} \tag{5.15}$$

which can be rearranged to give

$$\mathbf{y}\mathcal{N}(\mathbf{y}; \mu, \Sigma) = \mu\mathcal{N}(\mathbf{y}; \mu, \Sigma) - \Sigma \frac{\mathrm{d}\mathcal{N}(\mathbf{y}; \mu, \Sigma)}{\mathrm{d}\mathbf{y}}. \tag{5.16}$$

Putting this back into the cell centroid gives

$$\begin{aligned}
\bar{\mathbf{y}}_t^{m,i} &= \frac{\mu}{G_t^{m,i}} \int_{W^i} \mathcal{N}\left(\mathbf{y}; h\left(\hat{\mathbf{x}}_t^m\right), \Sigma\right) \mathrm{d}\mathbf{y} - \frac{\Sigma}{G_t^{m,i}} \int_{W^i} \frac{\mathrm{d}\mathcal{N}(\mathbf{y}; \mu, \Sigma)}{\mathrm{d}\mathbf{y}} \mathrm{d}\mathbf{y}, \\
&= \mu - \frac{\Sigma}{G_t^{m,i}} \int_{W^i} \frac{\mathrm{d}\mathcal{N}(\mathbf{y}; \mu, \Sigma)}{\mathrm{d}\mathbf{y}} \mathrm{d}\mathbf{y}.
\end{aligned} \tag{5.17}$$

In the one-dimensional case where $W^i \equiv [W_L^i, W_U^i)$ this becomes

$$\bar{\mathbf{y}}_t^{m,i} = \mu - \frac{\Sigma}{G_t^{m,i}} \left[\mathcal{N}\left(W_U^i; \mu, \Sigma\right) - \mathcal{N}\left(W_L^i; \mu, \Sigma\right)\right]. \tag{5.18}$$

The only term that needs to be numerically approximated is $G_t^{m,i}$, which we have already determined earlier. In the next section we will see how to extend this to higher dimensions.

5.3 Vectorised Two-Dimensional Case

The H-PMHT derivation developed in the previous chapter uses single indexing of the sensor cells. Single indexing does not limit the sensor dimensionality, but it makes it more difficult to decouple the sensor dimensions if they are independent because the index does not intuitively map spatially. One of the most common measurement types is a two-dimensional image, for example a video camera. In this case it is possible to write much of the H-PMHT maths as matrix-vector algebra if we can assume independence between the two dimensions. This representation was developed in [8]. Dimensional independence has two requirements: firstly, the pixels must be a Cartesian product of two one-dimensional spaces, that is a rectangular grid, although not necessarily a uniform one; secondly, the target appearance function should be separable into two one-dimensional functions. Mathematically, the Cartesian product condition means that

$$G_t^{m,l} = \int_{W_l} g\left(\mathbf{y}|\mathbf{x}_t^m\right) \mathrm{d}\mathbf{y} \equiv \int_{W_X^i} \int_{W_Y^j} g\left(u, v|\mathbf{x}_t^m\right) \mathrm{d}v\mathrm{d}u, \tag{5.19}$$

where the arbitrary index l corresponds to the same cell as the arbitrary double index i, j. The separable appearance function means that

$$g\left(u, v|\mathbf{x}_t^m\right) \equiv g_X\left(u|\mathbf{x}_t^m\right) g_Y\left(v|\mathbf{x}_t^m\right). \tag{5.20}$$

The standard H-PMHT equations derived in Chap. 4 apply equally well to this problem, but it is desirable to exploit the structure of the target function to achieve a factorised algorithm. The factorised algorithm will be easier to read, easier to implement and efficient. We now derive a matrix-vector implementation of H-PMHT for the separable problem above. Pakfiliz and Efe [17] described a mathematically equivalent form for the two-dimensional case but their representation did not arrive at the compact intuitive expressions we will present here. The form of the matrix-vector expressions points towards efficient implementation. In the following discussion, the time subscripts are suppressed to simplify notation.

The per-target cell probability (4.9) directly factorises into a product of two integrals:

$$\begin{aligned} G^{m,i,j}\left(\mathbf{x}^m\right) &= \left\{\int_{W_X^i} g_X(u|\mathbf{x}^m)\mathrm{d}u\right\}\left\{\int_{W_Y^j} g_Y(v|\mathbf{x}^m)\mathrm{d}v\right\} \\ &\equiv G_X^m(i)G_Y^m(j). \end{aligned} \tag{5.21}$$

Define the stacked vectors

$$\boldsymbol{G}_X^m = \left[G_X^m(1), G_X^m(2), \ldots G_X^m(I_X)\right]^\mathsf{T}, \tag{5.22}$$

$$\boldsymbol{G}_Y^m = \left[G_Y^m(1), G_Y^m(2), \ldots G_Y^m(I_Y)\right]^\mathsf{T}, \tag{5.23}$$

where I_X and I_Y are the numbers of cells in the X and Y dimensions respectively. So $\boldsymbol{G}_X^m$ is a $I_X \times 1$ vector and $\boldsymbol{G}_Y^m$ is a $I_Y \times 1$ vector. Let G_X be a $I_X \times M$ matrix of the X per-cell contributions

$$\mathsf{G}_X = \left[\boldsymbol{G}_X^0, \boldsymbol{G}_X^1, \ldots \boldsymbol{G}_X^M\right], \tag{5.24}$$

and similarly define G_Y, then

$$\begin{aligned}\mathsf{G} &\equiv \left\{G^{i,j}\left(\mathbf{X};\Pi\right)\right\} = \left\{\sum_{m=0}^{M} \pi^m G^{m,i,j}\left(\mathbf{x}^m\right)\right\}, \\ &= \mathsf{G}_X \,\Lambda\, \mathsf{G}_Y{}^{\mathsf{T}}, \end{aligned} \tag{5.25}$$

where Λ is a $M \times M$ diagonal matrix of the π^m. Note that (5.25) is simply a matrix-vector version of (4.8), which is

$$G_t^i = \sum_{m=0}^{M} \pi_t^m G_t^{m,i}.$$

They are equivalent, but (5.25) represents an explicit factorisation of (4.8) exploiting the separable point spread function.

For consistency of notation, let $\bar{\mathsf{Z}}$ denote the extrapolated measurement image arranged as a matrix. The mixing proportion update is given by Eq. (4.62)

$$\hat{\pi}^m := \frac{\sum_{i=1}^{I+I^U} \bar{\mathbf{z}}^i w^{i,m}}{\sum_{s=0}^{M} \sum_{i=1}^{I+I^U} \bar{\mathbf{z}}^i w^{i,s}},$$

or equivalently using associated images (4.69)

$$\hat{\pi}_t^m = \frac{||\mathfrak{z}_t^m||}{||\bar{\mathbf{z}}_t||}.$$

The associated image power $||\mathfrak{z}_t^m||$ can be expressed as an inner product of the vectors $\boldsymbol{G}_X$ and $\boldsymbol{G}_Y$

$$\tilde{\pi}^m = \hat{\pi}^m \sum_{i,j} \frac{\bar{\mathbf{z}}^{ij}}{G^i} G_X^m(i) G_Y^m(j) = \hat{\pi}^m \{\boldsymbol{G}_X^m\}^{\mathsf{T}} \left(\bar{\mathsf{Z}}./\mathsf{G}\right) \boldsymbol{G}_Y^m, \tag{5.26}$$

where the *MATLAB* style notation. denotes Hadamard element-wise division.

The cell-level centroids are given by

$$\begin{aligned}\bar{\mathbf{y}}_t^{m,i,j} &= \frac{1}{G_X^m(i) G_Y^m(j)} \int_{W_X^i} \int_{W_Y^j} \begin{bmatrix} u \\ v \end{bmatrix} g_X(u|\mathbf{x}^m) g_Y(v|\mathbf{x}^m) dv du, \\ &= \frac{1}{G_X^m(i) G_Y^m(j)} \begin{bmatrix} \int_{W_X^i} \int_{W_Y^j} u g_X(u|\mathbf{x}^m) g_Y(v|\mathbf{x}^m) dv du \\ \int_{W_X^i} \int_{W_Y^j} v g_X(u|\mathbf{x}^m) g_Y(v|\mathbf{x}^m) dv du \end{bmatrix},\end{aligned}$$

$$= \frac{1}{G_X^m(i)G_Y^m(j)} \begin{bmatrix} G_Y^m(j) \int_{W_X^i} u g_X(u|\mathbf{x}^m) du \\ G_X^m(i) \int_{W_Y^j} v g_Y(v|\mathbf{x}^m) dv \end{bmatrix},$$

$$= \begin{bmatrix} 1/G_X^m(i) \int_{W_X^i} u g_X(u|\mathbf{x}^m) du \\ 1/G_Y^m(j) \int_{W_Y^j} v g_Y(v|\mathbf{x}^m) dv \end{bmatrix}, \tag{5.27}$$

From this we define one-dimensional centroids

$$\tilde{\mathbf{y}}_X^m(i) = \int_{W_X^i} u g_X(u|\mathbf{x}^m) du, \tag{5.28}$$

$$\tilde{\mathbf{y}}_Y^m(j) = \int_{W_Y^j} v g_Y(v|\mathbf{x}^m) dv, \tag{5.29}$$

that combine to form the overall vector centroid

$$\bar{\mathbf{y}}_t^{m,i,j} = \left[\tilde{\mathbf{y}}_X^m(i)/G_X^m(i), \quad \tilde{\mathbf{y}}_Y^m(j)/G_Y^m(j)\right]^{\mathsf{T}}. \tag{5.30}$$

As intuitively expected, the centroids are independent in the sensing dimensions. The synthetic measurements are formed by the weighted combination of the vector centroids

$$\bar{\mathbf{y}}_t^m = \frac{\pi^m}{||\mathfrak{z}_t^m||} \sum_i \sum_j \bar{z}^{i,j} G_X^m(i) G_Y^m(j) \bar{\mathbf{y}}_t^{m,i,j},$$

$$= \frac{\pi^m}{||\mathfrak{z}_t^m||} \begin{bmatrix} \sum_i \sum_j \bar{z}_{ij} \; G_X^m(i) \; G_Y^m(j) \; \tilde{\mathbf{y}}_X^m(i) \; / \; G_X^m(i) \\ \sum_i \sum_j \bar{z}_{ij} \; G_X^m(i) \; G_Y^m(j) \; \tilde{\mathbf{y}}_Y^m(j) \; / \; G_Y^m(j) \end{bmatrix},$$

$$= \frac{\pi^m}{||\mathfrak{z}_t^m||} \begin{bmatrix} \sum_i \tilde{\mathbf{y}}_X^m(i) \left\{ \sum_j \bar{z}^{i,j} G_Y^m(j) \right\} \\ \sum_j \tilde{\mathbf{y}}_Y^m(j) \left\{ \sum_i \bar{z}^{i,j} G_X^m(i) \right\} \end{bmatrix},$$

$$= \frac{\pi^m}{||\mathfrak{z}_t^m||} \begin{bmatrix} \left\{ \bar{\mathbf{y}}_t^{m,i} X^m \right\}^{\mathsf{T}} \bar{\mathsf{Z}} \, \boldsymbol{P}_Y^m \\ \{\boldsymbol{P}_X^m\}^{\mathsf{T}} \, \bar{\mathsf{Z}} \, \bar{\mathbf{y}}_t^{m,i} Y^m \end{bmatrix}, \tag{5.31}$$

where

$$\bar{\mathbf{y}}_t^{m,i} X^m = \left[\tilde{z}_X^m(1), \ldots \tilde{z}_X^m(N_X)\right]^\mathsf{T}, \tag{5.32}$$

and similarly for $\bar{\mathbf{y}}_t^{m,i} Y^m$.

The synthetic measurement and process covariances remain the same, and the state estimates may now be determined using a Kalman filter.

The matrix-vector representation leads to a more efficient implementation because the per-target cell probabilities and the synthetic measurements can be calculated using marginal expressions rather than joint calculations across the whole image. If the number of pixels influenced by a target is N in each dimension, the per-target cell probabilities can be determined using $2N$ one-dimensional integrals instead of N^2 two-dimensional integrals. This represents a significant simplification. The integrals required for the synthetic measurements experience the same benefit. Further, if the algorithm is implemented in a higher level computer language, it is likely that expressing the mathematics this way makes the algorithm amenable to the language's intrinsic pipelining and parallelization capabilities. It is the authors' experience that within the Matlab environment the speed difference is significant.

5.4 Single-Target Chip Processing

The H-PMHT is a multi-target algorithm that performs joint target data association. This is a good thing from the perspective of estimation performance when targets are closely spaced, but the coupled nature of the pixel association weights is undesirable from the perspective of parallel implementation. Further, a direct implementation of the pixel weights in (4.24)

$$w_t^{i,m} = \frac{\hat{\pi}_t^m G_t^{m,i}}{\sum_{s=0}^{M} \hat{\pi}_t^s G_t^{s,i}}.$$

requires the calculation of the integrals $G_t^{m,i}$ discussed above for every target and every pixel. It is simple enough to implement a gate and avoid unnecessary calculations of this kind, but the denominator of (4.24) still needs to be defined over all of the pixels. This can be excessive when the image size is large and again it inhibits parallel implementation. We now propose an approximate method that allows for independent target calculations.

The EM iterations of H-PMHT mean that the pixel weights are determined by the state estimates from the previous iteration and the new estimates are obtained independently. So the interaction between targets is indirect through the weights. The direct implementation of this process is shown in Fig. 5.2a. We refer to this architecture as centralised association since the implementation is built around a single monolithic pixel association processor. Assuming gating restricts target calculations to I_t^m pixels, then the computation complexity of the data association scales approximately with $\sum_m I_t^m$. The memory requirement of the data association scales with

$I + \sum_m I_t^m$ where the second term is due to the denominator of (4.24). When the image is very large, the I term can become a dominant factor. A decentralised alternative implementation scheme is to have M processors, each of which performs first pixel association and then estimation for a single target. Between EM iterations the processors share their state estimates. This data flow is illustrated in Fig. 5.2b. In this scheme the pixel association blocks are redundant copies of each other. However, since each feeds into a single-target estimator it needs only to determine the pixel weights for the pixels close enough to that target to be significant. Due to redundancy, the decentralised computation cost is greater; the worst case is $M \sum_m I_t^m$ when all of the targets overlap completely, but this is very unlikely to occur in practice. The memory requirement becomes $2M \sum_m I_t^m$, which can be either more than the central architecture or less depending on the target density and the image size. This expression arises because each pixel association processor only needs to retain the numerator terms due to its target and the corresponding denominator terms. Note also that these pixel association blocks do not require access to the whole sensor image, they only need an image chip gated around the target state estimate for their particular target.

The most efficient architecture is an approximation to the decentralised implementation we refer to as single-target chip processing. The difference here is that the communication between the single-target processors between EM iterations is removed, as shown in Fig. 5.2c. Each single-target processor knows the initial value of the state estimates for the other targets but the refined estimates due to the EM iterations are not available. This scheme is an approximation to the proper EM solution because the estimates are not shared but it completely decouples the association-estimation cycle. There are two advantages: first the pixel probabilities due to the clutter and other targets do not change from one EM iteration to the next because the state estimates of the other targets are not changing. This means that these terms need only be calculated once and the cost of each EM iteration is reduced. Second, since the processing of each target is now truly independent, the single-target processors can be run in parallel and multiple threads can be used.

We now illustrate the effectiveness of single-target chip processing on the canonical multi-target scenario. Three versions of the H-PMHT[1] are compared: the core algorithm, here denoted H-PMHT-c, that uses central association as in Fig. 5.2a; a bank of independent single-target filters, H-PMHT-i; and the single-target chip described above, H-PMHT-s. The bank of independent filters will be expected to have low quality output when the targets are close together, but it is a performance bound: no other approach can be quicker. For both the H-PMHT-i and H-PMHT-s the local chip around each track was defined as a box of size $3\sigma_X \times 3\sigma_Y$ where σ_X and σ_Y are the standard deviations of $g_X(\cdot)$ and $g_Y(\cdot)$, in this case both unity.

---

[1] *MakeHPMHTParams*
The H-PMHT toolbox contains two main functions for running H-PMHT. *MakeHPMHTParams* defines tracking parameters, including which version of H-PMHT to execute, and *HPMHTTracker* actually does the tracking. The default tracking parameters for the single-target chip H-PMHT are created by *MakeHPMHTParams('STC')*. For details on how to use this function, refer to the H-PMHT toolbox documentation.

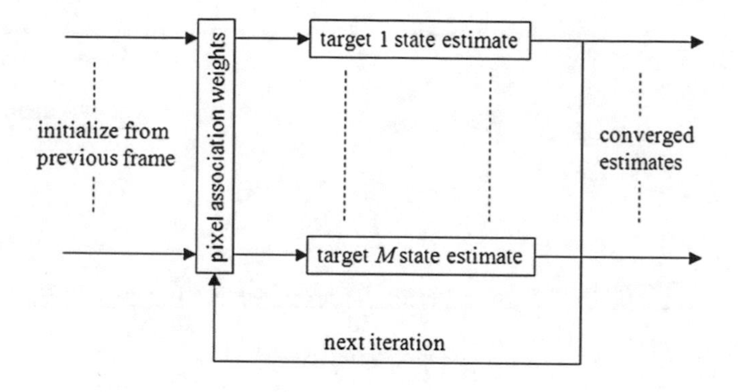

(a) Central association

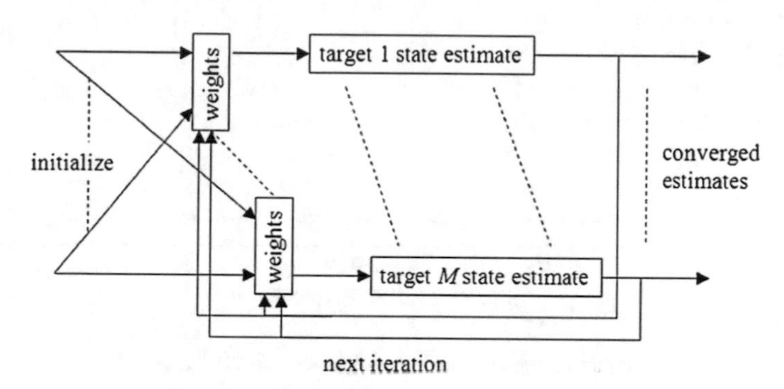

(b) Decentralized association

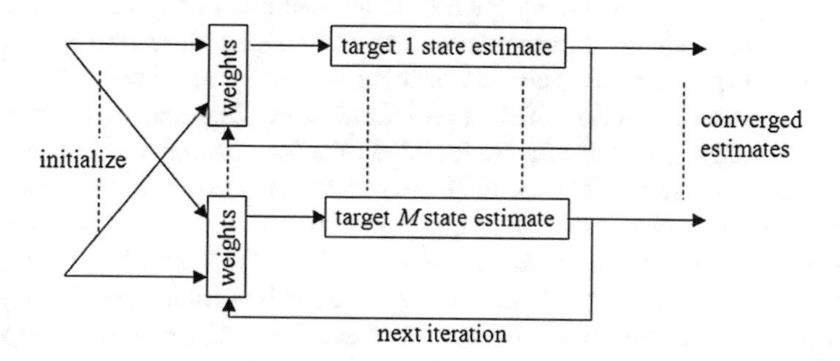

(c) Single target chip processing

Fig. 5.2 Implementation architectures

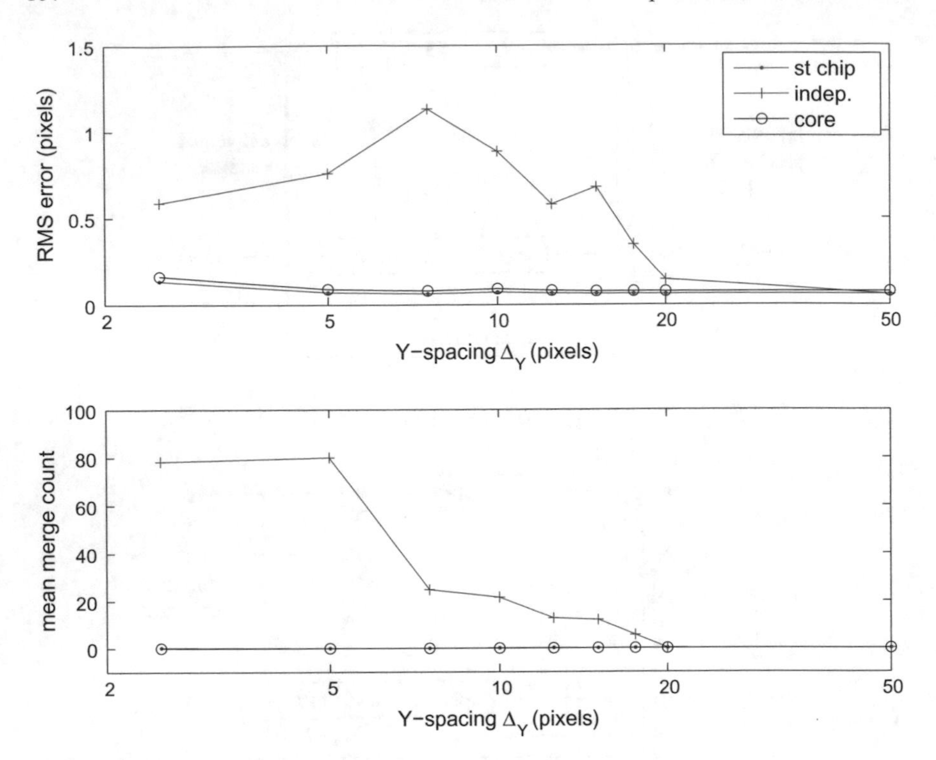

Fig. 5.3 Estimation error and merge metrics as a function of spacing Δ_Y

We first demonstrate that the approximation does not significantly impact on the track output quality by measuring the estimation and cardinality accuracy through the RMS error and redundant track count. Redundant tracks cause extra computation overhead for no performance gain and in this analysis we are chiefly interested in resource scaling as a function of the problem size. For this reason, the redundant tracks are merged and new tracks are initiated by a track manager that is described in detail in the next section. The cardinality plot shows the mean merge count, which is equivalent to the mean number of redundant tracks. The measures are shown as a function of the target spacing parameter Δ_Y for 20 targets in Fig. 5.3. As expected, the independent H-PMHT-i performs well when Δ_Y is large and all of the targets are a long way apart but does poorly when they interact. For $\Delta_Y \geq 20$ the independent H-PMHT-i does just as well as the coupled versions. The mean number of merges increases monotonically as the spacing is reduced but the estimation error peaks at around $\Delta_Y \approx 8$ and actually improves for very close target spacing even though there are more merges. This happens because association errors result in lower estimation error when the targets become very close. The single-target chip H-PMHT-s shows no significant performance degradation over the central H-PMHT-c: both track all of the targets well even for very dense scenarios.

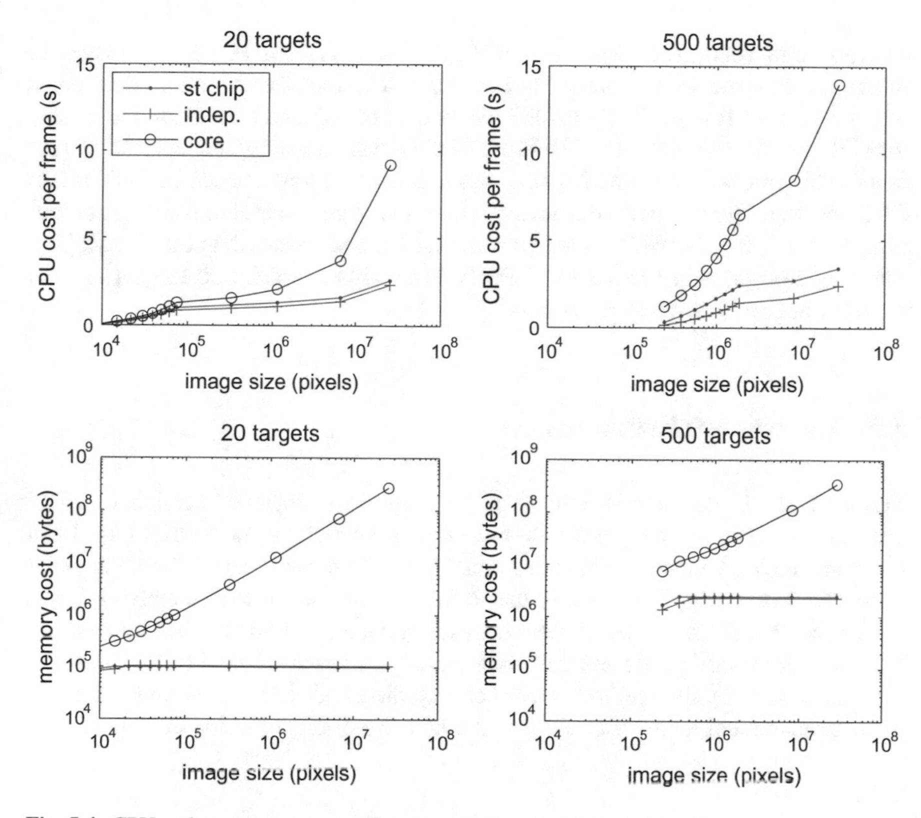

Fig. 5.4 CPU and memory costs as a function of M and total image size N

The scaling performance was investigated by measuring the algorithm resource requirements as the image size was increased for 20 and for 500 targets. The 500 target case was realised by tiling 25 copies of the 20 target canonical multi-target scenario in a 5×5 grid; the trajectories are repeated with offsets so that there are clusters of 20 targets and the clusters do not interact. Resource requirements were measured in CPU seconds and bytes allocated per frame. The parameter Δ_Y controls how closely the targets are spaced in Y and was used to adjust the total image size for a prescribed number of targets. For each combination of target count and spacing, 100 Monte Carlo trials were used.

Figure 5.4 shows the CPU and memory costs of the H-PMHT variants as a function of target count and image size: the left column is for 20 targets and the right 500 targets. As predicted, the H-PMHT-s and H-PMHT-i implementations had a constant memory footprint independent of the image size and the memory cost scaled linearly with the target count. Note that the figures do not include the memory required to store the sensor image itself. In contrast, the central H-PMHT-c memory footprint grew with sensor image size: it showed less dependence on the number of targets because the memory cost was dominated by terms that are a function of input image size, not target count. The CPU costs for H-PMHT-s and H-PMHT-i grew slowly with image size because the cost of the track manager searching for new tracks dominates when the image is large. This cost is independent of the target count, which is why the CPU cost for H-PMHT-s and H-PMHT-i did not significantly increase with target count. This process is described in the next section. In contrast, the central H-PMHT-c CPU cost grew quickly with sensor image size and target count. For the largest input images tested, the H-PMHT-c cost was around 4 times greater than the H-PMHT-s. The single-target chip version of H-PMHT was able to track 500 targets in a 28 Mpixel image at around 3 s per frame.

5.5 Covariance Estimates

Earlier in this chapter we raised the issue of covariance estimates. Recall that the EM process gives a point estimate for the kinematic state and that the matrices used in a Kalman filter implementation are not a reflection of the mean square error of this point estimate. One strategy is to simply use these Kalman filter matrices anyway, but it can be shown that they underestimate the estimation error. Two different approaches have been developed to address this problem for point measurement PMHT, they are summarised in [6]. No methods have been applied to H-PMHT, although the two point measurement methods could be extended. We briefly describe them here.

5.5.1 Observed Information

The first approach is based on estimating the Fisher Information matrix for the PMHT measurement model. This was first developed by Walsh [23]. To build intuition, consider a scalar Gaussian likelihood $p(x) = \mathcal{N}(x; \mu, \sigma^2)$. The second derivative of the log-likelihood is

$$\begin{aligned} \frac{\mathrm{d}^2 \log\{p(x)\}}{\mathrm{d}x^2} &= \frac{\mathrm{d}^2}{\mathrm{d}x^2}\left(-0.5\log\{2\pi\} - \log\{\sigma\} - \frac{(x-\mu)^2}{2\sigma^2}\right), \\ &= -0.5\sigma^{-2}. \end{aligned} \tag{5.33}$$

Thus the curvature of the log-likelihood is inversely proportional to the covariance. In the multi-variate case, this second derivative is the Hessian of the log-likelihood and the analogous relationship is

$$\frac{\mathrm{d}^2 \log\{p(\mathbf{x})\}}{\mathrm{d}\mathbf{x}^2} = -0.5\Sigma^{-1} \equiv -\mathsf{I}, \tag{5.34}$$

and the inverse covariance matrix is referred to as the Fisher Information matrix. In a multi-dimensional sense, the covariance matrix of a multi-variate Gaussian specifies the curvature of the log-likelihood. When the pdf is not Gaussian, then in general, this Hessian is not constant. In this case, one way to approximate the likelihood is as a Gaussian with a mean at the peak of the likelihood and a covariance matrix chosen to match the local curvature at the mean. For example, if the likelihood is a sinc function, we approximate it as a Gaussian matched to the central main peak. Under this approximation, the covariance is the inverse of the Hessian of the log-likelihood evaluated at the peak of the likelihood.

We present here the expressions from the summary in [6] and simplify notation by assuming constant matrices F, H, Q and R. The observed information matrix is the Hessian of the joint likelihood of the states and measurements evaluated at the EM state estimate and is the sum of a kinematic state term and a measurement term

$$\mathsf{I}(\hat{\mathbb{X}}, \mathbb{Y}) = \mathsf{I}^{\text{prior}}(\hat{\mathbb{X}}) + \mathsf{I}^{\text{data}}(\mathbb{Y}|\hat{\mathbb{X}}). \tag{5.35}$$

The state prior term $\mathsf{I}^{\text{prior}}(\hat{\mathbb{X}})$ is block diagonal

$$\mathsf{I}^{\text{prior}}(\hat{\mathbb{X}}) = \begin{bmatrix} \mathsf{I}_1^{\text{prior}} & & \\ & \ddots & \\ & & \mathsf{I}_{T-1}^{\text{prior}} \end{bmatrix}, \quad \mathsf{I}_t^{\text{prior}} = \begin{bmatrix} \chi_t & -\delta_t \\ -\delta_t^{\mathsf{T}} & \chi_{t+1} \end{bmatrix} \tag{5.36}$$

where χ_t and δ_t are block diagonal matrices with M identical blocks χ_t^m and δ_t^m given by

$$\chi_t^m = \begin{cases} \mathsf{P}_0^{-1} + \mathsf{F}^{\mathsf{T}}\mathsf{Q}^{-1}\mathsf{Q} & t = 1, \\ \mathsf{Q}^{-1} + \mathsf{F}^{\mathsf{T}}\mathsf{Q}^{-1}\mathsf{Q} & 1 < t < T, \\ \mathsf{Q}^{-1} & t = T, \end{cases} \tag{5.37}$$

$$\delta_t^m = \mathsf{F}^{\mathsf{T}}\mathsf{Q}^{-1}. \tag{5.38}$$

The data-dependent term is also block diagonal

$$\mathsf{I}^{\text{data}}(\mathbb{Y}|\hat{\mathbb{X}}) = \begin{bmatrix} \mathsf{I}_1^{\text{data}} & & \\ & \ddots & \\ & & \mathsf{I}_{T-1}^{\text{data}} \end{bmatrix}, \quad \mathsf{I}_t^{\text{data}} = \mathsf{B}_t - \mathsf{C}_t + \mathsf{D}_t \tag{5.39}$$

where B_t and C_t are again block diagonal with M blocks given by

$$\mathsf{B}_t^m = \mathsf{H}^\mathsf{T}\left(\tilde{\mathsf{R}}^m\right)^{-1}\mathsf{H}, \tag{5.40}$$

$$\mathsf{C}_t^m = \mathsf{H}^\mathsf{T}\mathsf{R}^{-1}\left[\sum_{r=1}^{n^t} w_t^{m,r}\nu_t^{m,r}\left(\nu_t^{m,r}\right)^\mathsf{T}\right]\mathsf{R}^{-1}\mathsf{H}, \tag{5.41}$$

with $\nu_t^{m,r} = \mathsf{H}\mathbf{x}_t^m - \mathbf{y}_t^r$. The final term is given by

$$\mathsf{D}_t = \sum_{r=1}^{n_t} \mathsf{D}_t^r\left(\mathsf{D}_t^r\right)^{-1}, \tag{5.42}$$

$$\mathsf{D}_t^r = \begin{bmatrix} w_t^{m,r}\mathsf{H}^\mathsf{T}\mathsf{R}^{-1}\nu_t^{1,r} \\ \vdots \\ w_t^{m,r}\mathsf{H}^\mathsf{T}\mathsf{R}^{-1}\nu_t^{M,r}. \end{bmatrix} \tag{5.43}$$

5.5.2 *Joint Probabilistic Data Association*

The second approach is less rigorous but is simpler. The method was originally presented by Blanding et al. [3] and uses the PMHT measurement to track assignment weights to define normalised probabilities under the single-measurement-per-target constraint. These probabilities are then used to evaluate the JPDA updated covariance matrices.

The PMHT weights $\tilde{w}_t^{m,r}$ give the probability that measurement r in frame t is associated with track m, and probability that measurement r is not due to the track is clearly $1 - \tilde{w}_t^{m,r}$. Since the PMHT measurement model is independent assignments, the probability that there is no measurement due to track m is

$$\beta_t^{m,0} = \prod_{r=1}^{n_t}\left(1 - \tilde{w}_t^{m,r}\right). \tag{5.44}$$

The normalised association probability is then given by

$$\beta_t^{m,r} = \left(1 - \beta_t^{m,0}\right)\frac{\tilde{w}_t^{m,r}}{\sum_{r=1}^{n_t}\tilde{w}_t^{m,r}}. \tag{5.45}$$

Although we have just called these $\beta_t^{m,r}$ quantities probabilities, it is important to note that they are really an ad hoc construct that sums to unity across the measurements for each track. The covariance estimate is now obtained by applying these to the PDA covariance update rule, which gives

$$\mathsf{P}_{t|t}^m = \beta_t^{m,0}\mathsf{P}_{t|t-1}^m + \left(1 - \beta_t^{m,0}\right)\mathsf{P}_{t|t}^m\{\mathrm{KF}\} + \tilde{\mathsf{P}}_t^m, \tag{5.46}$$

where $\mathsf{P}^m_{t|t}\{\mathrm{KF}\}$ is the updated covariance matrix for a Kalman filter with a single known associated measurement, and the spread term is given by

$$\tilde{\mathsf{P}}^m_t = \mathsf{K}^m_t \left[\sum_{r=1}^{n_t} \beta^{m,r}_t \nu^{m,r}_t \left(\nu^{m,r}_t\right)^\mathsf{T} - \nu^m_t \left(\nu^m_t\right)^\mathsf{T} \right] \left(\mathsf{K}^m_t\right)^\mathsf{T}, \tag{5.47}$$

where

$$\nu^{m,r}_t = \mathsf{HF}\hat{\mathbf{x}}^m_{t-1} - \mathbf{y}^r_t, \qquad \nu^m_t = \sum_{r=1}^{n_t} \nu^{m,r}_t. \tag{5.48}$$

This PDA covariance is increased when the track associates multiple measurements because of the scatter term $\tilde{\mathsf{P}}^m_t$. In contrast, the covariances within the PMHT Kalman filter are reduced when the track associates multiple measurements because the synthetic covariance is inversely proportional to the weights sum. When the true measurement model follows the single-measurement constraint, these ad hoc PDA covariances have better consistency properties than the Kalman filter matrices [3].

5.6 Track Management

The core H-PMHT algorithm was derived earlier in this book under the assumption that the number of targets in the scene is fixed and known. This is rarely the case in practice. More often targets enter and leave the sensor field of view over time and the tracking system needs to be capable of automatically initiating new tracks when new targets appear and terminating old tracks when existing targets disappear. These functions are essential for an operationally relevant tracker, especially since the benefit of H-PMHT is espoused as high detection sensitivity. This section describes how track initiation and termination can be incorporated into the algorithm. Several methods for track management for PMHT and H-PMHT have been described in the literature, for example [7, 10, 15, 24]. Here we follow the PMHT framework of Davey [7] that was extended to H-PMHT in [8]. This was further extended in [25] to allow for merging and splitting targets but we will leave these complications for later.

The tasks of initiating and terminating tracks will be collectively referred to as track management. One can view track management as a model order estimation problem where the model order is a Markov chain over time. Model order estimation is the general problem of fitting parameters to a set of observations when the number of parameters is unknown. For a known model order, the parameters can be estimated by optimising a statistical cost function, such are the likelihood or the mean squared error. The complication with model order estimation is that increasing the model order, that is introducing more parameters, will generally lead to a statistically superior cost function. A simple expression of this is fitting a polynomial to a collection of points: as the order of the polynomial is increased, the error between the curve and the points will decrease until the order is high enough for the fitted curve to pass exactly through every point. Sometimes this is called over-fitting the data. To avoid this problem a penalty term can be introduced so that a higher order model is only chosen if it reduces the residual error by more than the cost of increasing the number of parameters. Different methods exist to determine what this penalty should be: the most well known are due to Akaike [1], Rissanen [18], and Schwartz [19].

The problem with general model order estimation methods is that they are based on fitting the highest likelihood parameters for a given model order. In the tracking context this really means optimizing over the joint target space of M targets, which is impractical. Instead, a common strategy is to over-model the data and then use a statistical test to reject tracks that make an insignificant contribution to the model. This amounts to starting a large number of tracks and then using a statistical track quality measure to make a decision between two hypotheses: this track really does correspond to a target; or this track is false. In the context of a mixture model, these two hypotheses are equivalent to: this track makes a significant contribution to the mixture; or this track is superfluous. In practice it is often desirable to defer the decision for unsure tracks. This leads naturally to two track lists: *established* tracks, are those that the algorithm has high confidence in; and *candidate* or *tentative* tracks, are those that correspond to potential targets, but do not have a high confidence. The tentative list contains the unsure tracks for which a decision has been deferred.

5.6.1 Track Quality Score

The track manager assumes that each track in the established list and the candidate list has an automatically assigned track quality score. Two common examples of this in point measurement tracking are the track probability of existence, which has many aliases [4, 5, 14, 16], and denotes the probability of the track corresponding to a real target; and the sequential track likelihood ratio that computes the ratio of the likelihoods of the associated measurements under the target-track and false-track hypotheses [2]. For H-PMHT in this book, we will use the estimated track SNR as the quality score. For the core H-PMHT this SNR can be expressed in terms of the mixing proportion π_t^m. We use the peak SNR, defined as the ratio of the peak level of the target spread function to the local noise floor. It will be expressed in decibels, namely

$$s_t^m \equiv 10 \log_{10} \frac{\pi_t^m}{\pi_t^0} \max_i \left\{ \frac{G_t^{m,i}}{G_t^{0,i}} \right\}, \tag{5.49}$$

For a rectangular image, a Gaussian point spread function, and uniform clutter, the SNR is simplified to

$$s_t^m = 10 \log_{10} \frac{\pi_t^m I_X I_Y}{\pi_t^0 |2\pi \mathsf{R}|}. \tag{5.50}$$

For radar problems where the spread of the target is narrow, point measurement tracking tends to break down at around 12 dB peak SNR whereas track-before-detect can typically function reliably down to around 6 dB, for example see [9, 11]. Peak SNR is a good measure of the detection difficulty for small targets but it can overestimate the difficulty for widely spread targets. For example, a narrow −3 dB target is not detectable but if the target is spread over hundreds of pixels then the small change to the mean level across this area is detectable. An alternative way to quantify SNR is to describe the total target power to the total noise power. This form tends to give numbers that are much smaller in absolute terms than the peak SNR and is often used in sonar literature. The overall SNR can be a better reflection of detection difficulty for spread targets but it is only ever useful in a relative sense because it depends on the image size. This is because the total noise power is a function of the number of noise pixels for a uniform background. Adding more noise pixels doesn't decrease the contrast between target and noise and so shouldn't make detection more difficult but it will increase total noise power and therefore reduce overall SNR. This is why we use peak SNR, but it is good to remember that it can be deceptive for spread targets.

In Chap. 6 an alternative assignment prior is developed based on modelling the number of shots from each component as Poisson. This results in a modified track score function that is discussed in Chap. 6. This Poisson version of the H-PMHT will be used for the application chapters later in this book.

5.6.2 Hierarchical Track Update

When the tracker automatically initializes new tracks it is possible for multiple duplicate tracks to be formed on a single object. This can happen when the tracker fails to associate all of the target data with its corresponding track, or it can be due to clutter energy that falls close to the target by chance. A common way to reduce the number of these duplicates is for the track manager to separate the established and candidate tracks and apply the measurement data to them sequentially. Each established track has equal access to the sensor image, but the candidate tracks are not allowed to use energy associated with an existing established track. Neither should we initialize new candidates where a track already exists.

Figure 5.5 illustrates a hierarchical flow diagram for the sequential processing of established and candidate tracks. This flow is generic in the sense that the association and update algorithms could be point measurement or TkBD methods. In this book, the two track-update blocks are separate instances of the same H-PMHT[2] algorithm. The measurements and residual measurements are sensor images.

---

[2] *HPMHTTracker*

The H-PMHT toolbox contains two main functions for running H-PMHT. *MakeHPMHTParams* defines tracking parameters, including which version of H-PMHT to execute, and *HPMHTTracker* actually does the tracking. *HPMHTTracker* is a direct implementation of the flow illustrated in Fig. 5.5. For details on how to use this function, refer to the H-PMHT toolbox documentation.

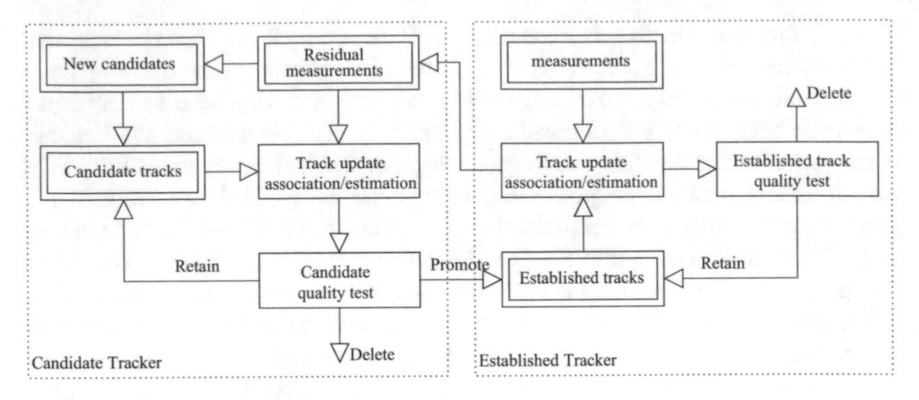

Fig. 5.5 Track management flow diagram

In order to realise the management structure in Fig. 5.5 the following elements are required: a set of rules for track decisions; a method for vetting the sensor image to enable hierarchical data access; and a method for forming new candidate tracks based on data. Each of these is now explained further.

5.6.3 Track Decisions

The track manager is required to make automated decisions about when to terminate tracks and when to promote candidate tracks. These decisions are based on the track quality, which is quantified through the estimated SNR described above.

In principle, the decision rules could be as simple as thresholds on the estimated track SNR. However, in practice our experience has been that simple thresholds are not adequate in applications where there is a large dynamic range on SNR. In this case there are some targets that are very strong and one expects the tracker to quickly promote tracks on these. There are also weaker targets that one expects the tracker to only recognise by their persistence over many frames. The track manager used in this book implements these different types of decisions based on multiple parallel composite decision rules.

A composite decision rule combines multiple simple decisions to produce a more detailed function. In our case, we apply a simple SNR threshold to each frame and then an M out of N decision over the resulting binary sequence. In words, this amounts to: promote a candidate if its estimated SNR is greater than s for at least M of the most recent N frames. A rule for detecting higher SNR targets quickly is implemented using a high SNR threshold value s and a low count threshold N. Conversely, a rule for detecting persistent weak targets uses a low SNR threshold and a high count threshold. We use the logical *OR* to combine three composite rules for strong, persistent and intermediate targets. The threshold values used are application dependent and provided for each individual example. Note that the SNR estimates produced by the core H-PMHT are only correlated because the previous estimate is used for initialisation: the maximisation for the current $\hat{\Pi}$ does not depend on the previous estimate. The result is that estimated SNR changes rapidly for fluctuating targets and this should be considered when designing track decision rules. In the next chapter we will introduce an extension of core H-PMHT that builds correlation into the model.

5.6.4 Image Vetting

The data vetting function is used to give the different status tracks hierarchical access to the sensor image. In a point measurement tracker, this is as simple as removing measurements that have been associated with established tracks from the measurement list before data association with the candidate tracks. The H-PMHT requires a vetting method that modifies the sensor image and removes target energy associated with existing tracks.

A natural way to do image vetting is through the associated images described in Chap. 4. In Sect. 4.4 we introduced the idea of H-PMHT as a data association algorithm that assigns energy to each track by defining a weight for every pixel. The associated image for track m is defined by scaling the sensor image by the pixel weight for that track and is given by Eq. 4.67:

$$\mathfrak{z}_t^{i,m} = w_t^{i,m} \bar{\mathbf{z}}_t^i .$$

In particular, note that we can determine the associated image for the background component, $\mathfrak{z}_t^{i,0}$. This image contains the residual image energy not associated with an existing track. Intuitively, this quantity is exactly what we want as the output of image vetting.

Earlier works on H-PMHT used a method referred to as *whitening* that was proposed in [21]. This whitening is mathematically exactly the same as the background associated image, we are just using a different intuition to motivate it. In [21] the association weight $w_t^{i,0}$ is described as a whitening mask that takes a value of unity far away from all tracks and a much smaller value close to the existing tracks.

This background associated image is provided as the input to subsequent stages. Namely, the background associated image from the established track mixture is the input to the candidate H-PMHT, and new candidates are formed from the background associated image from the candidate track mixture.

Figure 5.6 shows a simplified illustration of the process. In the example, there are two targets: one broad target near the centre of the image and a more compact one above the broad target. The background noise is uniform and Gaussian. Figure 5.6a shows the sensor image. Suppose that the tracker has an established track on the

(a) Original sensor image (b) Existing tracks (c) Residual image

Fig. 5.6 Sensor image vetting

broad target. Figure 5.6b shows the mixture model corresponding to the established track list. Figure 5.6c shows the background associated image that is effectively the pixel-wise ratio of the image in (a) and the image in (b). The second target is preserved but the first target, for which a track already exists, is suppressed. In this example it would be relatively easy to then form a new candidate track on the narrow target because both targets are quite high in SNR.

5.6.5 New Candidate Formation

The final element of the track manager is the process to create new candidate tracks. One way to do this is to use a birth prior and create new tracks based on this, for example, new candidates could be uniformly distributed in the state space. This can work but is rather inefficient. Instead, it is better to use a data-driven strategy that places new candidates in regions where they are more likely to be assigned a higher track quality score. As discussed in the previous section, the background associated image is the sensor energy not associated with an existing track. Placing new candidates in areas of high energy in the background associated image are going to result in higher quality candidates as these areas are more likely to correspond to undetected targets.

The simplest way to exploit the higher energy regions of the background associated image is to apply a single-frame detector and use the resulting point measurements to seed new candidate tracks. More complicated schemes can be developed but their success is application dependent. For example, two-point differencing [2] can improve initialisation if the target speed is high but relies on relatively high SNR, conversely incoherent integration can give reliable initialisation for lower SNR targets but relies on low target speed.

An obvious criticism of this method for track management is that the detection of targets now hinges on using single-frame detections to seed tracks. The motivation of track-before-detect was to get rid of the single-frame detector! However, this detector can be run at a very low threshold and the sensitivity difference between track-before-detect and detect-then-track is affected significantly more by the track-update procedure than candidate seeding [9].

5.6.6 *Integrated Track Management*

Some methods in the tracking literature make a fuss about being *integrated* track management methods, by which the authors mean that the tracking algorithm intrinsically decides for itself when to start new tracks and which candidates are valid. Intellectually this sounds appealing and perhaps the approach described above may seem a little ad hoc. However, it is important to recognise that these integrated methods are really just a slightly different model: they do not reflect the real physics of the sensor system more accurately, so any claims of optimality are not valid. In practice, the implementation arrives at essentially the same point as we have already described. We have decided to adopt a pragmatic approach and use a system that is highly effective. The two main differences are the use of a target birth density and the block treatment of all of the tracks rather than the candidate and established method described above. Here we briefly outline how one might build this into core H-PMHT. A little later in the book we will introduce a more powerful approach based on the extensions developed in the next few chapters.

The multinomial process allows us to subdivide the assignment sequentially. This is essentially what the hierarchical assignment above implements. We can apply one further assignment layer between the background and a density of newly born targets. The energy associated with the new targets can be associated using H-PMHT. If both the background and the birth density are uniform, then this is a constant scaling factor. Otherwise it becomes

$$\mathfrak{z}_t^{i,\text{birth}} = \frac{G_t^{i,\text{birth}}}{G_t^{i,\text{birth}} + G_t^{i,0}} \bar{\mathbf{z}}_t^i. \tag{5.51}$$

What remains is to fit a number of components to the density of new targets. This can be done by choosing a fixed number of components and then learning the best fit to the data. Alternatively we can use point measurement methods, such as starting a new candidate on every location with anomalously high energy, which amounts to a single-frame detector.

5.7 Summary

This chapter has discussed a number of practical issues for H-PMHT that included: efficient software implementation strategies; arranging the algorithm in a parallel-ready form; estimating track covariances; and track management. These are all important to consider in an operational context. Our hope has been to share a little of the black magic and dispel some of the misconceptions about perceived limitations of the method.

References

1. Akaike, H.: A new look at the statistical model identification. IEEE Trans. Autom. Control **19**(6), 716–723 (1974)
2. Blackman, S.S., Popoli, R.: Design and Analysis of Modern Tracking Systems. Artech House, Norwood (1999)
3. Blanding, W.R., Willett, P., Steir, R.L., Dunham, D.: Consistent covariance estimation for PMHT. In: Proceedings of SPIE: Signal and Data Processing of Small Targets, vol. 6699, San Diego, USA (2007)
4. Colegrove, S.B., Davis, A.W., Ayliffe, J.K.: Track initiation and nearest neighbours incorporated into probabilistic data association. J. Electr. Electron. Eng. Australia **6**, 191–198 (1986)
5. Colegrove, S.B., Davey, S.J.: PDAF with multiple clutter regions and target models. IEEE Trans. Aerosp. Electron. Syst. **39**, 110–124 (2003)
6. Crouse, D.F., Guerriero, M., Willett, P.: A critical look at the PMHT. ISIF J. Adv. Inf. Fusion **4**, 93–116 (2009)
7. Davey, S.J.: Extensions to the Probabilistic Multi-Hypothesis Tracker for Improved Data Association. School of Electrical and Electronic Engineering, The University of Adelaide (2003)
8. Davey, S.J.: Detecting a small boat with histogram PMHT. ISIF J. Adv. Inf. Fusion **6**, 167–186 (2011)
9. Davey, S.J.: SNR limits on kalman filter Detect-Then-Track. IEEE Signal Process. Lett. **20**, 767–770 (2013)
10. Davey, S.J., Gray, D.A.: Integrated track maintenance for the PMHT via the hysteresis model. IEEE Trans. Aerosp. Electron. Syst. **43**(1), 93–111 (2007)
11. Davey, S.J., Rutten, M.G., Gordon, N.J.: Track before detect techniques. In: Integrated Tracking, Classification and Sensor Management: Theory and Applications. Wiley, London (2012)
12. Davis, P.J., Rabinowitz, P.: Methods of Numerical Integration. Dover Publications, New York (2007)
13. Gaetjens, HX, Davey, SJ, Luginbuhl, TE: Robust state models for the Histogram-PMHT (in preparation)
14. Li, N., Li, X.R.: Target perceivability and its application. IEEE Trans. Signal Process. **49**, 2588–2604 (2001)
15. Luginbuhl, T.E., Sun, Y., Willett, P.: A Track Management System for the PMHT Algorithm. In: Proceedings of the 4th International Conference on Information Fusion (2001)
16. Musicki, D., Evans, R., Stankovic, S.: Integrated probabilistic data association. IEEE Trans. Autom. Control **39**, 1237–1241 (1994)
17. Pakfiliz, A.G., Efe, M.: Multi-target tracking in clutter with histogram probabilistic multi-hypothesis tracker. In: IEEE Conference on Systems Engineering, pp. 137–142 (2005)
18. Rissanen, J.: Modeling by shortest data description. Automatica **14**, 465–471 (1978)
19. Schwartz, G.: Estimating the dimension of a model. Ann. Stat. **6**(2), 461–464 (1978)
20. Streit, R.L.: Tracking on intensity-modulated data streams. Technical report 11221, NUWC, Newport, Rhode Island, USA (2000)
21. Streit, R.L., Graham, M.L., Walsh, M.J.: Multitarget tracking of distributed targets using histogram-PMHT. Digit. Signal Process. **12**(2), 394–404 (2002)
22. Vu, H.X.: Track-before-detect for active sonar. Ph. D. thesis, The University of Adelaide (2015)
23. Walsh, M.J.: Computing the observed information matrix for dynamic mixture models. Technical report 11768, NUWC, Newport, Rhode Island, USA (2006)

24. Wieneke, M., Willett, P.: On track-management within the PMHT framework. In: Proceedings of the 11th International Conference on Information Fusion (2008)
25. Wieneke, M., Davey, S.J.: Histogram-PMHT for extended targets and target groups in images. IEEE Trans. Aerosp. Electron. Syst. **50**(3) (2014)

Chapter 6
Poisson Scattering Field

The PMHT and the H-PMHT algorithms reviewed in Chaps. 3 and 4 both assume that the source of each point measurement is an independent draw from a probability mass function. The individual component probability masses are denoted π_t^m and are hyperparameters of the measurement function. When these hyperparameters are unknown they are estimated using a weighted relative frequency. This limits the dynamics of π_t^m: it must be either time independent or constant. As described in Chap. 5, these assignment prior estimates can be a useful metric for track management because they specify how much influence a particular track has over the frame at time t. However, if the target amplitude fluctuates then the estimates of the prior will be pelagic. Figure 6.1 shows two examples of a fluctuating target following a Swerling II model [7]. The plots show the instantaneous amplitude of the target and the mixing proportion estimate $\hat{\pi}_t^m$ resulting from applying H-PMHT to images with this target. Two different average SNRs are shown. For a 10 dB SNR, the estimated mixing proportion $\hat{\pi}_t^m$ is so highly correlated with the instantaneous amplitude that it is hard to pick them apart. Even for the 0 dB average SNR, there is a high degree of correlation: the estimate follows the target fluctuations very closely. This would be desirable if the aim was to characterise the temporal variations but the SNR, in this case, has a stationary statistical distribution, a better characterisation would be to estimate the mean or another appropriate sufficient statistic. An alternative perspective is to treat the instantaneous SNR as a noisy time series. The underlying signal component is contained in the low-frequency part of the spectrum and the higher frequencies are noise.

The variability of the estimate occurs because the mixing proportion has been modelled as an unknown constant. Although the target strength can evolve with time, the underlying mean is usually time correlated and its estimate would be less noisy if it incorporated a history of data. This can be done by modelling the mixing proportion as a hidden random process. Davey and Gray introduced a hidden Markov chain that dictated the mixing proportions for the point measurement PMHT [1, 2] and referred to this as the PMHT with Hysteresis. They showed that by using this hidden Markov model over the π it was possible to significantly improve track initiation

S. J. Davey and H. Gaetjens, *Track-Before-Detect Using Expectation Maximisation*, Signals and Communication Technology,
https://doi.org/10.1007/978-981-10-7593-3_6

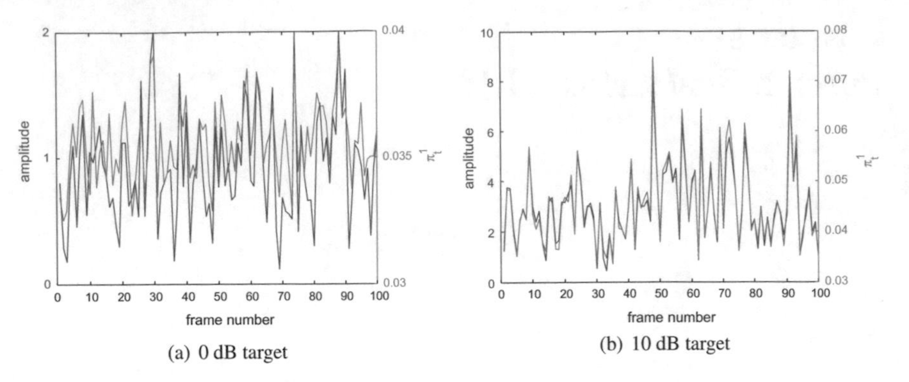

(a) 0 dB target

(b) 10 dB target

Fig. 6.1 Fluctuating target SNR and estimated π_t^1

performance, that is the discrimination between valid target tracks and false tracks. However, the hysteresis model comes with a cost. Since the mixing proportion is a normalised probability vector, it is constrained to sum to unity and this constraint means that individual elements of the vector π_t^m are not independent: if one decreases then the others must increase to maintain normalisation. The dependence between the different elements demands a joint estimator and the complexity of that estimator is exponential in the number of targets.

An alternative approach was adopted by Gaetjens et al. [3, 4, 9, 10] where the link between Poisson and multinomial processes was exploited to avoid this complexity issue. The PMHT measurement model uses a multinomial process to describe the association of the measurements, each measurement is an independent draw from the mixing vector π, and the H-PMHT inherits the same model. The alternative versions of PMHT and H-PMHT use a Poisson process to model the arrival of measurements. If the measurements arrive according to a Poisson process, then the conditional distribution of the assignments given the total measurement count $p(k|n)$ follows a multinomial distribution; we will prove this a little later in Sect. 6.2. The key difference is that the Poisson processes for each target are independent and do not require any normalisation, which means that the estimators for the parameters of these processes are also independent and thus the complexity is linear.

This chapter reviews the hysteresis model for the assignment prior and shows how placing a state directly on the priors results in exponential computation complexity. It then demonstrates how H-PMHT can be built up from a Poisson measurement model instead of a multinomial model. Simulations are used to illustrate how the use of this Poisson model improves the estimation of target amplitude and track initiation.

The Poisson H-PMHT derived in this chapter is a superior version of the algorithm that provides flexible modelling for the mixing proportions without incurring a significant overhead and will be used as the base for the further extensions in the subsequent chapters of this book.

6.1 Hysteresis

The PMHT with hysteresis modifies the PMHT measurement model by adding a hyperparameter for the assignment prior [1, 2]. Each target state is augmented with an assignment state d_t^m which follows a Markov chain and determines the mixing proportion for that target. The clutter can itself be a mixture but for the discussion here we consider a single uniform clutter component. The hysteresis model in [1] is applied to the point measurement PMHT. It would be relatively straightforward to apply the same model to the H-PMHT but as stated already, the complexity of this approach ultimately makes it undesirable: we will stay with the simpler PMHT.

The point measurement PMHT, which uses point measurements $\mathbb{Y}$ to estimate target states $\mathbf{X}$ by introducing an assignment variable $\mathbb{K}$ and treating it as missing data in the EM framework. The hysteresis model extends this model by adding a hyperparameter to the assignment distribution. Following our usual convention, the stacked vector of all of the assignment states at frame t is $\mathbf{D}_t$ and the collection of all of the assignment states over the batch is $\mathbb{D}$. The hysteresis states are assumed to be independent from the kinematic states and each d_t^m follows an independent Markov chain, so

$$p\left(\mathbb{D}\right) = \prod_{m=1}^{M} \left\{ d_0^m \prod_{t=1}^{M} p\left(d_t^m | d_{t-1}^m\right) \right\}. \tag{6.1}$$

Each assignment is still assumed to be an independent discrete random variable. The prior probability that $k_t^r = m$ is again denoted by π_t^m and is the mixing proportion for target m. Assume a common mapping function from d_t^m to k_t^r across all targets such that

$$\pi_t^m = \begin{cases} \phi\left(d_t^m\right) & 1 \leq m \leq M \\ 1 - \sum_s \phi\left(d_t^s\right) & m = 0. \end{cases} \tag{6.2}$$

Note that the function $\phi(d)$ must map to a probability, so $0 \leq \phi(d) \leq 1\ \forall d$ and the additive normalisation for the clutter term assumes that $\sum_s \phi\left(d_t^s\right) < 1$. The implication is that the target priors are relatively small, which is the case when the clutter density is relatively high.

Since the assignments remain independent the probability of $k_{1:T}$ remains a double product

$$p\left(\mathbb{K}|\mathbb{X}, \mathbb{D}\right) = p\left(\mathbb{K}|\mathbb{D}\right) = \prod_{t=1}^{T} \prod_{r=1}^{n_t} \pi_t^{k_t^r}. \tag{6.3}$$

Although we have not acknowledged it through the notation, the probability $\pi_t^{k_t^r}$ depends on the full hysteresis state vector $\mathbf{D}_t$ due to the clutter term.

The measurement pdf term does not change because it is conditionally independent of the hysteresis variables and the complete data likelihood is

$$p_{\mathsf{comp}}\left(\mathbb{X},\mathbb{D},\mathbb{K},\mathbb{Y}\right)=p\left(\mathbb{X}\right)p\left(\mathbb{D}\right)p\left(\mathbb{K}|\mathbb{D}\right)p\left(\mathbb{Y}|\mathbb{X},\mathbb{K}\right)$$
$$=\prod_{m=1}^{M}\left[\left(p\left(\mathbf{x}_0^m\right)\prod_{t=1}^{T}p\left(\mathbf{x}_t^m|\mathbf{x}_{t-1}^m\right)\right)\left(p\left(d_0^m\right)\prod_{t=1}^{T}p\left(d_t^m|d_{t-1}^m\right)\right)\right]$$
$$\times\prod_{t=1}^{T}\prod_{r=1}^{n_t}\pi_t^{k_t^r}p\left(\mathbf{y}_t^r;h\left(\mathbf{x}_t^{k_t^r}\right),\Sigma\right). \quad (6.4)$$

There are two fundamental ways that the hysteresis state could be treated in the EM process: either as additional missing data or as an extra parameter to estimate. Both options were explored by Davey [1]; the option preferred for the application of track initiation in [2] is to treat the hysteresis state as missing data because this allows for a smaller state space. Assuming a discrete state space for the hysteresis parameter, the modified auxiliary function is

$$\mathscr{Q}\left(\mathbb{X}|\hat{\mathbb{X}}\right)=E_{\mathbb{D},\mathbb{K}|\hat{\mathbb{X}},\mathbb{Y}}\Big[\log\Big\{p_{\mathsf{comp}}\left(\mathbb{X},\mathbb{D},\mathbb{K},\mathbb{Y}\right)\Big\}\Big],$$
$$=\sum_{\mathbb{K}}\sum_{\mathbb{D}}\log\Big\{p_{\mathsf{comp}}\left(\mathbb{X},\mathbb{D},\mathbb{K},\mathbb{Y}\right)\Big\}p_{\mathsf{miss}}\left(\mathbb{K},\mathbb{D}|\mathbb{Y},\hat{\mathbb{X}}\right). \quad (6.5)$$

The missing data probability is now the joint probability of the hysteresis states and the assignments given the measurements and a kinematic state estimate. This can be decomposed into two terms in a similar way to the standard PMHT missing data conditional probability,

$$p_{\mathsf{miss}}\left(\mathbb{K},\mathbb{D}|\mathbb{Y},\hat{\mathbb{X}}\right)=\frac{p_{\mathsf{comp}}\left(\hat{\mathbb{X}},\mathbb{D},\mathbb{K},\mathbb{Y}\right)}{\sum_{\mathbb{K}}\sum_{\mathbb{D}}p_{\mathsf{comp}}\left(\hat{\mathbb{X}},\mathbb{D},\mathbb{K},\mathbb{Y}\right)}$$
$$=\frac{p\left(\hat{\mathbb{X}}\right)p\left(\mathbf{D}_0\right)\prod_{t=1}^{T}\left[p\left(\mathbf{D}_t|\mathbf{D}_{t-1}\right)\prod_{r=1}^{n_t}\pi_t^{k_t^r}p\left(\mathbf{y}_t^r;h\left(\mathbf{x}_t^{k_t^r}\right),\Sigma\right)\right]}{p\left(\hat{\mathbb{X}}\right)\sum_{\mathbb{D}}\left\{p\left(\mathbf{D}_0\right)\prod_{t=1}^{T}\left[p\left(\mathbf{D}_t|\mathbf{D}_{t-1}\right)\prod_{r=1}^{n_t}\sum_{s=0}^{M}\pi_t^s p\left(\mathbf{y}_t^r;h\left(\mathbf{x}_t^s\right),\Sigma\right)\right]\right\}}$$
$$=\frac{\prod_{t=1}^{T}\left[p\left(\mathbf{D}_t|\mathbf{D}_{t-1}\right)\prod_{r=1}^{n_t}\pi_t^{k_t^r}p\left(\mathbf{y}_t^r;h\left(\mathbf{x}_t^{k_t^r}\right),\Sigma\right)\right]}{\sum_{\mathbb{D}}\prod_{t=1}^{T}\left[p\left(\mathbf{D}_t|\mathbf{D}_{t-1}\right)\prod_{r=1}^{n_t}\sum_{s=0}^{M}\pi_t^s p\left(\mathbf{y}_t^r;h\left(\mathbf{x}_t^s\right),\Sigma\right)\right]}, \quad (6.6)$$

where in the final step the initial hysteresis state d_0^m is assumed constant and known (chosen). This probability expression has insurmountable complexity. For the independent π_t^m case, it was possible to factorise both the numerator and denominator into products of single measurement terms. In this case, it isn't possible because every term in the denominator depends on the entire sequence of joint hysteresis state vectors which has a cardinality of $||d||^{MT}$. Davey mitigated this complexity by making d_t^m binary and setting $\pi_t^m\left(d_t^m=0\right)=0$. This reduces the size of the hysteresis state space and renders half of this limited space uninteresting. The complexity could also be mitigated by realising that the auxiliary function does not explicitly use the joint probability above but instead depends on marginals of it. The approach of Williams and Lau [11] that uses belief propagation to efficiently approximate association marginal probabilities could be adapted to this problem. However, it is not necessary since we will soon see that the Poisson model removes this complexity entirely. Low complexity is the main benefit of PMHT and for this reason, the hysteresis approach is not considered any further.

6.2 Poisson and Multinomial Equivalence

The PMHT measurement model considered so far in this book has assumed that the source of each measurement is an IID realisation of a discrete probability mass and hence the count of measurements across the sensor bins follows a multinomial distribution. We will now show that this multinomial distribution is equivalent to the conditional distribution of Poisson measurements. This is a well-known result, e.g. [5], and is repeated here because of its importance in context.

Recall from Chap. 4 that the likelihood for the measurement counts across the sensor pixels is given by the multinomial distribution in (4.5):

$$p\left(\mathbf{N}|\mathbf{X};\Pi\right) = \frac{n!}{\prod_{i=1}^{I} n^i!} \prod_{i=1}^{I} \left(G^i\right)^{n^i},$$

where n^i is the measurement count in pixel i; $n = \sum_i n^i$ is the total measurement count across the image; $\mathbf{N} = \{n^i\}$ is a vector of the pixel counts; and the time index has been hidden to simplify notation. From the outset in Sect. 3.1 a subtle assumption has been made that the total number of measurements in the frame is known, for example, we immediately proceed from $p\left(\mathbf{Y},\ldots\right)$ to a product of n conditionally independent terms without stopping to consider $p(n)$. In fact, the core PMHT, and hence core H-PMHT, offer no guidance on $p(n)$ because n is implicitly assumed fixed.

Consider now the case where the number of point measurements is random and follows a Poisson distribution with mean Λ. The Poisson distribution is given by

$$p(n) = \exp\{-\Lambda\}\frac{\Lambda^n}{n!}, \tag{6.7}$$

for nonnegative n. The Poisson distribution has two important properties that make it well suited to the mixture problem: first, it follows superposition, and second, it can be subdivided into Poisson child distributions. The first property means that the sum of multiple Poisson distributed variables is Poisson distributed. If n^m is Poisson distributed with mean Λ^m for $m = 1\ldots M$ then $n = \sum_m n^m$ is Poisson distributed with mean $\Lambda = \sum_m \Lambda^m$. This is simple to prove: for the two-component case one can convolve together the two Poisson distributions and obtain the result directly. Alternatively, applying characteristic functions gives an even faster result. Extending this to an arbitrary number of terms can be done by iteratively repeating the two-component result $M-1$ times. The superposition result means that a Poisson mixture is Poisson distributed.

The second Poisson property is somewhat like superposition in reverse. Consider the case where calls arrive at a call centre and the switchboard randomly assigns each call to an operator. There are I operators and the probability that a call is assigned to operator i is G^i. If the number of calls in an hour follows a Poisson distribution with mean Λ then the number of calls in an hour assigned to operator i follows a Poisson distribution with mean $G^i \Lambda$. What is unexpected is that the I Poisson random variables created in this way are all independent of each other!

These are standard properties of Poisson distributions and proofs can readily be found in the literature, for example, [8].

Consider the subdivision example above: the total number of calls is n and the number of calls to each operator is n^i. Due to the subdivision independence property, the joint probability of the counts is

$$p\left(\mathbf{N}\right) = \prod_{i=1}^{I} \exp\{-G^i \Lambda\} \frac{\left(G^i \Lambda\right)^{n^i}}{n^i!} \tag{6.8}$$

$$= \Lambda^n \exp\{-\Lambda\} \prod_{i=1}^{I} \frac{\left(G^i\right)^{n^i}}{n^i!}. \tag{6.9}$$

Using Bayes rule the conditional probability of the counts is

$$\begin{aligned} p\left(\mathbf{N}|n\right) &= \frac{p\left(\mathbf{N}, n\right)}{p\left(n\right)}, \\ &= \frac{\delta\left(n - \sum_i n^i\right) p\left(\mathbf{N}\right)}{p\left(n\right)}, \\ &= \delta\Big(n - \sum_i n^i\Big) \Lambda^n \exp\{-\Lambda\} \prod_{i=1}^{I} \frac{\left(G^i\right)^{n^i}}{n^i!} \exp\{\Lambda\} \frac{n!}{\Lambda^n}, \\ &= \delta\Big(n - \sum_i n^i\Big) \frac{n!}{\prod_{i=1}^{I} n^i!} \prod_{i=1}^{I} \left(G^i\right)^{n^i}, \end{aligned} \tag{6.10}$$

where $\delta(x)$ is the dirac delta function, which is unity when $x = 0$ and zero everywhere else. This is clearly the multinomial distribution in (4.5) and repeated above. The delta function simply asserts that the total sum of all of the counts must equal n. What this illustrates is that assuming a Poisson number of measurements from each mixture component is consistent with the existing mathematical framework.

6.3 Dynamic Non-homogeneous Poisson Mixture Model

As described above, the fundamental modelling assumption underlying this chapter is that the histogram shots are realisations of a dynamic non-homogeneous Poisson Point Process (PPP), which is parameterised by its intensity function $\lambda_t(\mathbf{y})$. Because the PPP is non-homogeneous the intensity is spatially varying and because it is dynamic it is also time varying. The intensity function is finite and non-negative over the whole observation space W and zero outside it. Let Λ_t denote the integral of the intensity function over the observation space, namely

$$\Lambda_t = \int_W \lambda_t(\mathbf{y})\mathrm{d}\mathbf{y}. \tag{6.11}$$

We will limit λ_t by asserting that Λ_t is finite and positive. In practice, the observation space W is usually a closed interval so this limitation is not particularly restrictive. Situations where λ_t is zero everywhere or unbounded are of no practical interest and so are ignored. The most common situation is that each pixel is a closed interval contained in λ_t and that the pixels do not overlap so a finite value to the integral implies a finite integral over every pixel. These pixel integrals are defined as

$$\Lambda_t^i = \int_{W^i} \lambda_t(\mathbf{y})\mathrm{d}\mathbf{y}, \tag{6.12}$$

and clearly

$$\Lambda_t = \sum_{i=1}^{I} \Lambda_t^i. \tag{6.13}$$

As before, we have assumed that the number and extent of pixels are constant with time. Relaxing this only requires more careful notation.

The fundamental properties of a PPP are that the number of measurements n_t is a Poisson distributed random variable with rate parameter (mean) Λ_t

$$p\,(n_t;\,\Lambda_t) = \exp\{-\Lambda_t\}\,\frac{\Lambda_t^{n_t}}{n_t!}, \tag{6.14}$$

and that the measurements are IID random variables with a spatial distribution given by

$$p(\mathbf{y};\,\lambda_t) = \frac{\lambda_t(\mathbf{y})}{\Lambda_t}. \tag{6.15}$$

These combine to give the probability of a set of measurements $\mathbf{Y}_t$

$$p(\mathbf{Y}_t; \lambda_t) = p(n_t; \Lambda_t) \prod_{r=1}^{n_t} \frac{\lambda_t(\mathbf{y}_t^r)}{\Lambda_t}$$
$$= \frac{\exp\{-\Lambda_t\}}{n_t!} \prod_{r=1}^{n_t} \lambda_t(\mathbf{y}_t^r) \qquad (6.16)$$

The system intensity function is now modelled as a mixture of M component Poisson distributions, the mth of which is given by

$$\lambda_t^m(\mathbf{y}) = \Lambda_t^m g^m(\mathbf{y}|\mathbf{x}_t^m), \qquad (6.17)$$

where Λ_t^m is a positive scalar-valued random variable and represents the overall measurement rate for the mth Poisson component, we will refer to this as the component rate. The function $g^m(\mathbf{y}|\mathbf{x}_t^m)$ is the pdf of measurements from the mth component, which is consistent with how we've used this notation in earlier chapters. Note that Λ_t and Λ_t^m have a discrete domain over pixels or mixture components whereas λ in its various guises is a function over the continuous domain W. The random variable Λ_t^m is assumed to follow an independent Markov chain $p(\Lambda_t^m|\Lambda_{t-1}^m)$. As before, $\mathbf{x}_t^m$ is a target state vector that determines a translation on $g^m(\cdot)$ and this also follows an independent Markov chain. The mixture is defined as

$$\lambda_t(\mathbf{y}) = \sum_{m=0}^{M} \lambda_t^m(\mathbf{y}). \qquad (6.18)$$

By inspection observe that the overall measurement rate is the superposition of the component rates,

$$\Lambda_t = \sum_{m=0}^{M} \Lambda_t^m. \qquad (6.19)$$

The collection of component rates at frame t is denoted by $\mathbf{L}_t \equiv \Lambda_t^{0:M}$ and the overall collection of component rates over the batch is $\mathbb{L} \equiv \mathbf{L}_{0:T} \equiv \Lambda_{0:T}^{0:M}$. One could choose to absorb these rates into an augmented state vector but we prefer to keep them explicitly separate.

6.3.1 *Point Measurement Data*

Consider first the point measurement case addressed by PMHT; the precise location of each measurement is available. The measurement data is the collection of point measurements $\mathbb{Y}$; the missing data is the source of each $\mathbb{K}$; and the variables to be estimated are the kinematic states $\mathbb{X}$ and the rates $\mathbb{L}$. The kinematic states $\mathbb{X}$ evolve according to independent Markov chains and the measurement rates $\mathbb{L}$ also follow independent Markov chains. The complete data likelihood is,

$$p_{\text{comp}}(\mathbb{X}, \mathbb{L}, \mathbb{K}, \mathbb{Y}) = p(\mathbb{X})p(\mathbb{L})p(\mathbb{K}|\mathbb{L})p(\mathbb{Y}|\mathbb{X}, \mathbb{K}), \tag{6.20}$$

and the missing data conditional likelihood is $p_{\text{miss}}\left(\mathbb{K}|\hat{\mathbb{X}}, \hat{\mathbb{L}}, \mathbb{Y}\right)$. The EM auxiliary function is

$$\begin{aligned}
\mathscr{Q}\left(\mathbb{X}, \mathbb{L}|\hat{\mathbb{X}}, \hat{\mathbb{L}}\right) &= E_{\mathbf{K}|\hat{\mathbb{X}},\hat{\mathbb{L}},\mathbf{Y}}\Big[\log\left\{p_{\text{comp}}(\mathbb{X}, \mathbb{L}, \mathbb{K}, \mathbb{Y})\right\}\Big] \\
&= \sum_{\mathbb{K}} \log\left\{p_{\text{comp}}(\mathbb{X}, \mathbb{L}, \mathbb{K}, \mathbb{Y})\right\} p_{\text{miss}}(\mathbb{K}|\hat{\mathbb{X}}, \hat{\mathbb{L}}, \mathbb{Y}), \\
&= \log\{p(\mathbb{X})\} + \log\{p(\mathbb{L})\} + \sum_{\mathbb{K}} \log\{p(\mathbb{K}|\mathbb{L})\}\, p_{\text{miss}}(\mathbb{K}|\hat{\mathbb{X}}, \hat{\mathbb{L}}, \mathbb{Y}) \\
&\quad + \sum_{\mathbb{K}} \log\{p(\mathbb{Y}|\mathbb{X}, \mathbb{K})\}\, p_{\text{miss}}(\mathbb{K}|\hat{\mathbb{X}}, \hat{\mathbb{L}}, \mathbb{Y}),
\end{aligned} \tag{6.21}$$

6.3.1.1 Expectation Step

The E-step consists of deriving the auxiliary function $\mathscr{Q}\left(\mathbb{X}, \mathbb{L}|\hat{\mathbb{X}}, \hat{\mathbb{L}}\right)$. The first term in (6.21) is the log-likelihood of the state sequence and is the same as introduced in Eq. (3.3) of Chap. 3, that is

$$p(\mathbb{X}) = \prod_{m=1}^{M}\left[p\left(\mathbf{x}_0^m\right)\prod_{t=1}^{T} p\left(\mathbf{x}_t^m|\mathbf{x}_{t-1}^m\right)\right].$$

The second term in (6.21) is the log-likelihood of the measurement rate sequence. The same general assumptions made about the kinematic state sequence apply to the measurement rate sequence: each component's rate is independent of all of the others and the rate sequence for a single component follows a first-order Markov chain. So the probability of $\mathbb{L}$ looks just like the likelihood of the state sequence.

$$p(\mathbb{L}) = \prod_{m=0}^{M}\left[p\left(\Lambda_0^m\right)\prod_{t=1}^{T} p\left(\Lambda_t^m|\Lambda_{t-1}^m\right)\right]. \tag{6.22}$$

Notice that we have defined a rate sequence for the clutter component whereas, so far, there is no clutter state.

The third term in (6.21) depends on the likelihood of the assignments $p(\mathbb{K}|\mathbb{L})$ and the missing data likelihood. The assignments are conditionally independent over time given the rates so their likelihood is

$$p(\mathbb{K}|\mathbb{L}) = \prod_{t=1}^{T} p\left(\mathbf{K}_t|\mathbf{L}_t\right). \tag{6.23}$$

The assignment list $\mathbf{K}_t$ implicitly defines the number of measurements from each component, which will be denoted as $\tilde{n}_t^m$. The collection of these counts across a single frame is denoted $\tilde{\mathbf{N}}_t$. Note that n_t^i is the measurement count in pixel i so we adopt the tilde to discriminate between the two. The counts are redundant given the assignments and we can write

$$p\left(\mathbf{K}_t|\mathbf{L}_t\right) = p\left(\mathbf{K}_t, \tilde{\mathbf{N}}_t|\mathbf{L}_t\right) = p\left(\mathbf{K}_t|\tilde{\mathbf{N}}_t\right) p\left(\tilde{\mathbf{N}}_t|\mathbf{L}_t\right). \tag{6.24}$$

The first term is the probability of a particular configuration of assignments given that the number of measurements due to each component is known. All combinations are equally likely and the value of this probability is then the reciprocal of the number of combinations

$$p\left(\mathbf{K}_t|\tilde{\mathbf{N}}_t\right) = \frac{\prod_{m=0}^{M} \tilde{n}_t^m!}{n_t!} \tag{6.25}$$

The second term is the probability of the counts, which is Poisson by construction

$$p\left(\tilde{\mathbf{N}}_t|\mathbf{L}_t\right) = \prod_{m=0}^{M} \exp\left\{-\Lambda_t^m\right\} \frac{\left(\Lambda_t^m\right)^{\tilde{n}_t^m}}{\tilde{n}_t^m!}. \tag{6.26}$$

Combining these gives.

$$p\left(\mathbf{K}_t|\mathbf{L}_t\right) = \frac{\exp\left\{-\Lambda_t\right\}}{n_t!} \prod_{m=0}^{M} \left(\Lambda_t^m\right)^{\tilde{n}_t^m} = \frac{\exp\left\{-\Lambda_t\right\}}{n_t!} \prod_{r=1}^{n_t} \Lambda_t^{k_t^r}. \tag{6.27}$$

The final term in (6.21) depends on the measurement likelihood of the assignments $p(\mathbb{Y}|\mathbb{X}, \mathbb{K})$ and the missing data likelihood. The measurements are conditionally independent of the measurement rates given the assignments, so this is exactly the same as developed in Chap. 3. Conditioned on the states and rates, the measurements are independent so this probability is a double product

$$p\left(\mathbb{Y}|\mathbb{X}, \mathbb{K}\right) = \prod_{t=1}^{T} \prod_{r=1}^{n_t} g^{k_t^r}\left(\mathbf{y}_t^r|\mathbf{x}_t^{k_t^r}\right). \tag{6.28}$$

The missing data likelihood, as usual, can be simplified using Bayes' rule

$$
\begin{aligned}
p_{\text{miss}}\left(\mathbb{K}|\mathbb{Y},\hat{\mathbb{X}},\hat{\mathbb{L}}\right) &= \frac{p\left(\mathbb{K},\mathbb{Y},\hat{\mathbb{X}},\hat{\mathbb{L}}\right)}{\sum_{\acute{\mathbb{K}}} p\left(\acute{\mathbb{K}},\mathbb{Y},\hat{\mathbb{X}},\hat{\mathbb{L}}\right)}, \\
&= \frac{p\left(\hat{\mathbb{X}}\right)p\left(\hat{\mathbb{L}}\right)p(\mathbb{K}|\mathbb{L})p(\mathbb{Y}|\mathbb{X},\mathbb{K})}{p\left(\hat{\mathbb{X}}\right)p\left(\hat{\mathbb{L}}\right)\sum_{\acute{\mathbb{K}}}\left\{p(\acute{\mathbb{K}}|\mathbb{L})p(\mathbb{Y}|\mathbb{X},\acute{\mathbb{K}})\right\}}, \\
&= \frac{\left[\prod_{t=1}^{T}\frac{\exp\left\{-\hat{\Lambda}_t\right\}}{n_t!}\prod_{r=1}^{n_t}\hat{\Lambda}_t^{k_t^r}\right]\left[\prod_{t=1}^{T}\prod_{r=1}^{n_t}g^{k_t^r}\left(\mathbf{y}_t^r|\hat{\mathbf{x}}_t^{k_t^r}\right)\right]}{\sum_{\acute{\mathbb{K}}}\left\{\left[\prod_{t=1}^{T}\frac{\exp\left\{-\hat{\Lambda}_t\right\}}{n_t!}\prod_{r=1}^{n_t}\hat{\Lambda}_t^{\acute{k}_t^r}\right]\left[\prod_{t=1}^{T}\prod_{r=1}^{n_t}g^{\acute{k}_t^r}\left(\mathbf{y}_t^r|\hat{\mathbf{x}}_t^{\acute{k}_t^r}\right)\right]\right\}}, \\
&= \frac{\prod_{t=1}^{T}\prod_{r=1}^{n_t}\hat{\Lambda}_t^{k_t^r}g^{k_t^r}\left(\mathbf{y}_t^r|\hat{\mathbf{x}}_t^{k_t^r}\right)}{\sum_{\acute{\mathbb{K}}}\left\{\prod_{t=1}^{T}\prod_{r=1}^{n_t}\hat{\Lambda}_t^{\acute{k}_t^r}g^{\acute{k}_t^r}\left(\mathbf{y}_t^r|\hat{\mathbf{x}}_t^{\acute{k}_t^r}\right)\right\}}, \\
&= \prod_{t=1}^{T}\prod_{r=1}^{n_t}\frac{\hat{\Lambda}_t^{k_t^r}g^{k_t^r}\left(\mathbf{y}_t^r|\hat{\mathbf{x}}_t^{k_t^r}\right)}{\sum_{m=0}^{M}\hat{\Lambda}_t^{m}g^{k_t^r}\left(\mathbf{y}_t^r|\hat{\mathbf{x}}_t^{m}\right)} \\
&= \prod_{t=1}^{T}\prod_{r=1}^{n_t}p\left(k_t^r|\mathbf{y}_t^r,\hat{\mathbf{X}},\hat{\Pi}_t\right).
\end{aligned}
\tag{6.29}
$$

The missing data probability is again given by the product of the marginal conditional probabilities of each measurement. This form continues to arise because of the fundamental assumption that the source of each measurement is an independent random variable rather than the more typical case in tracking where each target can only make at most one measurement, which inherently makes the measurement sources dependent. The modified data association weight $\tilde{w}^{m,r}$ is, therefore,

$$
\tilde{w}_t^{m,r} = p\left(k_t^r = m|\mathbf{y}_t^r,\hat{\mathbf{X}},\hat{\Pi}_t\right) = \frac{\hat{\Lambda}_t^m g^m\left(\mathbf{y}_t^r|\hat{\mathbf{x}}_t^m\right)}{\sum_{j=0}^{M}\hat{\Lambda}_t^j g^j\left(\mathbf{y}_t^r|\hat{\mathbf{x}}_t^j\right)}. \tag{6.30}
$$

If the numerator and denominator of the weight ratio are divided by $\hat{\Lambda}_t$ then the expression becomes

$$
\tilde{w}_t^{m,r} = \frac{\hat{\Lambda}_t^m/\hat{\Lambda}_t g^m\left(\mathbf{y}_t^r|\hat{\mathbf{x}}_t^m\right)}{\sum_{j=0}^{M}\hat{\Lambda}_t^j/\hat{\Lambda}_t g^j\left(\mathbf{y}_t^r|\hat{\mathbf{x}}_t^j\right)}, \tag{6.31}
$$

$$
= \frac{\hat{\pi}_t^m g^m\left(\mathbf{y}_t^r|\hat{\mathbf{x}}_t^m\right)}{\sum_{j=0}^{M}\hat{\pi}_t^j g^j\left(\mathbf{y}_t^r|\hat{\mathbf{x}}_t^j\right)}. \tag{6.32}
$$

where $\hat{\pi}_t^m = \hat{\Lambda}_t^m / \hat{\Lambda}_t$; it is straightforward to demonstrate that these form a probability vector. This is the same expression as used in the original multinomial version of PMHT, which was defined by (3.56). So the difference between the multinomial data association weights and the Poisson data association weights is that the multinomial version uses a normalised prior whereas the Poisson uses an unnormalised prior. Since the weight expression is a ratio, constant scale factors are irrelevant.

Now that the missing data likelihood has been defined we can return to the auxiliary function. The third term in the auxiliary function (6.21) is the expectation of the logarithm of the likelihood of the assignments $p(\mathbb{K}|\mathbb{L})$ over the missing data. Using (6.27) and the weights above, this becomes

$$\begin{aligned}
&\sum_{\mathbb{K}} \log\{p(\mathbb{K}|\mathbb{L})\}\, p_{\text{miss}}(\mathbb{K}|\hat{\mathbb{X}}, \hat{\mathbb{L}}, \mathbb{Y}) \\
&= \sum_{\mathbb{K}} \left(-\sum_{t=1}^{T} \Lambda_t - \sum_{t=1}^{T} \log\{n_t!\} + \sum_{t=1}^{T}\sum_{r=1}^{n_t} \log\left\{\Lambda_t^{k_t^r}\right\}\right) \prod_{t=1}^{T}\prod_{r=1}^{n_t} \tilde{w}^{k_t^r,r}, \\
&= C - \sum_{t=1}^{T} \Lambda_t + \sum_{\mathbb{K}}\sum_{t=1}^{T}\sum_{r=1}^{n_t} \log\left\{\Lambda_t^{k_t^r}\right\} \prod_{t=1}^{T}\prod_{r=1}^{n_t} \tilde{w}^{k_t^r,r}, \\
&= C - \sum_{t=1}^{T} \Lambda_t + \sum_{t=1}^{T}\sum_{r=1}^{n_t}\sum_{m=0}^{M} \log\{\Lambda_t^m\}\, \tilde{w}_t^{m,r} \sum_{\mathbb{K}\setminus k_t^r} p\left(\mathbb{K}\setminus k_t^r | \mathbb{Y}, \hat{\mathbb{X}}, \hat{\mathbb{L}}\right), \\
&= C - \sum_{t=1}^{T} \Lambda_t + \sum_{t=1}^{T}\sum_{r=1}^{n_t}\sum_{m=0}^{M} \log\{\Lambda_t^m\}\, \tilde{w}_t^{m,r}, \\
&= C - \sum_{t=1}^{T} \Lambda_t + \sum_{t=1}^{T}\sum_{m=0}^{M} \log\{\Lambda_t^m\}\, \bar{n}_t^m,
\end{aligned} \tag{6.33}$$

where $\bar{n}_t^m = \sum_{r=1}^{n_t} \tilde{w}_t^{m,r}$ is the mean number of measurements due to component m at frame t conditioned on the current estimates of the kinematic state and measurement rate. The term C is constant with respect to the kinematic states and the measurement rates so its details are irrelevant. Its contents may change as the algebra develops but we need not find this troublesome. The auxiliary function term can be arranged into single component terms recalling that $\Lambda_t = \sum_{m=0}^{M} \Lambda_t^m$.

$$\begin{aligned}
&\sum_{\mathbb{K}} \log\{p(\mathbb{K}|\mathbb{L})\}\, p_{\text{miss}}(\mathbb{K}|\hat{\mathbb{X}}, \hat{\mathbb{L}}, \mathbb{Y}) \\
&= C - \sum_{t=1}^{T}\sum_{m=0}^{M} \Lambda_t^m + \sum_{t=1}^{T}\sum_{m=0}^{M} \log\{\Lambda_t^m\}\, \bar{n}_t^m, \\
&= C + \sum_{m=0}^{M}\sum_{t=1}^{T} \log\left\{\exp\{-\Lambda_t^m\} \left(\Lambda_t^m\right)^{\bar{n}_t^m}\right\}.
\end{aligned} \tag{6.34}$$

The third term from the auxiliary function now looks like a single rate observation for each component at each time. The observation pdf is almost Poisson: what is missing is a $\tilde{n}_t^m!$ term in the denominator. Such a term would be constant with respect to the states and rates so we have the liberty to introduce it, but $\tilde{n}_t^m$ is not an integer so the factorial isn't defined and indeed the Poisson distribution is only defined over integers. This is not really a problem because there is no requirement to construct an equivalent measurement pdf. During the M-step we will see that this measurement function can readily be accommodated.

The final term in the auxiliary function (6.21) is the expectation of the logarithm of the measurement likelihood over the conditional probability of the assignments. This is the same as with the standard PMHT with a multinomial prior and simplifies as shown in (3.57). The result is

$$\sum_{\mathbb{K}} \log \{p(\mathbb{Y}|\mathbb{X}, \mathbb{K})\} p_{\text{miss}}(\mathbb{K}|\hat{\mathbb{X}}, \hat{\mathbb{L}}, \mathbb{Y}) = \sum_{t=1}^{T}\sum_{m=0}^{M}\sum_{r=1}^{n_t} \log \left\{g^m\left(\mathbf{y}_t^r|\mathbf{x}_t^m\right)\right\} \tilde{w}_t^{m,r}, \tag{6.35}$$

where the weight is now defined by (6.30).

6.3.1.2 Maximisation Step

Once again, the independence assumptions have lead to an auxiliary function that separates into sub-functions for the assignment prior hyperparameters and the kinematic states

$$\mathcal{Q}\left(\mathbb{X}, \mathbb{L}|\hat{\mathbb{X}}, \hat{\mathbb{L}}\right) = \sum_{m=0}^{M} \mathcal{Q}_\Lambda^m + \sum_{m=1}^{M} \mathcal{Q}_x^m, \tag{6.36}$$

where the kinematic state part is the same as the core PMHT, given by (3.57)

$$\mathcal{Q}_x^m = \log\left\{p\left(\mathbf{x}_0^m\right)\right\} + \sum_{t=1}^{T} \log\left\{p\left(\mathbf{x}_t^m|\mathbf{x}_{t-1}^m\right)\right\} + \sum_{t=1}^{T}\sum_{r=1}^{n_t} \log\left\{g^m\left(\mathbf{y}_t^r|\mathbf{x}_t^m\right)\right\} \tilde{w}_t^{m,r}.$$

It is maximised in the same way as before, if $g^m(\cdot)$ is Gaussian then a Kalman filter can be used.

The Poisson rate term is what is new; it is given by

$$\mathcal{Q}_\Lambda^m = \log\left\{p\left(\Lambda_0^m\right)\right\} + \sum_{t=1}^{T} \log\left\{p\left(\Lambda_t^m|\Lambda_{t-1}^m\right)\right\} + \sum_{t=1}^{T} \log\left\{\exp\left\{-\Lambda_t^m\right\}\left(\Lambda_t^m\right)^{\tilde{n}_t^m}\right\}. \tag{6.37}$$

Maximising this expression requires one to adopt a specific form for the transition probability $p\left(\Lambda_t^m|\Lambda_{t-1}^m\right)$ and the prior $p\left(\Lambda_0^m\right)$. The conjugate prior for a Poisson measurement is a gamma distribution. Consider a single time step instead of a batch.

The gamma prior is defined as

$$p_{\text{gamma}}(\Lambda_t; \alpha_t, \beta_t) = \frac{\beta_t^{\alpha_t}}{\Gamma(\alpha_t)} \Lambda_t^{\alpha_t - 1} \exp\{-\beta_t \Lambda_t\}, \tag{6.38}$$

and the posterior with a Poisson measurement is, therefore,

$$\begin{aligned} \frac{\beta_t^{\alpha_t}}{\Gamma(\alpha_t)} \Lambda_t^{\alpha_t - 1} \exp\{-\beta_t \Lambda_t\} \frac{\exp\{-\Lambda_t\}}{n_t!} \Lambda_t^{n_t} &= C \Lambda_t^{\alpha_t - 1 + n_t} \exp\{-(\beta_t + 1)\Lambda_t\} \\ &= p_{\text{gamma}}(\Lambda_t; \alpha_{t|t}, \beta_{t|t}), \end{aligned} \tag{6.39}$$

where C is constant and

$$\alpha_{t|t} = \alpha_t + n_t \qquad \beta_{t|t} = \beta_t + 1. \tag{6.40}$$

Substituting these results back into the auxiliary function gives

$$\mathscr{Q}_\Lambda^m = \log\left\{p\left(\Lambda_0^m\right)\right\} + \sum_{t=1}^{T} \log\left\{p_{\text{gamma}}(\Lambda_t^m; \alpha_{t|t}^m, \beta_{t|t}^m)\right\}. \tag{6.41}$$

What remains is to describe a process noise model that transforms a gamma distribution at time $t-1$ to another gamma distribution at time t. It turns out that it is not possible to write this process noise in closed form so instead we follow an intuitive approach. The process noise should preserve the mean of the rate distribution but increases its uncertainty. This is achieved using a decay parameter $\exp\{-(\tau_t - \tau_{t-1})/\bar{\tau}\}$, where τ_t is the collection time of measurement t and $\bar{\tau}$ is a time constant. This time constant can be different for each target but we feel the notation is dense enough already. These time constants then give

$$p(\Lambda_t^m | n_1 \ldots n_t) \sim p_{\text{gamma}}(\Lambda_t^m; \alpha_{t|t-1}^m, \beta_{t|t-1}^m), \tag{6.42}$$

where

$$\alpha_{t|t-1}^m = \exp\{-(\tau_t - \tau_{t-1})/\bar{\tau}\}\, \alpha_{t-1|t-1}^m, \tag{6.43}$$

$$\beta_{t|t-1}^m = \exp\{-(\tau_t - \tau_{t-1})/\bar{\tau}\}\, \beta_{t-1|t-1}^m, \tag{6.44}$$

and

$$\alpha_{t|t}^m = \alpha_{t|t-1}^m + \bar{n}_t^m \qquad \beta_{t|t} = \beta_{t|t-1} + 1. \tag{6.45}$$

6.3.2 Image Measurement Data

It should be fairly clear that the treatment in the previous section applies to the case of a quantised count obtained from an image. The extra complication is that the specific location of each point measurement is missing data but this is no different to the core H-PMHT case. The complication is that the number of shots we generate through quantisation depends on the increment size $\hbar^2$. Since the Poisson model directly describes the count and not the relative proportion of measurements, we need to accommodate the quantisation in the Poisson model. The number of quantised shots is inversely proportional to the increment size, so the Poisson rate needs to also be inversely proportional to the increment size. Further, the mean of the gamma prior should be inversely proportional to the increment size as well.

Let the scaled target measurement rate be

$$\bar{\Lambda}_t^m = \frac{\Lambda_t^m}{\hbar^2} \tag{6.46}$$

and similarly for expressions such as $\bar{\mathbb{L}}$. We extend the Poisson model by assuming that the number of quantised shots from target m at time t is Poisson with rate $\bar{\Lambda}_t^m$.

To achieve the desired mean behaviour in the prior the precision parameter β remains unchanged and the α parameter is replaced with $\bar{\alpha} = \alpha/\hbar^2$ so that the prior becomes

$$p(\bar{\Lambda}_t^m|n_1 \dots n_t) \sim p_{\text{gamma}}(\bar{\Lambda}_t; \alpha_t, \beta_t), \tag{6.47}$$

$$\log\left\{p(\bar{\Lambda}_t^m|n_1 \dots n_t)\right\} = C + \frac{\alpha_t - 1}{\hbar^2}\log\{\Lambda_t\} - \frac{\beta_t}{\hbar^2}\Lambda_t. \tag{6.48}$$

Recall that the point measurement Poisson H-PMHT auxiliary function (6.36) is composed of two parts

$$\mathcal{Q}^n\left(\mathbb{X}, \mathbb{L}|\hat{\mathbb{X}}, \hat{\mathbb{L}}\right) = \sum_{m=0}^{M} \mathcal{Q}_\Lambda^m + \sum_{m=1}^{M} \mathcal{Q}_x^m.$$

In the image case, we need to consider the limit of this function

$$\mathcal{Q}\left(\mathbb{X}, \mathbb{L}|\hat{\mathbb{X}}, \hat{\mathbb{L}}\right) = \lim_{\hbar^2 \to 0} \hbar^2 \sum_{m=0}^{M} \mathcal{Q}_\Lambda^m + \lim_{\hbar^2 \to 0} \hbar^2 \sum_{m=1}^{M} \mathcal{Q}_x^m. \tag{6.49}$$

The second term in this limit is exactly the same as the core H-PMHT and we won't dwell on it. The first term can be expanded out by substituting (6.41)

$$\lim_{\hbar^2 \to 0} \hbar^2 \mathcal{Q}_\Lambda^m = \lim_{\hbar^2 \to 0} \hbar^2 \sum_{t=0}^{T} \log\left\{p_{\text{gamma}}(\Lambda_t^m; \alpha_{t|t}^m, \beta_{t|t}^m)\right\}. \tag{6.50}$$

where

$$\bar{\alpha}^m_{t|t-1} = \exp\{-(\tau_t - \tau_{t-1})/\bar{\tau}\}\,\bar{\alpha}^m_{t-1|t-1}, \tag{6.51}$$
$$\bar{\beta}^m_{t|t-1} = \exp\{-(\tau_t - \tau_{t-1})/\bar{\tau}\}\,\bar{\beta}^m_{t-1|t-1}, \tag{6.52}$$

and

$$\bar{\alpha}^m_{t|t} = \bar{\alpha}^m_{t|t-1} + \bar{n}^m_t \qquad \bar{\beta}_{t|t} = \bar{\beta}_{t|t-1} + 1. \tag{6.53}$$

Combining these with (6.48), the Poisson rate auxiliary sub-function becomes

$$\begin{aligned}
\lim_{\hbar^2 \to 0} \hbar^2 \mathscr{Q}^m_{\Lambda} &= \lim_{\hbar^2 \to 0} \hbar^2 \sum_{t=0}^{T} \left\{ C + \frac{(\alpha^m_{t|t-1} - 1)}{\hbar^2} \log\{\Lambda_t\} + \bar{n}^m_t - \frac{\beta_t}{\hbar^2}\Lambda_t. \right\}, \\
&= \sum_{t=0}^{T} \left\{ C + \left(\alpha^m_{t|t-1} - 1 + \lim_{\hbar^2 \to 0} \hbar^2 \bar{n}^m_t \right) \log\{\Lambda_t\} - \beta_t \Lambda_t \right\}, \\
&= \sum_{t=0}^{T} \left\{ C + \alpha^m_{t|t-1} - 1 + ||\mathfrak{z}^m_t|| - \beta_t \Lambda_t \right\}.
\end{aligned} \tag{6.54}$$

This is again equivalent to a gamma distribution with

$$\alpha^m_{t|t} = \alpha^m_{t|t-1} + ||\mathfrak{z}^m_t|| \mathfrak{z}^{i,m}_t. \tag{6.55}$$

The Poisson rate auxiliary sub-function is a sum of gamma distributions that are maximised at the mode of the gamma, which is given by

$$\hat{\Lambda}^m_t = \frac{\alpha^m_{t|t}}{\beta^m_t}. \tag{6.56}$$

Since the gamma parameters are linked over time the estimates can be expressed recursively. For compactness, we temporarily define $\rho = \exp\{-(\tau_t - \tau_{t-1})/\bar{\tau}\}$. Then

$$\begin{aligned}
\hat{\Lambda}^m_t &= \frac{\rho\,\alpha^m_{t|t-1} + ||\mathfrak{z}^m_t||}{\rho\,\beta^m_{t-1} + 1}, \\
&= \frac{\rho\,\beta^m_{t-1}}{\rho\,\beta^m_{t-1} + 1} \frac{\alpha^m_{t|t-1}}{\beta_{\tau-1}} + \frac{||\mathfrak{z}^m_t||}{\rho\,\beta^m_{t-1} + 1}, \\
&= A^m_t \hat{\Lambda}^m_{t-1} + \left(1 - A^m_t\right) ||\mathfrak{z}^m_t||,
\end{aligned} \tag{6.57}$$

where

$$A^m_t = \frac{\rho\,\beta^m_{t-1}}{\rho\,\beta^m_{t-1} + 1}, \tag{6.58}$$

and clearly $0 \leq A_t^m \leq 1$. The form of (6.57) is a single pole filter where the placement of the pole is a function of the precision of the estimate, β, which varies with time. This filter will achieve a steady state gain when

$$\beta = \rho\,\beta + 1, \tag{6.59}$$

$$\beta = \frac{1}{1-\rho}, \tag{6.60}$$

that is the rate of decay of information equals the rate of learning new information. At the steady state β, the coefficient A_t^m becomes

$$\begin{aligned} A &= \frac{\beta - 1}{\beta}, \\ &= 1 - (1-\rho) = \rho. \end{aligned} \tag{6.61}$$

The driving term on the Poisson rate filter is the total image power associated with the component, which is consistent with our initial assertion that Λ_t^m gives the mean power from component m in frame t.

The steady state gain of the Poisson rate filter is a function of the decay term $\rho = \exp\{-(\tau_t - \tau_{t-1})/\bar{\tau}\}$. When the time constant $\bar{\tau}$ is very large, then this exponential is close to $\exp\{0\} = 1$, the steady state β is very high, and the single pole is close to unity. That is, the filter integrates energy over a long time. Such a filter will provide a very low variance estimate of the true Λ_t^m if it is constant, but if it changes, the filter will have an overdamped response and lag behind. Conversely, if the time constant $\bar{\tau}$ is very small then the exponential term approaches zero and the steady state β approaches unity. In the limit, the filter completely ignores past data and the rate estimate is simply the current associated energy. This is as we would expect since this limiting case is when the Poisson rate is independent over time.

Note that this single pole form arises because of the dynamics we imposed on the gamma parameters in (6.44). The exponential correlation introduced by that relationship leads to a single pole estimator. This leads to the intuition that we can shape the step response of the rate estimator by modifying the recurrence relationship. We won't explore that in this book: we find that a single pole offers enough flexibility since we expect the true target cross section to vary smoothly over time.

Overall, the difference between Poisson H-PMHT and core H-PMHT is that core H-PMHT uses priors $\hat{\pi}_t^m$ that are independent from one time step to the next. The Poisson H-PMHT uses priors $\hat{\Lambda}_t^m$ that are smoothed over time using a recursive filter. The temporal response of this filter is controlled by the dynamic model we impose on the Poisson prior. This means that the Poisson H-PMHT offers the flexibility to tailor the estimate to trade off noise suppression and responsiveness to change.

6.4 Examples

We will now investigate the performance of the Poisson H-PMHT and Poisson PMHT by comparing them with the core H-PMHT and core PMHT through some simple examples. In this case we do not use the canonical multi-target scenario and instead, simulate a single constant velocity target. We will show examples of both point measurement and image measurement tracking. The point measurement examples given here were first presented in [3] and the image measurement examples were presented in [4, 9, 10].

6.4.1 Measurement Rate Estimation

Our first example will be point measurement tracking where the true number of measurements from the target follows a Poisson distribution. In this case, the Poisson PMHT derived in this chapter is matched to the truth.

Consider a single target moving with almost constant velocity and process noise 0.001 $\text{units}^2/\text{frame}^2$. The surveillance region is 100×100 units with high density uniformly distributed clutter. Both the target and the clutter generate a Poisson distributed number of measurements on each scan; the average number of clutter measurements is fixed at $\Sigma_t^0 = 50$ and the average number of target measurements is varied. Target measurements consist of the true target position plus unit variance Gaussian noise. A hundred Monte Carlo trials, each with 20 scans of data are simulated. The track is initialised using a single measurement on the first scan with a zero mean velocity prior. Each algorithm performs 20 EM iterations. We now present the accuracy of the estimated target measurement rate for various true Λ^1 values. The Poisson PMHT naturally produces an estimate of this parameter, the core PMHT estimate is obtained as $\hat{\Lambda} = N_t\, \hat{\pi}_{20}^1$.

The Poisson PMHT rate parameters α_0^1 and β_0^1 were initialised using the steady state values derived in the previous section and $\hat{\Lambda}_0^m = 1$. The decay constant $\rho = \exp\{-(\tau_t - \tau_{t-1})/\bar{\tau}\}$ was varied between 0.5, 0.8 and 0.9. Figure 6.2 shows the bias for each value of ρ determined by finding the mean error between $\hat{\Lambda}_{20}^1$ and the true Λ and for the core PMHT. It is clear from the bias curves that the core PMHT overestimates the target measurement rate, especially when the true rate is small. This is because there are clutter measurements that sometimes fall close to the target. The core PMHT assigns these measurements to the target and consequently assigns a value to π_{20}^1 that is too big. The Poisson PMHT still assigns the measurements to the target since they are physically close, but due to the imposed smoothness in the Λ_t^m estimates, it is less influenced by the incorrect association. For low values of ρ the initialisation bias in the Poisson PMHT is forgotten by the end of the batch, but for $\rho = 0.9$ the λ estimates do not yet converge after 20 scans. This translates to a bias in Fig. 6.2. If the sequence length had been longer, this bias would have been reduced.

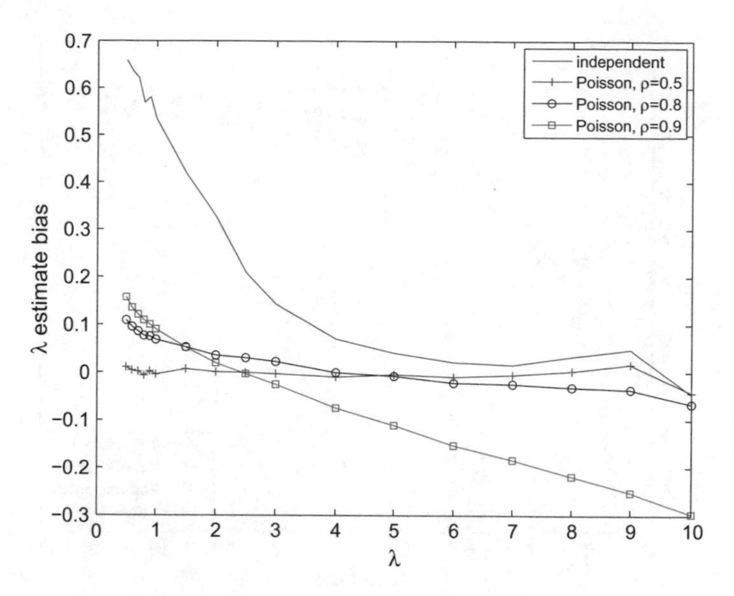

Fig. 6.2 Bias in estimated measurement rate $\hat{\lambda}$

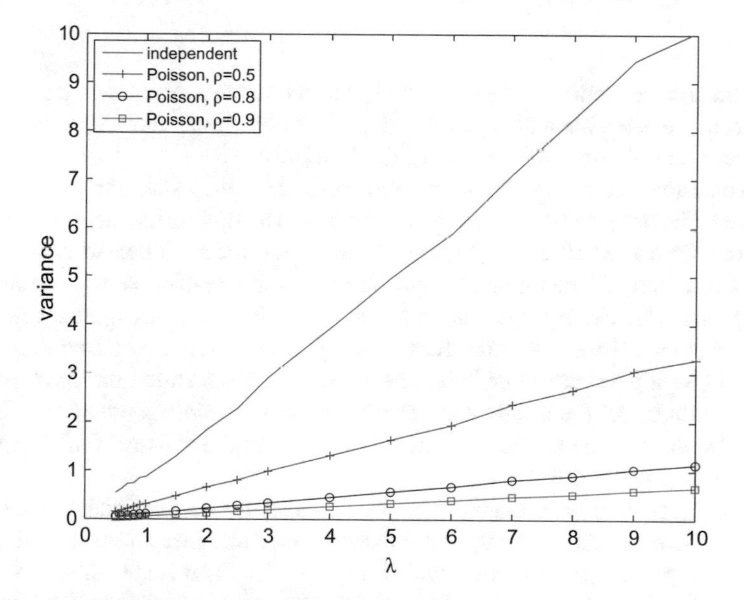

Fig. 6.3 Variance in estimated measurement rate $\hat{\lambda}$

Figure 6.3 shows the variance of the error between the Poisson PMHT $\hat{\lambda}^1_{20}$ and the truth λ^1 for the same values of ρ and for the core PMHT. The core PMHT variance is approximately equal to the true measurement rate λ: this is the variance of the driving Poisson process. Notwithstanding the incorrect associations described

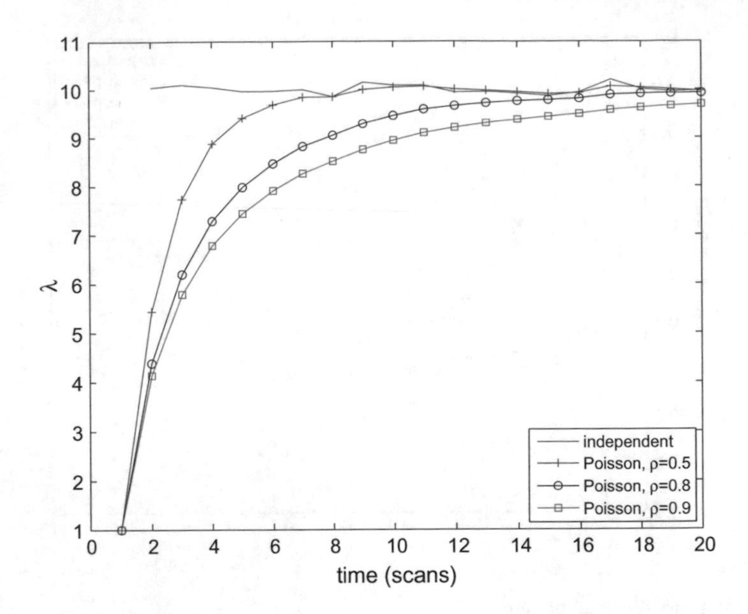

Fig. 6.4 Estimate step response for $\lambda = 10$

above, the independent prior PMHT mostly correctly identified which measurements come from the target but with no temporal model it could not reduce the measurement variance. The Poisson PMHT variance is much lower.

The bias and variance results above summarise the steady state performance of the estimates. The transient performance was measured by finding the mean and variance over time for a $\lambda = 10$ target. Figure 6.4 shows the mean estimated measurement rate as a function of time for an average measurement rate of $\lambda^1 = 10$. The achieved step responses are clearly exponential for the idealised known-assignment case. The curve for $\rho = 0.9$ illustrates the bias effect discussed above: the step response is slow and has still not converged after 20 scans. Figure 6.5 shows the corresponding variance values. As the steady state results showed, the independent prior PMHT has the same variance as the measurements whereas the Poisson PMHT variance decreases as ρ is increased.

We now repeat the experiment for targets that obey the conventional tracking measurement assumptions: each target produces exactly one measurement with probability P_D or no measurement with probability $1 - P_D$. As it is customary to assume a Poisson distributed number of false alarm measurements, this is retained from the first scenario. The performance measures used for the Poisson target case are repeated here. Figure 6.6 shows the mean estimated measurement rate as a function of the probability of detection. In this case the independent core PMHT significantly overestimates the measurement rate for the same reason as discussed with the Poisson targets: it incorrectly associates clutter measurements. The Poisson PMHT mitigates

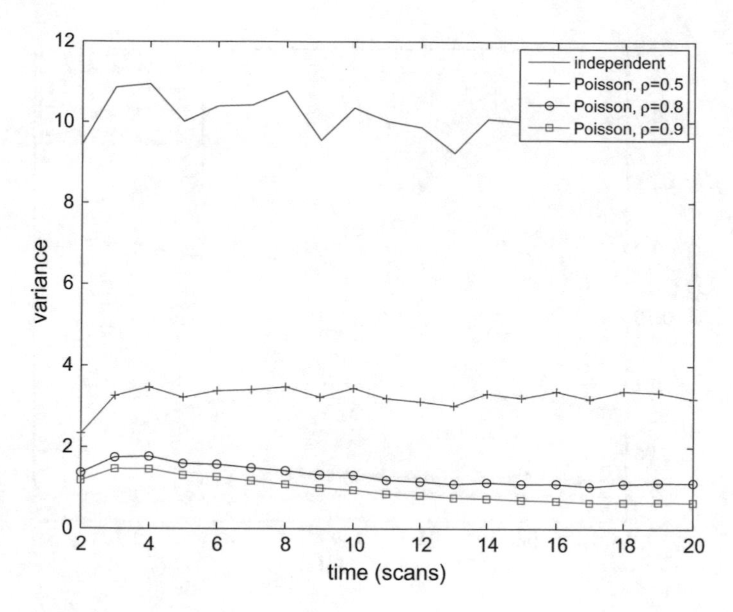

Fig. 6.5 Step response variance for $\lambda = 10$

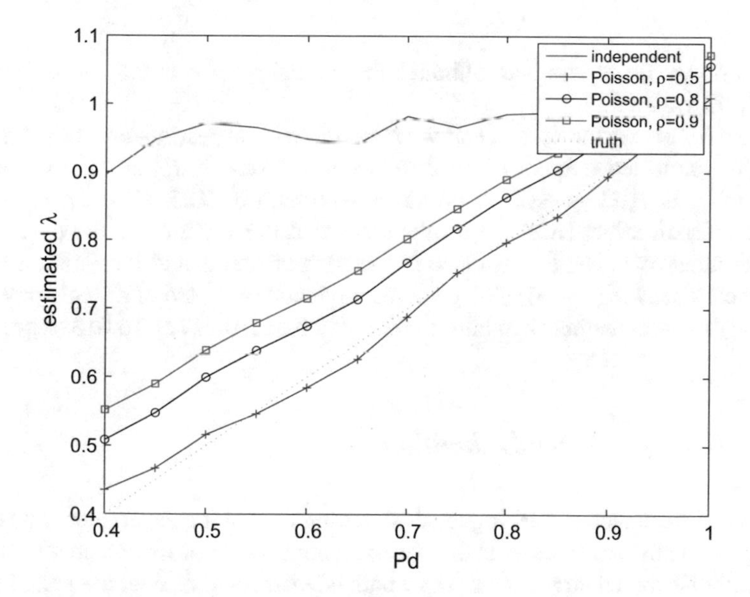

Fig. 6.6 Estimated measurement rate for Bernoulli targets

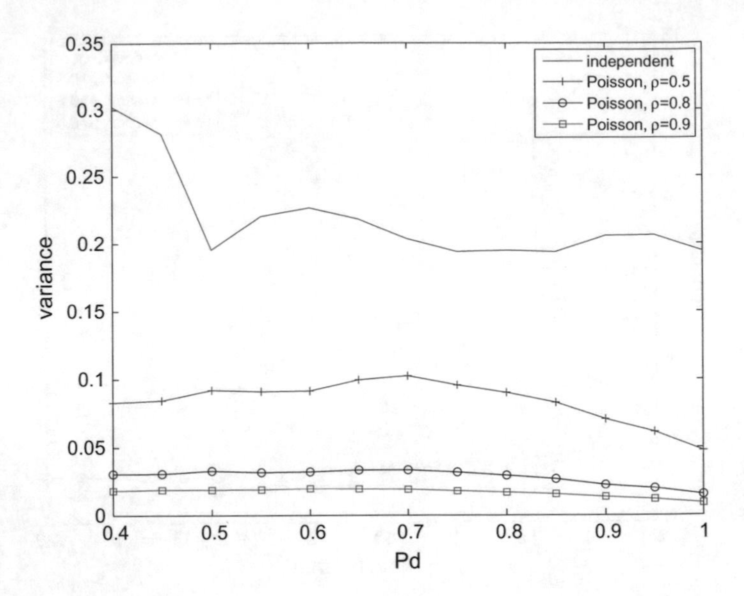

Fig. 6.7 Estimate variance for Bernoulli targets

this and shows much less bias, although the estimates for $\rho = 0.8$ and $\rho = 0.9$ are still a little too high.

Figure 6.7 shows the variance of the measurement rate estimates. For a Bernoulli target, the expected number of measurements in a scan is P_D and the variance in this number is $P_D(1 - P_D)$, which has a maximum of 0.25 when $P_D = 0.5$. As for the Poisson target case, the results show that the variance of the core PMHT approximately matches the variance of the measurements and the Poisson PMHT closely matches the $(1 - \rho)/(1 + \rho)$ reduction factor expected of a single pole filter, which in this case, predicts maximum variance of 0.083, 0.027 and 0.013 respectively.

6.4.2 Average Power Estimation

In track before detect, the analogue of estimating the point measurement rate is estimating the mean target power. In this case, it is not appropriate to consider Poisson or Bernoulli distributed target power since both are distributions over integers. Instead, we consider two deterministic but changing power situations and also a Swerling I case where the instantaneous power follows an exponential distribution. We note that there is a degree of mismatch here between the Poisson H-PMHT assumptions and the model used to generate data.

Figure 6.8 compares the output of the core H-PMHT and the Poisson H-PMHT for a smoothly varying target power with two different values of the Poisson rate forgetting factor ρ. The plots show a single realisation of the random elements. The

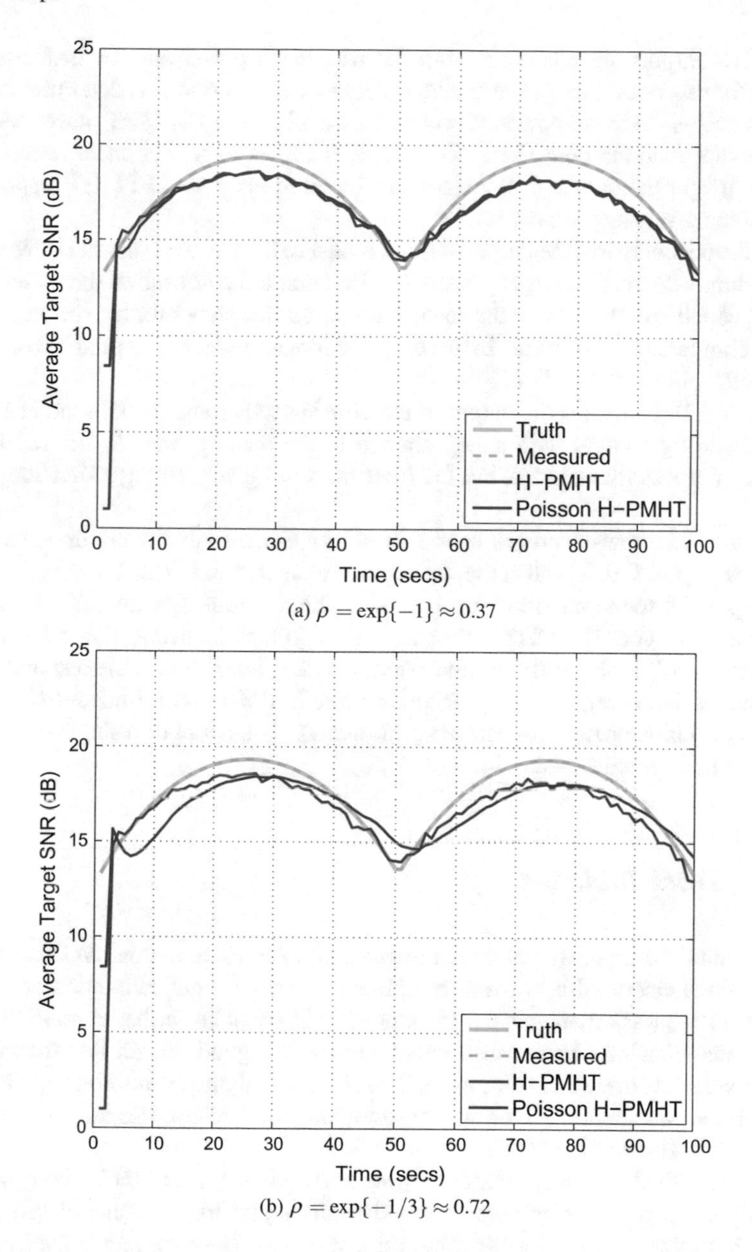

(a) $\rho = \exp\{-1\} \approx 0.37$

(b) $\rho = \exp\{-1/3\} \approx 0.72$

Fig. 6.8 Power estimation for a smoothly varying target amplitude

H-PMHT outputs are compared with the true mean power and the instantaneous power for this scan. For this example, the instantaneous power is deterministic and is the same as the mean power. It is clear that the Poisson H-PMHT power estimate is smoother than the core H-PMHT, that is, it shows lower variance. In this case, using a longer memory $\rho = 0.7$ leads to a lag in the Poisson H-PMHT response to the changing true target power.

Figure 6.9 compares the output of the core H-PMHT and the Poisson H-PMHT for a Swerling I fluctuating target power. The line marked *truth* shows the mean target power, which is constant, and the dashed *measured* line shows the true instantaneous power simulated on this frame. In this case, the longer memory ρ is preferred because it smooths out the fluctuations.

Figure 6.10 compares the output of the core H-PMHT and the Poisson H-PMHT for a Swerling I target with a step change in the mean power. Again, the longer memory reduces fluctuations, but the filter does not give a strong indication of the step.

Figure 6.11 shows averages over 100 Monte Carlo trials for the three example targets with $\rho = 0.9$. It is clear that the higher value of ρ makes the Poisson H-PMHT less responsive to the smoothly varying case, which is quite dynamic. In contrast, the variance in the core H-PMHT estimates is so high that the average over 100 Monte Carlo trials still looks noisy. In this average sense, both algorithms respond to the step change. However, referring back to the core H-PMHT performance in Fig. 6.10, there is very little prospect of detecting this average behaviour on a single realisation, so it's of little practical use.

6.4.3 Track Initiation

For our final example we consider the problem of track initiation. As discussed in the previous chapter this is really a problem of model order estimation and in the context of the track management structure in place can be further simplified to the idea of deciding whether or not a candidate track is good. Recall that for the core PMHT we made these decisions using the estimated mixing proportions $\hat{\pi}_t^m$. For this comparison, we use just the point measurement case because this allows us to also consider the Hysteresis PMHT.

In the original hysteresis work of Davey and Gray [2], the PMHT with hysteresis was shown to give performance approximately equal to the standard PMHT for uniform clutter but superior for non-uniform clutter. Here we repeat the initiation comparison for the same uniform and non-uniform clutter examples. The measure of performance is a Receiver Operating Characteristic (ROC) curve that is defined as the locus of the probability of detection as a function of the probability of false alarm: refer to Chap. 1 for a more detailed discussion. The ROC curve was constructed by performing 10,000 Monte Carlo trials with a target present and 10,000 with no target. When the target was present, the track was initialised on the first target measurement and the detection statistics were derived from these tracks. When no target was

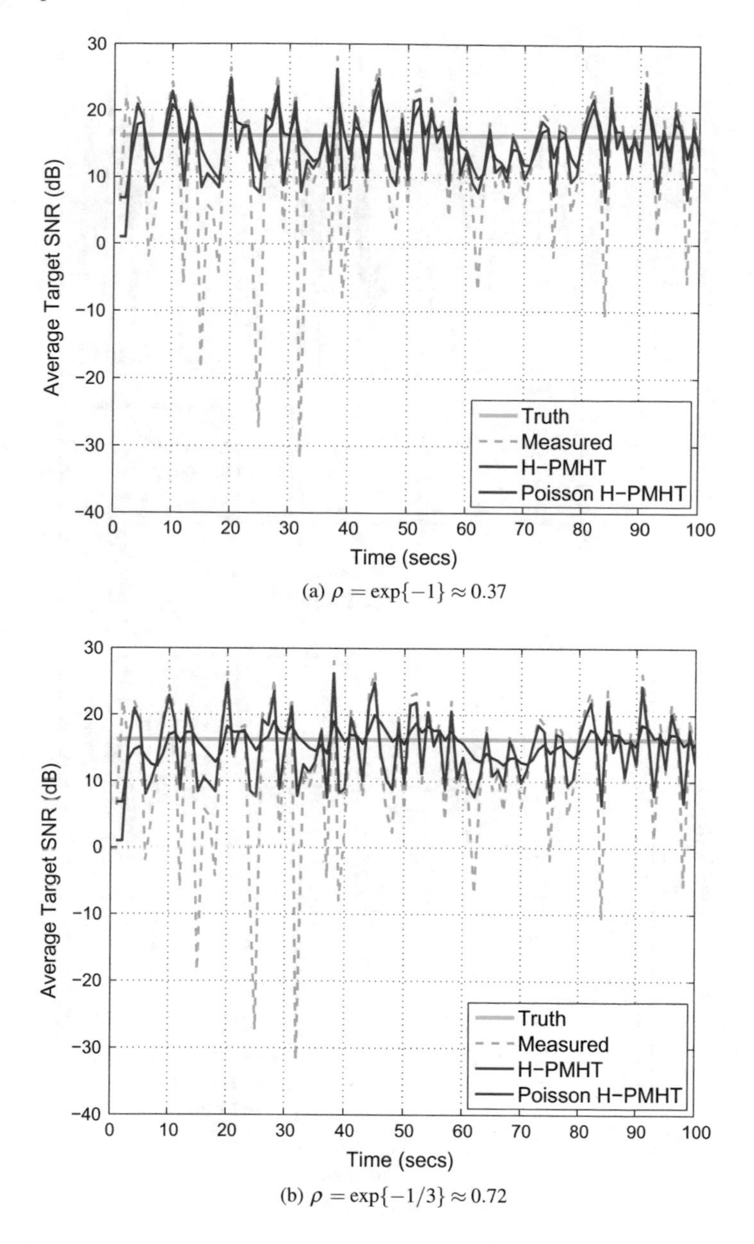

(a) $\rho = \exp\{-1\} \approx 0.37$

(b) $\rho = \exp\{-1/3\} \approx 0.72$

Fig. 6.9 Power estimation for a Swerling I target amplitude

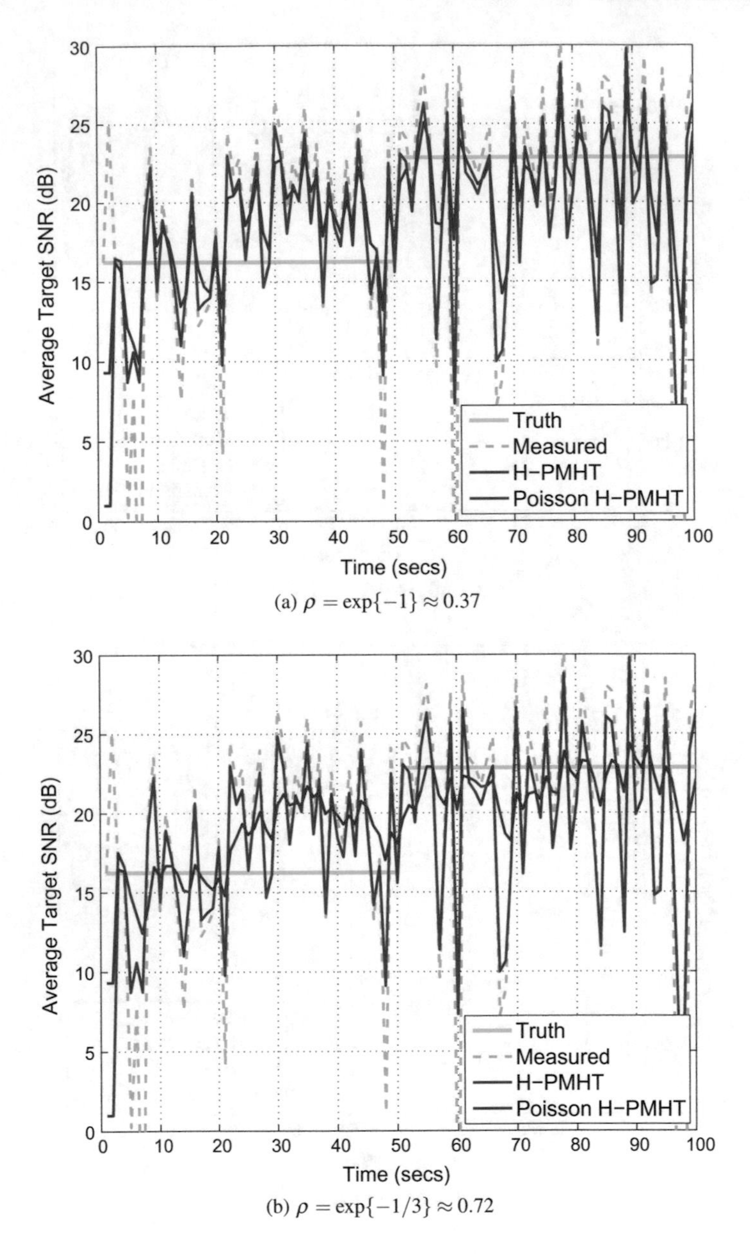

(a) $\rho = \exp\{-1\} \approx 0.37$

(b) $\rho = \exp\{-1/3\} \approx 0.72$

Fig. 6.10 Power estimation for a step-change target amplitude

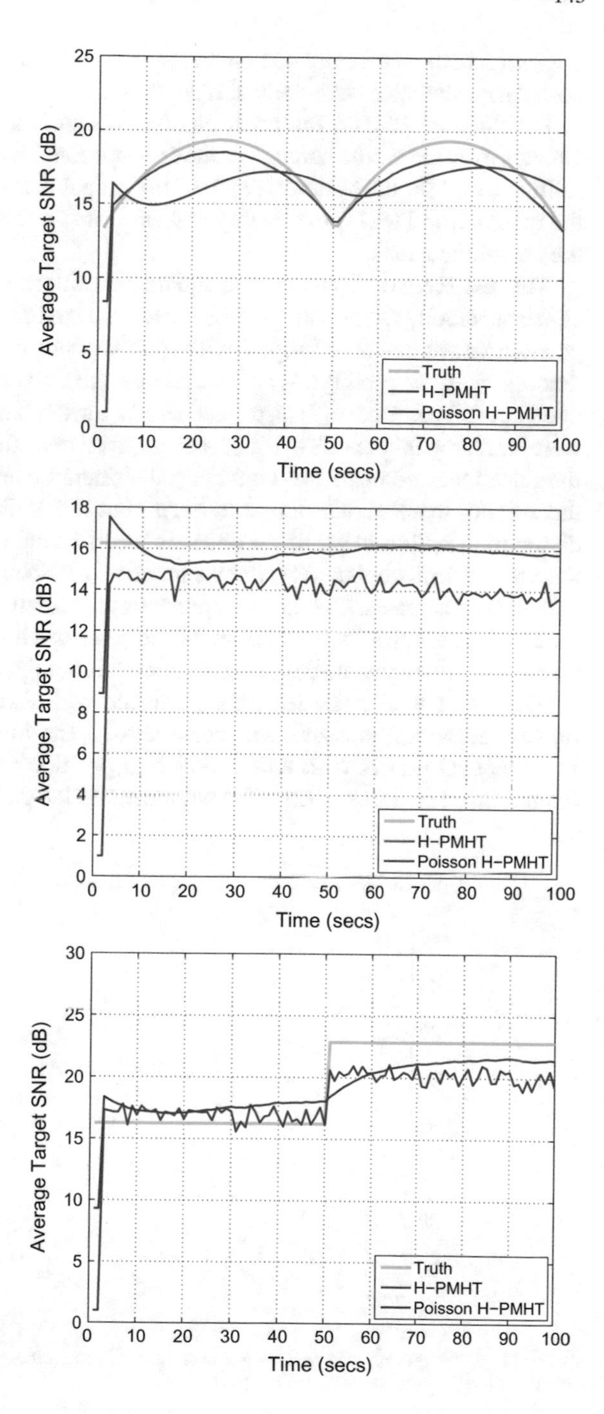

Fig. 6.11 Power estimation for a step-change target amplitude

present, a track was initialised on a single random clutter point and the false track acceptance statistics were derived from these.

For the core PMHT, the track quality statistic was the sum of the assignment weights over the whole batch, which is proportional to the batch mixing proportion estimate, $\hat{\pi}^m$; for the hysteresis PMHT the probability of a target existing was used; for the Poisson PMHT, the quality statistic was the estimated measurement rate at the end of the batch.

The test scenario consists of a square surveillance region with a single almost constant velocity target with process noise 0.001 units^2/frame^2. The sensor provides position measurements corrupted with additive white noise with unit variance that are detected with probability $P_\text{D} = 0.6$ and also false alarm measurements whose average rate is 50 per frame; [2] also considered a lower clutter density but all approaches were similar in this case. Two false alarm spatial densities are considered: in the first, the false measurements are uniformly distributed over the surveillance region; in the second, the X-dimension is uniform, but the Y-dimension has an exponential distribution with a mean of 20 units. Figure 6.12 shows an example realisation of the non-uniform clutter. Target measurements are shown as circles and clutter measurements as crosses. Due to the exponential distribution in Y, there are more clutter measurements towards the bottom of the plot. In all cases, the tracker knows the true-false measurement spatial distribution but not the true-false measurement rate.

The target state estimates were produced with Kalman filters that use the true process noise and measurement noise above. The initial state covariance was set to a diagonal matrix with unit variance in position and variance 0.1 in speed, i.e. diag(1, 0.1, 1, 0.1), as in [2]. The parameters of the gamma prior for the assignment

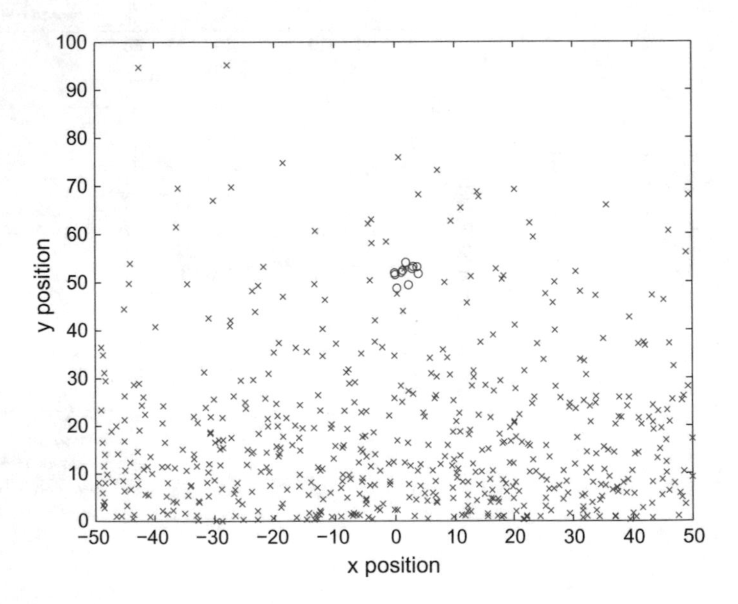

Fig. 6.12 Non-uniform clutter, example realisation

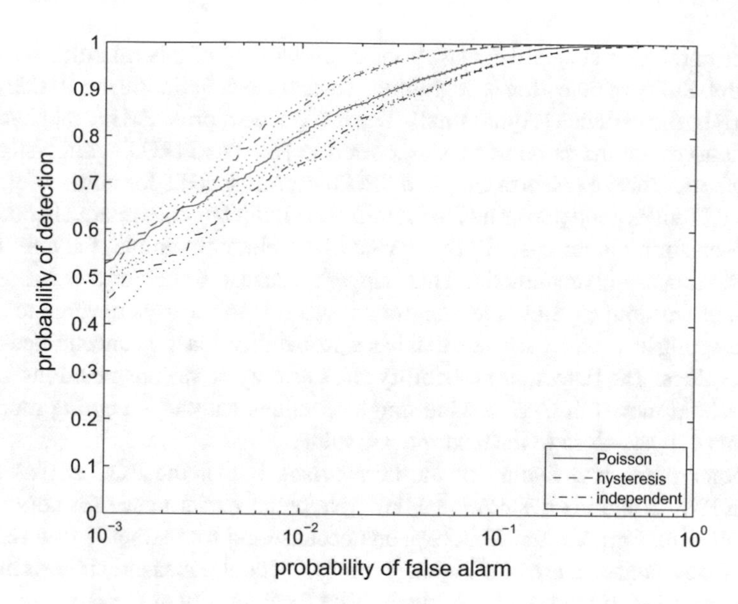

Fig. 6.13 Uniform clutter

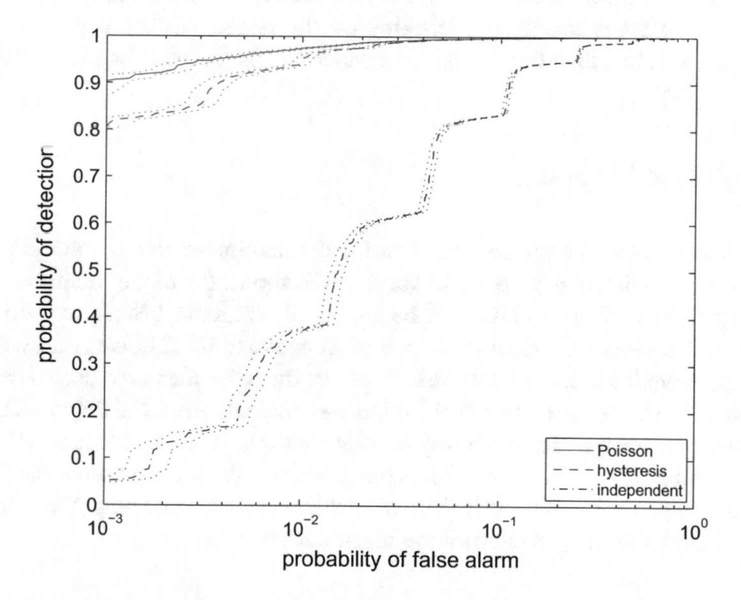

Fig. 6.14 Non-uniform clutter

rate was set to $\alpha_0^m = 0.3$ and $\beta_0^m = 10$ with $\exp\{-(\tau_t - \tau_{t-1})/\bar{\tau}\} = 0.9$, values which were empirically found to work well.

Figure 6.13 shows the ROC curves derived from these Monte Carlo trials for uniform clutter and Fig. 6.14 for non-uniform clutter. For each algorithm dotted lines mark the one-sigma standard deviation in the probability of false alarm; the variance in the probability of detection is negligible. Since the simulations used 10,000 Monte Carlo trials, the variance is quite small. The independent prior PMHT and hysteresis PMHT curves are the same as those reported in [2]. The PMHT with hysteresis is slightly better for the uniform case and the Poisson H-PMHT for non-uniform. The core PMHT gives good performance for uniform clutter but does relatively poorly for the non-uniform clutter case. In the non-uniform clutter example, the core PMHT shows a staircase characteristic. This happens because the algorithm has enough degrees of freedom to allow the assignment weights to mostly converge to zero or unity resulting in a quality statistic that has a probability density concentrated around integer values. The detection probability rises steeply at various positions because the threshold moves through a wide range of values without accepting more false tracks until it gets close to the next integer value.

Another interesting feature of the comparison is that the ROC curves for the Poisson PMHT and the hysteresis PMHT give better performance for non-uniform clutter than uniform clutter. This happens because the true target is in a region of relatively low clutter. The overall average number of clutter measurements is the same in both cases but in the non-uniform case, the local density at the target location is lower than the uniform density because the concentrated clutter is at a lower Y value. The overall effect is to improve tracking on the target. For the core PMHT this improvement is less than the degradation due to the increased false alarm rate.

6.5 Clutter Mapping

The development so far has assumed that the distribution of clutter and background noise is either uniform or known. In some applications, the clutter distribution will not be spatially uniform and will not be temporally stationary. Since we will not be told what this clutter distribution is, we must estimate it. This is actually a fairly small step from the mechanics already in place: the basic idea is to use some of the components to track the clutter distribution and others to track the targets. Tracking the clutter components requires a way to estimate the appearance function g^m, which has so far been assumed known. That is the subject of Chap. 8. However, the Poisson model developed here is also an important enabler for clutter mapping. We will revisit clutter mapping in Chap. 8 once all the machinery is in place.

6.6 Target Life Cycle

The implementation discussion in Chap. 5 presented a method for automatic track maintenance decisions, namely track formation and termination. In the context of Chap. 5 this was a rather ad hoc external layer of software that essentially amount to database operations on the track list. Within the context of a Poisson mixture this can now be couched in a mathematical formalism. The derivation in this chapter has developed an algorithm to perform inference over the parameters of a Poisson mixture with a fixed number of components. Track maintenance adds the ability to remove old components and add new components.

Track termination can be achieved in two ways in the Poisson mixture framework. This first way is to explicitly build it in to the rate transition probability by defining

$$p\left(\Lambda_t^m|\Lambda_{t-1}^m\right) = \begin{cases} p_{\text{death}}\left(\Lambda_{t-1}^m\right) & \Lambda_t^m = 0, \\ \left[1 - p_{\text{death}}\left(\Lambda_{t-1}^m\right)\right] p\left(\Lambda_{t-1}^m \to \Lambda_t^m\right) & \text{otherwise.} \end{cases} \tag{6.62}$$

Using this one could estimate the probability that each component transitions to zero rate, i.e. vanishes. This is the sort of thing usually used in many theoretical tracking papers. This may be problematic when applied to some of the fluctuating target scenarios presented in this chapter where targets drop down to almost zero rate; Swerling I targets have an exponential distribution on the instantaneous amplitude, which means that zero amplitude is the peak of the distribution even for arbitrarily high SNR. The alternative is to be more pragmatic and recognise that the filter already provides an estimate of the measurement rate Λ_t^m. Rather than impose a model that explicitly drives this to zero based on a Bernoulli coin toss, we can allow the algorithm to implicitly discard a component by ascribing a very small Λ_t^m estimate. Once this value gets small enough, the peak of the component distribution will be below the clutter floor, in which case future updates can only drive it to smaller values. Intuitively the algorithm has the capacity to intrinsically reject components without even building in the Bernoulli death process. For the sake of efficiency we can pre-empt this rejection by pruning components with rates below a prescribed threshold. Providing that the value of this threshold is sufficiently low it is arbitrary and simply controls computation efficiency.

Track birth requires the algorithm to automatically add new components when new targets appear. We can model the overall Poisson mixture as the superposition of a birth mixture and a continuing mixture where the first is the collection of new targets at this time and the second is the collection of components (target and clutter) that were present previously and still remain at this time. By Poisson superposition this is

$$\lambda_t^{\text{all}}\left(\mathbf{y}|\mathbb{Z}, \lambda_{t-1}^{\text{all}}(\cdot)\right) = \lambda_t^{\text{continuing}}\left(\mathbf{y}|\mathbf{Z}_t, \lambda_{t-1}^{\text{all}}(\cdot)\right) + \lambda_t^{\text{birth}}\left(\mathbf{y}|\mathbb{Z}, \lambda_{t-1}^{\text{all}}(\cdot)\right). \tag{6.63}$$

The previous overall Poisson mixture is a sufficient statistic for the past data for the continuing intensity and its components are estimated using the Poisson H-PMHT derived in this chapter with low rate components pruned as described above. The birth intensity is the intensity of new targets and is conditioned on the measurement sequence because we interpret *new* to mean anything not already contained in the prior mixture: there may be loitering targets that appeared before the current frame but are yet to be recognised by the estimator. In unrealistic simulations, this intensity might be given to you by a genie. In practice it is data driven and resembles a data-driven proposal for particle filtering TkBD [6], that is, components are placed in locations of high residual energy. An astute reader may notice that this birth intensity looks rather like the new tentative track collection with the serial numbers filed off. In fact this discussion does not provide any new machinery for dealing with track maintenance, instead the Poisson mixture model gives us a 'principled' way to interpret the track creation and termination methods already introduced.

6.7 Summary

This chapter has introduced a variation on H-PMHT that models the absolute amount of energy from each mixture component, not the relative proportion of energy like the core H-PMHT. To achieve this, it assumes a Poisson model for the number of shot measurements in the quantised interim working space whereas the core H-PMHT deals with a multinomial distribution. We showed that the two are compatible and that the Poisson absolute model implies a multinomial relative model for the number of shot measurements conditioned on knowing the total number.

The Poisson H-PMHT[1] is an important enabler for introducing a dynamic model on the assignment prior Π because its parameters can be estimated independently between components. In the core H-PMHT, the assignment parameters are coupled through a normalisation constraint and this leads to a requirement for joint estimation.

Through simulated examples of point measurements and imagery, the use of a dynamic model over the assignment prior was shown to give more robust performance when the amplitude of targets fluctuates and it gave improved performance for track initiation.

References

1. Davey, S.J.: Extensions to the probabilistic multi-hypothesis tracker for improved data association. School of Electrical and Electronic Engineering, The University of Adelaide (2003)
2. Davey, S.J., Gray, D.A.: Integrated track maintenance for the PMHT via the hysteresis model. IEEE Trans. Aerosp. Electron. Syst. **43**(1), 93–111 (2007)
3. Davey, S.J.: Probabilistic multihypothesis trackerwith an evolving poisson prior. IEEE Trans. Aerosp. Electron. Syst. **51**, 747–759 (2015)
4. Gaetjens, H.X., Davey, S.J., Arulampalam, S., Fletcher, F.K., Lim, C.C.: Histogram - PMHT for fluctuating target models. IET Radar Sonar Navig. **11**(8), 1292–1301 (2017)
5. Gelman, A., Carlin, J.B., Stern, H.S., Rubin, D.B.: Bayesian Data Analysis. Chapman and Hall/CRC, Boca Raton (2004)

[1] *MakeHPMHTParams*
The H-PMHT toolbox contains two main functions for running H-PMHT. *MakeHPMHTParams* defines tracking parameters, including which version of H-PMHT to execute, and *HPMHTTracker* actually does the tracking. The default tracking parameters for the Poisson H-PMHT are created by *MakeHPMHTParams('Poisson')*. For details on how to use this function, refer to the H-PMHT toolbox documentation.

6. Ristic, B., Arulampalam, S., Gordon, N.J.: Beyond the Kalman Filter: Particle Filters for Tracking Applications. Artech House, London (2004)
7. Skolnik, M.I.: Introduction to Radar Systems. McGraw-Hill, New York (2001)
8. Streit, R.L.: Poisson Point Processes: Imaging, Tracking, and Sensing. Springer, Berlin (2010)
9. Vu, H.X.: Track-before-detect for active sonar. Ph.D. thesis, The University of Adelaide (2015)
10. Vu, H.X., Davey, S.J., Arulampalam, S., Fletcher, F., Lim, C.C.: Histogram-PMHT with an evolving poisson prior. In: Proceedings of ICASSP, (2015)
11. Williams, J.L., Lau, R.: Approximate evaluation of marginal association probabilities with belief propagation. IEEE Trans. Aerosp. Electron. Syst. **50**, 2942–2959 (2014)

Chapter 7
Known Non-Gaussian Target Appearance

The mathematical development used to derive the H-PMHT used a general function notation for the appearance model $g^m\left(\mathbf{y}|\mathbf{x}_t^m\right)$ and a generic transition probability $p\left(\mathbf{x}_t^m|\mathbf{x}_{t-1}^m\right)$. However, the maximisation was significantly simplified by assuming a Gaussian appearance function and a linear Gaussian process model on the kinematic state. The numerical examples so far have all used these additional assumptions and the implementations have been based on the Kalman filter. At this point, we pause to emphasise that these Gaussian assumptions are not a necessary part of H-PMHT. They lead to convenient implementations but more general functions are possible. In this chapter, we consider how to implement the EM maximisation step when the appearance function $g^m\left(\mathbf{y}|\mathbf{x}_t^m\right)$ is not Gaussian. Importantly, we still assume that the target dynamics and appearance are *known*, they just are not Gaussian.

At one level the less constrained problem can be viewed as a nonlinear optimization and we could dismiss implementations as a simple matter of numerical programming. However, in practice the matter is not so simple and many have made the mistake of believing that H-PMHT requires the more restrictive Gaussian assumptions.

There are two parts of the estimation problem: the target dynamics and the measurement process. Chapter 3 demonstrated that the synthetic measurement approach depends on an assumed Gaussian appearance function, not on linearity. This means that H-PMHT can be implemented as a point measurement estimator provided the appearance function is Gaussian. If the mean location of the target response is nonlinear in the target state, or the target dynamics are nonlinear then a point measurement particle filter can be implemented without much difficulty. The bigger problem arises when the appearance is not Gaussian.

Four different approaches have been published for H-PMHT with known non-Gaussian appearance. The first two of these are generic numerical methods that can be applied to various non-Gaussian problems: particle filtering [2]; and grid-based approximation through the Viterbi algorithm [5, 9, 10]. Both the particle filter and the Viterbi algorithm are briefly reviewed in a more generic context in

S. J. Davey and H. Gaetjens, *Track-Before-Detect Using Expectation Maximisation*, Signals and Communication Technology,
https://doi.org/10.1007/978-981-10-7593-3_7

Appendix A. The third approach uses an analytic transformation to represent the appearance function as a Gaussian with non-constant parameters [3]. This amounts to generalising the Gaussian used so far to a broader parameter space, and then finding a good approximation to the non-Gaussian shape in this broad parameter space. The fourth approach uses a Gaussian mixture to represent the appearance function. In effect, this treats the family of shifted and scaled Gaussians as a basis for the appearance. In practice, this is only an approximation. The published work with Gaussian mixtures makes the further assumption that the appearance is unknown [4]. The extension of H-PMHT to learn the appearance function will be addressed in Chap. 8, in this chapter, we consider the case that the appearance is known but inconvenient and a mixture approximation is used to facilitate implementation.

7.1 Grid-Based Maximisation for Non-Gaussian Appearance

The Viterbi algorithm is a general method for finding the optimal sequence estimate in a discrete state system. The method is briefly reviewed in the Appendix A.5. Here we show how the Viterbi algorithm can be used as an optimisation tool for the H-PMHT auxiliary function if we approximate the state space by a grid.

In the general case the state dependent auxiliary sub-function in the core H-PMHT is given by (4.64):

$$\mathcal{Q}_x^m = \sum_{t=1}^{T} \frac{||\mathbf{z}_t||}{\sum_{i=1}^{I} G_t^i} \log\left\{p\left(\mathbf{x}_t^m|\mathbf{x}_{t-1}^m\right)\right\} + \sum_{t=1}^{T}\sum_{i=1}^{I+I^U} \bar{\mathbf{z}}_t^i w_t^{i,m} \frac{\int_{W^i} \log\left\{g^m\left(\mathbf{y}|\mathbf{x}_t^m\right)\right\} g^m\left(\mathbf{y}|\hat{\mathbf{x}}_t^m\right) d\mathbf{y}}{\int_{W^i} g^m\left(\mathbf{y}|\hat{\mathbf{x}}_t^m\right) d\mathbf{y}},$$

which is modified to (5.2):

$$\mathcal{Q}_x^m = \sum_{t=1}^{T}\sum_{i=1}^{I+I^U} \bar{\mathbf{z}}_t^i w_t^{i,m} \log\left\{p\left(\mathbf{x}_t^m|\mathbf{x}_{t-1}^m\right)\right\} + \sum_{t=1}^{T}\sum_{i=1}^{I+I^U} \bar{\mathbf{z}}_t^i w_t^{i,m} \frac{\int_{W^i} \log\left\{g^m\left(\mathbf{y}|\mathbf{x}_t^m\right)\right\} g^m\left(\mathbf{y}|\hat{\mathbf{x}}_t^m\right) d\mathbf{y}}{\int_{W^i} g^m\left(\mathbf{y}|\hat{\mathbf{x}}_t^m\right) d\mathbf{y}}, \tag{7.1}$$

when the alternative resampled prior is used. The first term in this function couples together the states over adjacent time steps and the second term combines a frame of data with the state at the same time instant. In this general case, there is no analytic solution to either of the integrals in the measurement dependent second term of the auxiliary sub-function. However, it is not restrictive to assume that we can point-wise evaluate $g^m\left(\mathbf{y}|\mathbf{x}_t^m\right)$ and its logarithm for a given state $\mathbf{x}_t^m$. This means that we can use the same numerical integration methods discussed in Chap. 5 to evaluate the measurement term for a given state. In effect, we can write

$$\sum_{t=1}^{T}\sum_{i=1}^{I+I^U} \bar{\mathbf{z}}_t^i w_t^{i,m} \frac{\int_{W^i} \log\left\{g^m\left(\mathbf{y}|\mathbf{x}_t^m\right)\right\} g^m\left(\mathbf{y}|\hat{\mathbf{x}}_t^m\right) d\mathbf{y}}{\int_{W^i} g^m\left(\mathbf{y}|\hat{\mathbf{x}}_t^m\right) d\mathbf{y}} = \log\left\{\phi\left(\mathbf{z}|\mathbf{x}_t^m\right)\right\}, \tag{7.2}$$

where in this case $\phi\left(\mathbf{z}|\mathbf{x}_t^m\right)$ is an inconvenient but not numerically insurmountable function. Using this representation the auxiliary sub-function becomes

$$\mathscr{Q}_x^m = \sum_{t=1}^{T}\left[A_t^m \log\left\{p\left(\mathbf{x}_t^m|\mathbf{x}_{t-1}^m\right)\right\} + \log\left\{\phi\left(\mathbf{z}|\mathbf{x}_t^m\right)\right\}\right], \tag{7.3}$$

which is clearly the log-likelihood for a nonlinear filtering problem. The EM process requires us to find the state value that maximises the auxiliary sub-function. A salient option is to throw a computer at it. We can view the sequence of states $\mathbf{x}_{0:T}^m$ as simply a high-dimension vector and use a numerical optimisation method, such as those available in commercial scientific software, to find the maximising value. Of course, this would work, but it is expensive and rather churlish. A better option is to exploit the structure of the problem. A good option is to optimise over a grid and use the Viterbi algorithm [1, 8]. When the state space is approximated as a discrete grid, then the possible states can be treated as a list enumerated by an index j, that is $\mathbf{x}_t^m \in \{\mathbf{x}^j\}_{j=1}^J$ and the auxiliary sub-function can be written as

$$\mathscr{Q}_x^m = \sum_{t=1}^{T}\left[A_t^m \log\left\{p\left(j_t^m|j_{t-1}^m\right)\right\} + \log\left\{\phi\left(\mathbf{z}|j_t^m\right)\right\}\right], \tag{7.4}$$

The general Viterbi algorithm is reviewed in Sect. A.5. The key idea with grid-based methods is that optimising (7.3) is easy if there is a finite number of possible values of $\mathbf{x}$. We can view this discrete approximation to the state space as a Reimann sum estimate of the integrals in (7.3). The types of function one is likely to encounter for the target appearance will generally be Reimann integrable, so this approximation is exact in the limit. The computational complexity of the direct maximisation is exponential in the number of frames, which can be very costly if the state space is large. Unfortunately, the estimation error is fundamentally limited by the granularity of the grid: good estimation performance requires a large state space. The Viterbi algorithm exploits the Markov property of the kinematic state to reduce this complexity to linear in time and quadratic in the size of the state space. If the transition probability $p\left(\mathbf{x}_t^m|\mathbf{x}_{t-1}^m\right)$ is compact and only has non-zero values in a small region then the complexity becomes linear in the size of the state space.

Algorithm 1 summarises Viterbi H-PMHT. The Viterbi implementation at the centre of H-PMHT looks almost the same as the Viterbi TkBD in Sect. A.5 except that the measurement likelihood is replaced by the H-PMHT measurement term. The H-PMHT also has no null-state because track management is addressed through the Poisson rate term. Note that the Viterbi algorithm is executed for every track inside every EM iteration. We might expect this to be a costly exercise. The initialisation before the EM iterations looks innocuous, but this is a batch estimator, so it amounts

Algorithm 1 Viterbi H-PMHT

1: Initialize the state estimates $\hat{\mathbb{J}}$ and rate estimates $\hat{\mathbb{L}}$
2: **while** not converged **do**
3: Calculate pixel probabilities G_t^m and G_t
4: Calculate the weights $w_t^{i,m}$
5: **for** each track $m = 1 \dots M$ **do**
6: Initialize $C_0(j_0) = -\log\left[p\left(j_0^m\right)\right]$ for all states
7: **for** each frame $t = 1 \dots T$ **do**
8: Calculate the normalized state costs

$$C_k(j_t^m) = -\log\left[\phi\left(\mathbf{z}|j_t^m\right)\right] + \min_{j_{t-1}^m}\left\langle C_{t-1}(j_{t-1}^m) - \log p\left(j_t^m|j_{t-1}^m\right)\right\rangle$$

9: The previous state in the most likely sequence leading to j_t is given by

$$\theta_t(j_t^m) = \arg\min_{j_{t-1}^m}\left\langle C_{t-1}(j_{t-1}^m) - \log p\left(j_t^m|j_{t-1}^m\right)\right\rangle$$

10: **end for**
11: The estimated state at time T is

$$\hat{j}_t^m = \arg\min_{j_t^m} C_t(j_t^m)$$

12: **for** each frame $t = T - 1 \dots 1$ **do**
13: The estimated state at time t is found by backtracking

$$\hat{j}_t^m = \theta_{t+1}\left(\hat{j}_{t+1}^m\right)$$

14: **end for**
15: **end for**
16: **end while**

to assuming a starting path through the whole batch. Since EM gives only local convergence, initialisation can be important in some applications. Here we simply predict the state forwards through the model.

7.2 Particle-Based Maximisation for Non-Gaussian Appearance

Using a grid-based numerical approximation allows us to exploit optimal and efficient algorithms over the grid. However, the estimation error is always fundamentally limited by the grid spacing: the best you might imagine to get is to choose the grid cell closest to the true state. Conversely, the computation cost scales at least quadratically with the resolution. If we use a constant velocity target model then the cost scales as twice the dimension of the position space, for example, resolution to the fourth power for constant velocity in the plane. Clearly, resolution is expensive! We can also see that much of the state space often has a very low likelihood. For a grid approximation the same resolution is used over the whole state space, so the majority of the samples are effectively wasting computer time. A better alternative would be to use an adaptive grid that has very fine resolution where we need it, close to the maximum of the likelihood, and coarse resolution in the low likelihood areas. A principled way to achieve this is the particle filter.

A brief review of the general particle filter is given in Sect. A.3. For H-PMHT we again use the SIR particle filter. There are some difficulties that need to be addressed to realise this combination. The EM iterations within H-PMHT require the estimator to supply the maximum-likelihood state estimate. This is not necessarily easy based on a random sample representation of the pdf: simply choosing the highest weight particle is not good enough. One possible solution is the method of Godsill et al. [6]. Also, care needs to be taken to ensure that the EM iterations converge: it is unlikely that the iterations would converge if a random sampling process were contained inside the iteration loop. So, it will not be possible to resample particles on each EM iteration. A detailed treatment of convergence is an important consideration but is not presented here.

The implementation described here uses a single estimator for the Poisson rate Λ_t^m. It is not sampled because the particles are used to optimise the state dependent part of the auxiliary function. The EM auxiliary has already separated the rate and state estimation tasks into independent optimisations. Sampling the rate would only be appropriate if we were trying to jointly estimate the two. Algorithm 2 illustrates the particle filter implementation of H-PMHT. This variant of H-PMHT was first published in [3].

Algorithm 2 Particle H-PMHT

1: **for** each frame $t = 1 \dots T$ **do**
2: **for** each particle j of each track m **do**
3: Propose a new state sample $\mathbf{x}_t^{m,j}$ at frame t by sampling the dynamics $p\left(\mathbf{x}_t^{m,j}|\mathbf{x}_{t-1}^{m,j}\right)$
4: **end for**
5: Initialize the state estimates $\hat{\mathbf{X}}_t$ from the particles and the rate estimates $\hat{\mathbf{L}}_t$ by predicting $\alpha_{t|t-1}^m$ and $\beta_{t|t-1}^m$
6: **while** not converged **do**
7: Calculate pixel probabilities G_t^m and G_t
8: Calculate the weights $w_t^{i,m}$
9: **for** each track m **do**
10: weight the particles using $w^j = \phi\left(\mathbf{Z}_t|\mathbf{x}_t^{m,j}\right)$
11: update the state estimate by taking the weighted mean of the particles

$$\hat{\mathbf{x}}_t^m = \frac{\sum_j w^j \mathbf{x}_t^{m,j}}{\sum_j w^j}$$

12: update the rate estimate $\hat{\Lambda}_t^m$
13: **end for**
14: **end while**
15: **end for**

7.3 Cell-Varying Point Spread Function

The two previous methods are numerical approximations to get around difficult integrals. The main problem with numerical approximations is usually the computation expense required to get sufficient accuracy. In the Viterbi case, resolution is the limitation, in the particle filter case the question is how many particles are required. In fact, the computation cost can be increased so much that standard TkBD based on particle filtering or Viterbi can be faster. We now present an alternative based on an unusual transformation that can allow us to implement non-Gaussian maximisation using a Kalman filter. The idea of using transforms to improve efficiency is common in signal processing: for example, convolution can be greatly simplified by transforming the arguments into their respective Fourier series. For H-PMHT we can use a different kind of transform on the appearance function to simplify the calculations. The fundamental idea here is to find an equivalent appearance function that gives the same cell probabilities $G_t^{m,i}$ as the true appearance but with a more convenient functional form.

Back in Chap. 4 we saw that it was possible to find an equivalent point measurement to replace the image measurement dependent part of the auxiliary function. The maths that shows us how to do this is essentially completing the square: the measurement dependent part of the auxiliary function is the sum of weighted log-likelihood terms. For the Gaussian case, the single time slice measurement contribution in (4.30) was written as

$$\sum_{i=1}^{I} \frac{n_t^i w_t^{i,m}}{G_t^{m,i}} \int_{W^i} \left(\mathbf{y} - h\left(\mathbf{x}_t^m\right)\right)^{\mathsf{T}} \Sigma^{-1} \left(\mathbf{y} - h\left(\mathbf{x}_t^m\right)\right) \mathcal{N}\left(\mathbf{y}; h\left(\hat{\mathbf{x}}_t^m\right), \Sigma\right) d\mathbf{y}. \qquad (7.5)$$

We now consider a slight generalisation of this form where the appearance function is $g^m\left(\mathbf{y}|\mathbf{x}_t^m\right) = \mathcal{N}\left(\mathbf{y}; h\left(\mathbf{x}_t^m\right) + \mu^i, \Sigma^i\right), \quad \forall \mathbf{y} \in W^i$. That is, within a single pixel the appearance is Gaussian but the mean and variance of this Gaussian vary as a function of the pixel index. We now show that the factorisation method still works and we can find an equivalent point measurement for this situation. The generalised measurement term is then

$$\sum_{i=1}^{I} \frac{n_t^i w_t^{i,m}}{G_t^{m,i}} \int_{W^i} \left(\mathbf{y} - h\left(\mathbf{x}_t^m\right) - \mu^i\right)^{\mathsf{T}} \left(\Sigma^i\right)^{-1} \left(\mathbf{y} - h\left(\mathbf{x}_t^m\right) - \mu^i\right) \times \mathcal{N}\left(\mathbf{y}; h\left(\hat{\mathbf{x}}_t^m\right) + \mu^i, \Sigma^i\right) d\mathbf{y}. \qquad (7.6)$$

Once again we expand the quadratic expression and then collect together terms in $\mathbf{x}^{\mathsf{T}}\mathbf{x}$ and terms in $\mathbf{x}^{\mathsf{T}}$. The other terms are constant with respect to the optimisation and are absorbing into an arbitrary term C.

$$\sum_{i=1}^{I} h\left(\mathbf{x}_t^m\right)^{\mathsf{T}} \left(\Sigma^i\right)^{-1} h\left(\mathbf{x}_t^m\right) \frac{n_t^i w_t^{i,m}}{G_t^{m,i}} \int_{W^i} \mathcal{N}\left(\mathbf{y}; h\left(\hat{\mathbf{x}}_t^m\right) + \mu^i, \Sigma^i\right) \mathrm{d}\mathbf{y}$$

$$- 2h\left(\mathbf{x}_t^m\right)^{\mathsf{T}} \sum_{i=1}^{I} \left(\Sigma^i\right)^{-1} \frac{n_t^i w_t^{i,m}}{G_t^{m,i}} \int_{W^i} (\mathbf{y} - \mu^i) \mathcal{N}\left(\mathbf{y}; h\left(\hat{\mathbf{x}}_t^m\right) + \mu^i, \Sigma^i\right) \mathrm{d}\mathbf{y} + C,$$

$$= h\left(\mathbf{x}_t^m\right)^{\mathsf{T}} \left[\sum_{i=1}^{I} \left(\Sigma^i\right)^{-1} n_t^i w_t^{i,m}\right] h\left(\mathbf{x}_t^m\right)$$

$$- 2h\left(\mathbf{x}_t^m\right)^{\mathsf{T}} \sum_{i=1}^{I} \left(\Sigma^i\right)^{-1} \frac{n_t^i w_t^{i,m}}{G_t^{m,i}} \int_{W^i} (\mathbf{y} - \mu^i) \mathcal{N}\left(\mathbf{y}; h\left(\hat{\mathbf{x}}_t^m\right) + \mu^i, \Sigma^i\right) \mathrm{d}\mathbf{y} + C. \tag{7.7}$$

Define the synthetic covariance as

$$\left(\tilde{\Sigma}_t^m\right)^{-1} = \sum_{i=1}^{I} \left(\Sigma^i\right)^{-1} n_t^i w_t^{i,m}. \tag{7.8}$$

This simplifies to the core H-PMHT expression we had in Chap. 4 if all of the Σ^i are the same, which is consistent with our notion that this is a generalisation of the Gaussian case. We can now write the measurement term as

$$h\left(\mathbf{x}_t^m\right)^{\mathsf{T}} \left(\tilde{\Sigma}_t^m\right)^{-1} h\left(\mathbf{x}_t^m\right) - 2h\left(\mathbf{x}_t^m\right)^{\mathsf{T}} \left(\tilde{\Sigma}_t^m\right)^{-1} \tilde{\mathbf{y}}_t^m + C, \tag{7.9}$$

where

$$\tilde{\mathbf{y}}_t^m = \tilde{\Sigma}_t^m \sum_{i=1}^{I} \left(\Sigma^i\right)^{-1} \frac{n_t^i w_t^{i,m}}{G_t^{m,i}} \int_{W^i} (\mathbf{y} - \mu^i) \mathcal{N}\left(\mathbf{y}; h\left(\hat{\mathbf{x}}_t^m\right) + \mu^i, \Sigma^i\right) \mathrm{d}\mathbf{y}, \tag{7.10}$$

$$= \tilde{\Sigma}_t^m \sum_{i=1}^{I} \left(\Sigma^i\right)^{-1} n_t^i w_t^{i,m} \frac{\int_{W^i} (\mathbf{y} - \mu^i) \mathcal{N}\left(\mathbf{y}; h\left(\hat{\mathbf{x}}_t^m\right) + \mu^i, \Sigma^i\right) \mathrm{d}\mathbf{y}}{\int_{W^i} \mathcal{N}\left(\mathbf{y}; h\left(\hat{\mathbf{x}}_t^m\right) + \mu^i, \Sigma^i\right) \mathrm{d}\mathbf{y}}, \tag{7.11}$$

$$= \tilde{\Sigma}_t^m \sum_{i=1}^{I} \left(\Sigma^i\right)^{-1} n_t^i w_t^{i,m} \bar{\mathbf{y}}_t^{m,i} \tag{7.12}$$

Again, this simplifies to the weighted sum of cell centroids that appears in core H-PMHT when the variances are the same and the offsets are zero.

The result in Eq. (7.9) says that we can find an equivalent point measurement and measurement covariance if the appearance is a Gaussian *with a different mean and covariance in every pixel*. This equivalent measurement can be used as the input to a Kalman filter. Our strategy now is to find functions for the means μ^i and covariances Σ^i so that the cell probabilities of the cell-varying Gaussian are the same as the cell probabilities of the true appearance. In this case, the cell-varying Gaussian 'looks the same' as the true appearance to the sensor but can be implemented using a point measurement Kalman filter. Each pixel in this cell-probability matching is controlled by two parameters: an offset term and a covariance matrix. The offset on its own is enough to match the pixel probabilities to the true appearance function. We could choose to make all of the covariances the same to simplify things. However, in some situations, the extra flexibility can be handy. For example, in a camera sensor sometimes individual pixels can become unresponsive. In this case, we could suppress such a pixel by choosing a larger covariance matrix for that one pixel than all of the others. It can also be desirable to suppress a pixel if we know that the true appearance function is identically zero in some places.

The problem with this approach is that we have not given the reader a method to actually define these cell-varying parameters. To be practically useful we would need a scheme to derive μ^i and Σ^i for a given appearance $g(\cdot)$. The cell-varying Gaussian was introduced in [3] and example parameters were determined there by inspection for the C appearance target that will be discussed a little later. However, in general, there is currently no method to derive these. An additional problem is that we will sometimes not know the appearance function at all. In Chap. 8 we will review methods to learn the target appearance, but there is no available method to learn these cell-varying Gaussian parameters. This cell-varying method would seem to be an example of modelling the appearance as a Gaussian process. So far this has not been explored wit H-PMHT but it is enticing.

7.4 Gaussian Mixture Appearance Approximation

The final approximate method we consider for non-Gaussian appearance is to use a Gaussian mixture to represent the appearance. The previous section described the cell-varying Gaussian approach that uses a single Gaussian with varying parameters. The method we describe now is similar but instead uses a mixture of several Gaussians, each with fixed parameters. The advantage of this method is that we will see it is much easier to fit the approximation and it readily extends to the unknown appearance case, which the cell-varying method cannot handle.

The Gaussian mixture appearance approximation is given by

$$\Lambda_t^m g^m(\mathbf{y}|\mathbf{x}_t^m) \approx \sum_{j=1}^{J} \Lambda_t^{m,j} \mathcal{N}\left(\mathbf{y}; h\left(\mathbf{x}_t^m\right) + \mu^j, \Sigma^j\right), \tag{7.13}$$

where the number of terms J is a parameter that controls a trade-off between accuracy in the appearance approximation and computation complexity. Intuitively one would expect the computation cost to scale linearly with J. Again we can trivially see that the mixture simplifies to a single Gaussian when $J = 1$.

In the context of a known true appearance function, the parameters J, μ^j and Σ^j are all known and fixed. Even in this case, we will allow the rate parameters to be unknown.

The assignment parameters $k_t^{i,r}$ usually describe the association of a single shot measurement to a single target. For the mixture target, we change that slightly and instead associate a shot with a single component of a target. It does not take much faith to see that this will result in an associated image for each component of each target. Since each component is Gaussian, we can also define a synthetic measurement and covariance for it using the same method as described in Chap. 4.

As usual, the auxiliary function is the sum of a kinematic state term and a measurement rate term. The difference in the rate term is that it now includes a sum over the appearance components

$$\mathscr{Q}\left(\mathbb{X}, \mathbb{L}|\hat{\mathbb{X}}, \hat{\mathbb{L}}\right) = \sum_{m=0}^{M}\sum_{j=1}^{J} \mathscr{Q}_{\Lambda}^{m,j} + \sum_{m=1}^{M} \mathscr{Q}_{x}^{m}. \tag{7.14}$$

Each appearance component rate term $\mathscr{Q}_{\Lambda}^{m,j}$ looks the same as the per-target term in Chap. 6 and is maximsed by the mode of a gamma distribution, given by

$$\hat{\Lambda}_t^{m,j} = \frac{\alpha_{t|t}^{m,j}}{\beta_t^{m,j}}, \tag{7.15}$$

with

$$\alpha_{t|t}^{m,j} = \exp\{-(\tau_t - \tau_{t-1})/\bar{\tau}\}\, \alpha_{t-1|t-1}^{m,j} + \sum_{i=1}^{I} \mathfrak{z}_t^{i,m,j}, \tag{7.16}$$

$$\beta_{t|t}^{m,j} = \exp\{-(\tau_t - \tau_{t-1})/\bar{\tau}\}\, \beta_{t-1|t-1}^{m,j} + 1, \tag{7.17}$$

The only difference between this model and the single-Gaussian case is that the notation has a double superscript. The total target mean power is the sum of the components since we have a Poisson model,

$$\hat{\Lambda}_t^{m} = \sum_{j=1}^{J} \hat{\Lambda}_t^{m,j}. \tag{7.18}$$

The target component looks a little different because there are J synthetic measurements at each time and only one state. The state-dependent auxiliary sub-function is then

$$\mathcal{Q}_x^m = \log\left\{p\left(\mathbf{x}_0^m\right)\right\}$$
$$+\sum_{t=1}^{T}\left[\log\left\{p\left(\mathbf{x}_t^m|\mathbf{x}_{t-1}^m\right)\right\}+\sum_{j=1}^{J}\log\left\{\mathcal{N}\left(\tilde{\mathbf{y}}_t^{m,j};h\left(\mathbf{x}_t^m\right)+\mu^j,\tilde{\Sigma}_t^{m,j}\right)\right\}\right]. \tag{7.19}$$

This is now the log-likelihood of an estimation problem with J point measurements per frame filter. One way to implement this is to stack the measurements together. Define

$$\nu_t^{m,j} = h\left(\mathbf{x}_t^m\right)+\mu^j-\tilde{\mathbf{y}}_t^{m,j}, \tag{7.20}$$

that is the innovation for component j, and the stacked innovation

$$\nu_t^m = \left[\left(\nu_t^{m,1}\right)^{\mathsf{T}},\ldots\left(\nu_t^{m,J}\right)^{\mathsf{T}}\right]^{\mathsf{T}}. \tag{7.21}$$

We need to also stack the measurement covariance matrix and the measurement matrix

$$\Sigma = \begin{bmatrix}\tilde{\Sigma}_t^{m,j} & & 0\\ & \ddots & \\ 0 & & \tilde{\Sigma}_t^{m,J}\end{bmatrix}, \qquad \mathsf{H} = \begin{bmatrix}\mathsf{H}\\ \vdots\\ \mathsf{H}\end{bmatrix}. \tag{7.22}$$

This stacked vector system then defines a high-dimension linear measurement vector that can be solved with a Kalman filter. An alternative implementation is to use the information form of the Kalman filter. The advantage of this implementation is that the update is linear in the number of components, which can be important if J is large. We have assumed the same J for every target. This can easily be relaxed at the cost of extra notation.

7.5 Simulated Examples

We now use some simplistic simulations to show situations where these more complex implementations can offer some advantage over the Kalman filter-based H-PMHT. Example experiments in [5] demonstrated the Viterbi version of H-PMHT and in [3] demonstrated the cell-varying Gaussian version. These variants were shown to give broadly similar to the particle filter H-PMHT, so here we focus on the particle version. The comparisons here were first presented in [2]. Four cases are considered: linear target motion with a Gaussian appearance; linear target motion with a non-Gaussian appearance; crossing linear targets with a non-Gaussian appearance; and initiation of closely spaced non-Gaussian targets. For brevity, the two algorithms will be referred to as H-PMHT-P and H-PMHT-K.

It is clear that the first case satisfies the assumptions leading to the Kalman filter H-PMHT so we would expect it to perform well. Of interest, in this case, is how many particles are required for the H-PMHT-P to give equivalent performance to H-PMHT-K, or if it is even possible to do so. The other simulation scenarios demonstrate simple examples where the particle method shows benefit.

For each of the scenarios, the sensor collected a 100×100 pixel image and a single target moved in the plane for 50 frames. The frames were collected at a uniform rate of one per second. The sensor noise was Rayleigh distributed with unit variance and so the pixels containing a target contribution followed a Rice distribution with unit variance and a mean dependent on the target location. These assumptions are typical for radar-based applications [4]. The H-PMHT-P algorithm was tested with 1000, 200 and 40 particles to loosely quantify the algorithm's dependence on this parameter.

7.5.1 *Linear Gaussian Appearance*

The first scenario consists of a target under constant velocity motion with a circular Gaussian appearance function. The variance of the appearance function was chosen to be 9 pixels^2. The target started near the corner of the image at approximately (20, 20) and moved with a constant speed of one pixel per frame at a heading of 45° (North-East).

Following our usual convention, both of the H-PMHT variants used a four element state consisting of position and velocity in the plane and an almost-constant-velocity target model, $p(\mathbf{x}_t | \mathbf{x}_{t-1}) \sim \mathcal{N}(\mathbf{x}_t; \mathsf{F}\mathbf{x}_{t-1}, \mathsf{Q})$ with

$$\mathsf{F} = \begin{bmatrix} 1 & 1 & 0 & 0 \\ 0 & 1 & 0 & 0 \\ 0 & 0 & 1 & 1 \\ 0 & 0 & 0 & 1 \end{bmatrix}, \quad \mathsf{Q} = 0.1 \begin{bmatrix} \frac{1}{3} & \frac{1}{2} & 0 & 0 \\ \frac{1}{2} & 1 & 0 & 0 \\ 0 & 0 & \frac{1}{3} & \frac{1}{2} \\ 0 & 0 & \frac{1}{2} & 1 \end{bmatrix}$$

Figure 7.1 shows the root mean square (RMS) position estimation accuracy for the two implementations of H-PMHT. For 1000 particles, the H-PMHT-P error curve is equivalent to the H-PMHT-K. As the number of particles was reduced, the estimation error degraded, as would be expected. This scenario verifies that the particle filter H-PMHT gives an appropriate answer for the simplest case.

7.5.2 *Linear Non-Gaussian Appearance*

In the second trial the target appearance function was changed to a highly non-Gaussian function that somewhat resembles the letter 'C'. Mathematically, the appearance function is specified in polar coordinates as

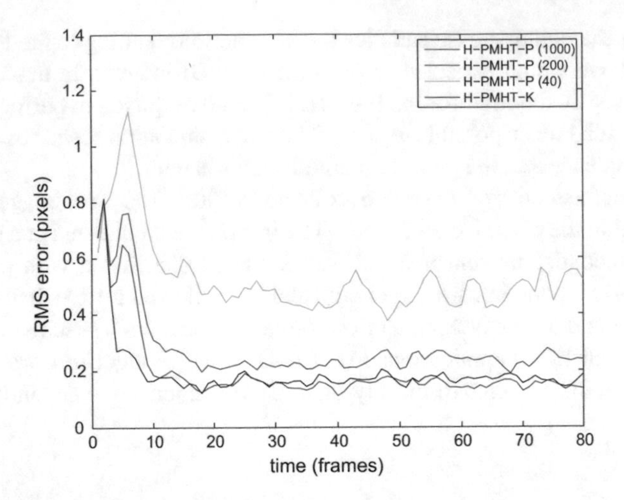

Fig. 7.1 Localisation accuracy, linear Gaussian scenario

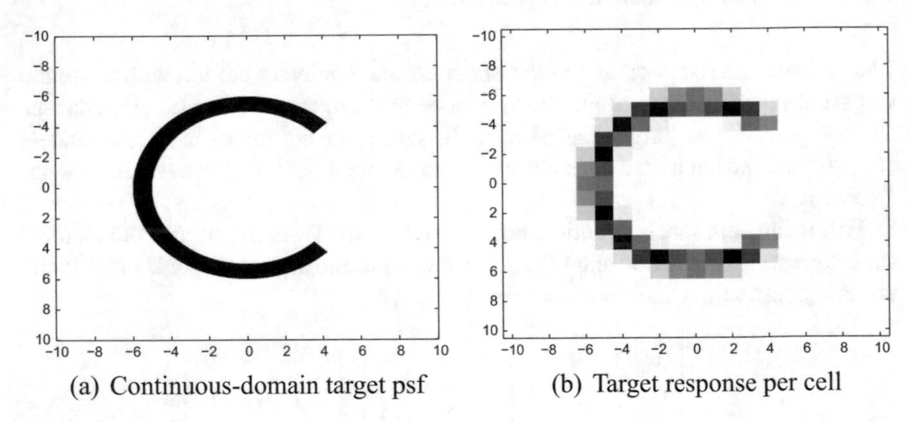

(a) Continuous-domain target psf

(b) Target response per cell

Fig. 7.2 Non-Gaussian point spread function

$$g(r, \theta) = \begin{cases} A & \text{if } 5 \geq r \geq 6 \text{ and } |\theta| > \frac{\pi}{4}, \\ 0 & \text{otherwise,} \end{cases} \tag{7.23}$$

where r is the distance from the appearance centre to the point $\mathbf{y}$ in the measurement space $r = |\mathsf{H}\mathbf{x} - \mathbf{y}|$; θ is the angle of the same direction vector; and A is a normalising constant. This appearance function is shown in Fig. 7.2a. The function is essentially binary: its value is either the constant A or zero. The response of the sensor to the appearance function is the cell-probability, that is, the integral of $g(r, \theta)$ over the extent of that pixel W^i. An example of this is shown in Fig. 7.2b for the pixel size used here. There is a large degree of blurring because the spatial sampling is relatively poor resolution compared with the features of the appearance function.

This artificial appearance was chosen because it is asymmetrical and the mean of the distribution is at a location of very low density. It is very different to a Gaussian with the same variance. For the H-PMHT-K implementation, the appearance was approximated by a Gaussian with covariance $\text{diag}(18.4, 9.2)$, which is the covariance of $g(r, \theta)$. Note that the mean of the appearance is not coincident with its centre at the projection of the target state because the appearance is asymmetrical. The H-PMHT-K can easily compensate for this since the offset is fixed and known: it can be treated as a bias. The mean of $g(r, \theta)$ was empirically found to be at approximately $(-1.66, 0)$ relative to the target state.

The H-PMHT-P has a more difficult problem to overcome. The appearance has a discontinuity and is also identically zero over most of the measurement space. This is not the sort of function that EM is suited to optimise because the log-likelihood is not bounded. Intuitively, it is attempting to use a hill-climbing method on a function with infinite slope and regions of infinite likelihood. The particle weights will contain $\log\{0\}$ for every particle that doesn't provide exactly the same pixel support as the previous EM iterated solution.

Happily, two simple regularisation steps alleviate this problem. First, the appearance is blurred by convolving it with a Gaussian with a relatively small variance and second a uniform pedestal is added to the appearance. The blurring function removes the discontinuity and the pedestal prevents numerical issues arising from attempts to evaluate $\log\{0\}$. This modified appearance function is then used by H-PMHT-P for the C-target: the simulations were created using the 'true' binary function above.

Figure 7.3 shows the position errors for an example realisation of this scenario. It is clear that the H-PMHT-K is still biased even with compensation for the asymmetry in the appearance. This bias is not present in the H-PMHT-P and leads to higher estimation error for H-PMHT-K.

RMS error curves for H-PMHT-P and H-PMHT-K are shown in Fig. 7.4. The error curve for the cases with fewer particles shows a saw-tooth characteristic. This is where the track diverges from the true target position and a new track is formed. This behaviour occurred for the cases with fewer particles because the filter was unable to adequately estimate the target velocity. This could possibly be alleviated by using a marginalised implementation of the particle estimator [7]. The H-PMHT-P with 1000 particles did not diverge and gave much lower estimation error than the H-PMHT-K.

7.5.3 Crossing Non-Gaussian Scenario

Next we consider a problem with two crossing targets, both with the 'C' appearance described above. The first target has the same trajectory as the previous scenario and a second target was introduced at a heading of $-45°$. For this scenario, the first target has a peak SNR of 10dB and the second target has a peak SNR of 7dB. Figure 7.5 shows an example realisation of the scenario; both of the targets move from left to right and cross over in the middle of the image, halfway through the scenario. The

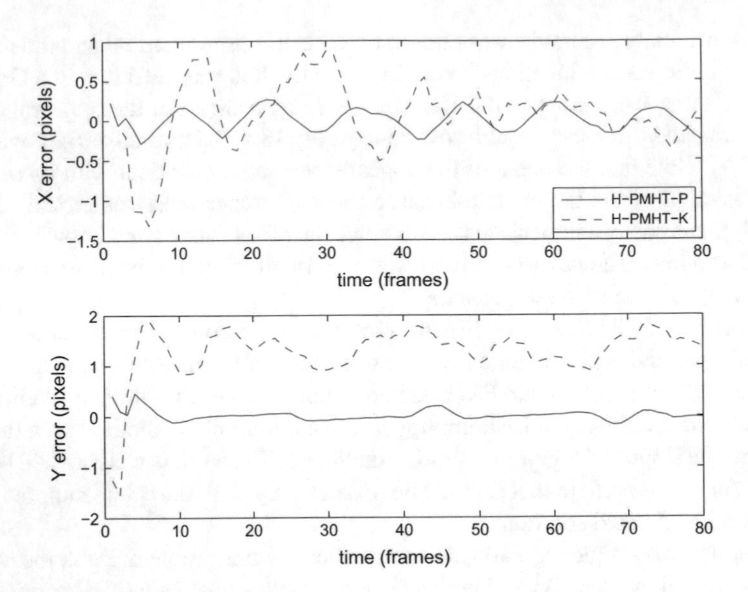

Fig. 7.3 Example realisation of scenario 2

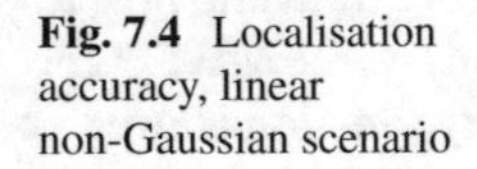

Fig. 7.4 Localisation accuracy, linear non-Gaussian scenario

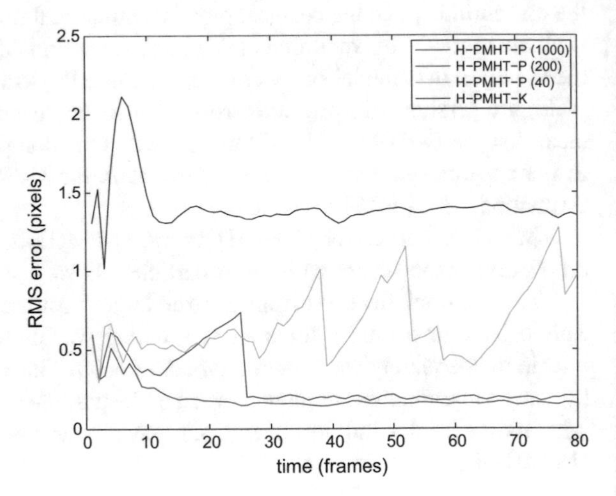

H-PMHT-P tracks are shown for 1000 particles and clearly follow the targets very closely. The H-PMHT-K tracks show the bias of the previous example and failed to navigate through the crossing event. The track on the weaker target incorrectly swapped onto the strong target and a new track was formed once the targets are sufficiently separated.

Each Monte Carlo trial for this scenario was classified as either: correctly tracked; single swap, such as the example in Fig. 7.5; double swap, which is the case where both tracks swap onto the other target; or uninitiated, where the tracker failed to establish a track on both targets prior to the crossing. The number of each of these is

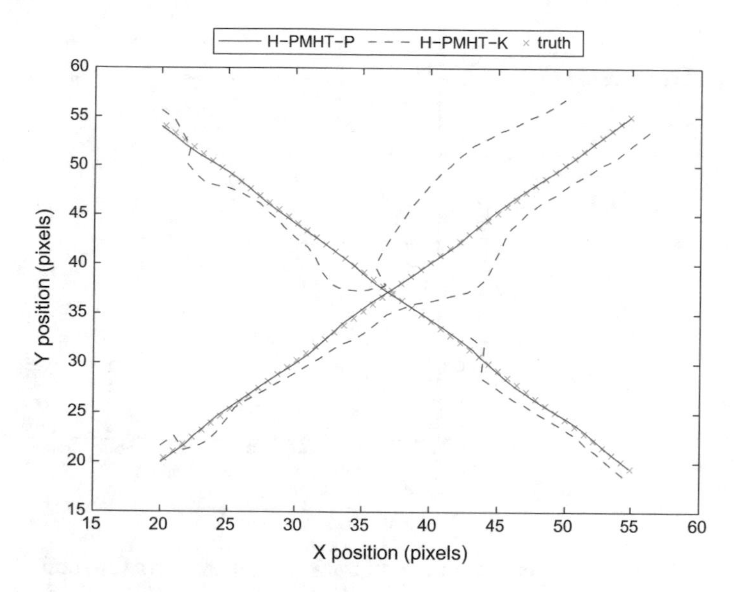

Fig. 7.5 Example realisation of crossing targets

Table 7.1 Crossing targets

	H-PMHT-P (1000)	H-PMHT-P (200)	H-PMHT-P (40)	H-PMHT-K
correct	100	99	87	2
1 swap	0	0	1	96
2 swap	0	0	0	2
uninit.	0	1	12	0

shown in Table 7.1. The results show that with 1000 particles H-PMHT-P correctly tracked every trial and that for 200 particles the loss in performance was minor. For 40 particles, the algorithm had difficulty establishing a track. This is related to the velocity estimation problem described above. In contrast, the H-PMHT-K failed on nearly every trial. Since the assumed Gaussian was poorly matched to the true appearance, the track on the weaker target swapped in 98 percent of the trials.

7.5.4 Diverging Target Scenario

The final scenario consisted of two targets that began co-located and then gradually spread apart with time. Of interest is how quickly the tracker establishes track on the second track. Again the first target was at 10 dB peak SNR and the second one at 7 dB. Both moved with unit speed, the first at a heading of 45° and the second at 60°.

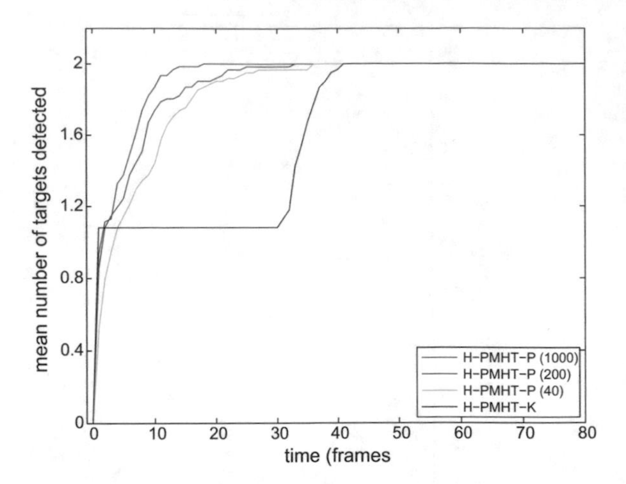

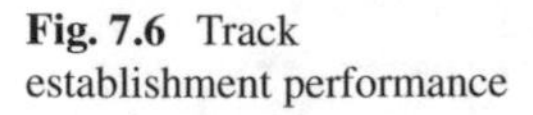
Fig. 7.6 Track establishment performance

Figure 7.6 shows the average number of tracks established as a function of time. In almost all cases, both algorithms established a track very quickly on the higher SNR target. The H-PMHT-P established a second track on the lower SNR target a little later. However, the H-PMHT-K was unable to establish a second track for 35 frames (on average). This was the time duration for the weaker target to be unambiguously separated from the stronger one.

There were a small number of trials where the H-PMHT-K established two tracks very quickly. This was a result of the track initiation logic which occasionally started multiple tracks on the strong target. A Gaussian is a poor approximation to true C appearance which means that the H-PMHT-K can sometimes establish duplicate tracks on a single target. In this scenario, a duplicate track can leave H-PMHT-K fortuitously placed when the targets diverged. The crossing target example in Fig. 7.5 illustrates an example of H-PMHT-K maintaining duplicate tracks on a single target and this behaviour was typical in the crossing scenario.

Once more the best performance for the H-PMHT-P was achieved with 1000 particles but the degradation was gradual as the number of particles decreased and is driven mainly by the filter's general initiation degradation rather than a specific loss of discrimination.

Figure 7.7 shows an example trial where both of the algorithms have established one track. Figure 7.7a shows the sensor frame in the region of the targets. There is a high degree of overlap, but the eye can resolve two targets. Figure 7.7b, c show the output of the frame vetting that is part of the track management described in Chap. 5 for H-PMHT-P and H-PMHT-K respectively. The existing first track in the particle H-PMHT is following the stronger target, but the residual frame clearly shows the response of the second target and the H-PMHT-P was subsequently able to establish a second track. Figure 7.7c shows the output of the frame vetting for H-PMHT-K. Since the algorithm models the target response using a broad Gaussian blob, the vetting suppresses all target information around the location of the extant track and

Fig. 7.7 Closely spaced non-Gaussian targets

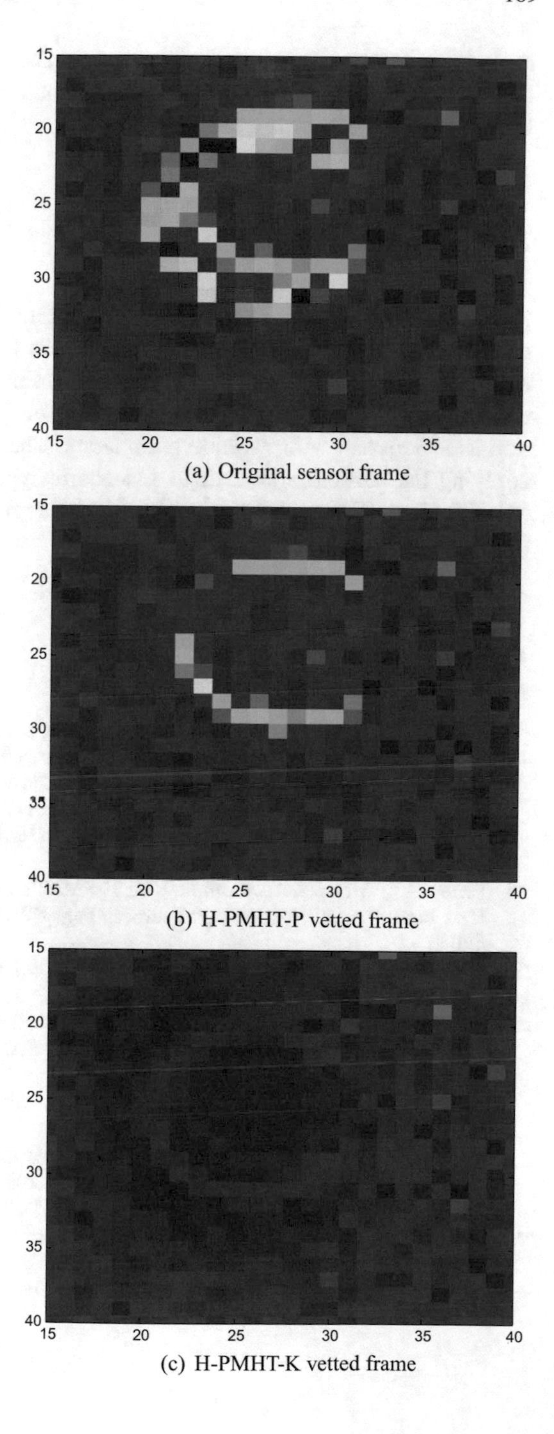

(a) Original sensor frame

(b) H-PMHT-P vetted frame

(c) H-PMHT-K vetted frame

there is almost no sign of the weaker target. The frame shows a dark region where the data has been over-suppressed and a faint outline of the stronger target which has been reduced to the mean background level.

7.6 Summary

This chapter has discussed various methods for implementing the maximisation stage of H-PMHT when the linear Gaussian Kalman filter assumptions are broken. The methods included deterministic and Monte Carlo sampling approaches, a Gaussian mixture approximation, and an unusual way of formulating a non-Gaussian function as a Gaussian with variable parameters. The simulations provide examples of applying the particle approach to a scenario where the appearance function was non-Gaussian. These methods could also be applied to situations where the target dynamics are nonlinear.

References

1. Barniv, Y.: Dynamic programming algorithm for detecting dim moving targets. Multitarget-Multisensor Tracking: Advanced Applications. Artech House (1990)
2. Davey, S.J.: Histogram PMHT with particles. In: Proceedings of the 14th International Conference on Information Fusion. Chicago, USA (2011)
3. Davey, S.J.: Detecting a small boat with histogram PMHT. ISIF J. Adv. Inf. Fusion **6**, 167–186 (2011)
4. Davey, S.J., Wieneke, M., Gordon, N.J.: H-PMHT for correlated targets. In: SPIE Signal and Data Processing of Small Targets, Proceedings SPIE, vol. 8393, 83930R. Baltimore, USA (2012)
5. Davey, S.J., Wieneke, M., Vu, H.X.: Histogram-PMHT unfettered. IEEE J. Sel. Top. Signal Process. **7**(3), 435–447 (2013)
6. Godsill, S., Doucet, A., West, M.: Maximum a Posteriori Sequence Estimation Using Monte Carlo Particle Filters Annals of the Institute of Statistical Mathematics, vol. 53. Springer, Netherlands (2001)
7. Ristic, B., Arulampalam, S., Gordon, N.J.: Beyond the Kalman Filter: Particle Filters for Tracking Applications. Artech House (2004)
8. Viterbi, A.J.: Error bounds for convolutional codes and an asymptotically optimum decoding algorithm. IEEE Trans. Inf. Theory **13**, 260–269 (1967)
9. Vu, H.X.: Track-before-detect for active sonar. Ph.D. thesis, The University of Adelaide (2015)
10. Vu, H.X., Davey, S.J., Fletcher, F., Arulampalam, S., Lim, C.C.: Track-before-detect using histogram PMHT and dynamic programming. In: Digital Image Computing: Techniques and Applications, pp. 1–8. Fremantle, Western Australia (2012)

Chapter 8
Adaptive Appearance Models

The treatment so far has assumed that the appearance function $g^m(\mathbf{y}|\mathbf{x}_t^m)$ is known. In general, the target appearance is affected by two factors. The first is the physical shape of the object itself and the second is the way that the sensor hardware and signal processing spreads energy across pixels. The second factor is often referred to as the point spread function of the sensor and depends on physical features of the sensor, such as the size of the aperture and the properties of the optics, and also software features, especially the tapers used in Fourier transforms to form range cells and beams. If an object is very small compared with the resolution of the sensor, that is the physical size of a pixel, then it can be approximated as a point and the appearance in the image is purely driven by the point spread function. This is typically the case for a sensor such as low bandwidth radar or a telescope looking at stars. Conversely, if the object is very large compared with the resolution, such as in the case of a video surveillance of people, then the appearance function is dominated by the physical characteristics of the object. Not only are these unknown, but they can also be time varying and highly complex, especially in the example of surveillance of people. For objects that occupy a relatively large number of sensor pixels, it is not sufficient to assume that the appearance is known.

In broad terms, objects are often classified as *rigid bodies* or *non-rigid bodies*. Rigid body objects are those whose intrinsic shape does not change, although their appearance can vary with aspect angle. Examples are vehicles, such as cars and buses. Non-rigid body objects change shape over time, such as people. For rigid bodies, it can be possible to actually estimate the object shape but for non-rigid bodies, a probabilistic description is usually more suitable. In this case, the appearance represents both the intensity of returns in a spatial area and the probability that the object will occupy that area at a particular time.

This chapter describes a range of different models that can be used to describe objects with unknown appearance. The majority of these assume that the shape

S. J. Davey and H. Gaetjens, *Track-Before-Detect Using Expectation Maximisation*, Signals and Communication Technology,
https://doi.org/10.1007/978-981-10-7593-3_8

is a randomly evolving property of the object and apply a Markov model to the parameters that describe the shape. The simplest representation is a Gaussian with unknown covariance matrix; the most complicated is a non-parametric grid of average intensities.

8.1 Deterministic Gaussian Appearance

The original derivation of the core H-PMHT [14] considered the case where the target appearance was Gaussian with an unknown covariance matrix and a linear mean, namely $g\left(\mathbf{y}|\mathbf{x}_t^m, \Sigma_t^m\right) \sim \mathcal{N}\left(\mathbf{y}; \mathsf{H}_t^m \mathbf{x}_t^m, \Sigma_t^m\right)$ with H_t^m known but Σ_t^m unknown. This is the simplest appearance estimator. Note that in this case, the covariance matrix is not a random variable and is treated in the same way as the mixing probability vector Π. We review this approach now and relax it slightly by allowing the mean to be a known nonlinear function $g\left(\mathbf{y}|\mathbf{x}_t^m, \Sigma_t^m\right) \sim \mathcal{N}\left(\mathbf{y}; h\left(\mathbf{x}_t^m\right), \Sigma_t^m\right)$.

There are now three parameters to estimate: the kinematic states, the Poisson rate parameters and the spreading matrices. Defining the collection of all spreading matrices as $\Sigma = \{\Sigma_t^m\}$, the modified auxiliary function is

$$\begin{aligned}
\mathcal{Q}\left(\mathbb{X}, \mathbb{L}, \Sigma|\hat{\mathbb{X}}, \hat{\mathbb{L}}, \hat{\Sigma}\right) &= E_{\mathbb{K}|\hat{\mathbb{X}},\hat{\mathbb{L}},\hat{\Sigma},\mathbb{Y}}\Big[\log\left\{p_{\mathsf{comp}}(\mathbb{X}, \mathbb{L}, \mathbb{K}, \mathbb{Y}; \Sigma)\right\}\Big] \\
&= \sum_{\mathbb{K}} \log\left\{p_{\mathsf{comp}}(\mathbb{X}, \mathbb{L}, \mathbb{K}, \mathbb{Y}; \Sigma)\right\} p_{\mathsf{miss}}(\mathbb{K}|\hat{\mathbb{X}}, \hat{\mathbb{L}}, \mathbb{Y}; \hat{\Sigma}), \\
&= \log\{p(\mathbb{X})\} + \log\{p(\mathbb{L})\} + \sum_{\mathbb{K}} \log\{p(\mathbb{K}|\mathbb{L})\}\, p_{\mathsf{miss}}(\mathbb{K}|\hat{\mathbb{X}}, \hat{\mathbb{L}}, \mathbb{Y}; \hat{\Sigma}) \\
&\quad + \sum_{\mathbb{K}} \log\{p(\mathbb{Y}|\mathbb{X}, \mathbb{K}; \Sigma)\}\, p_{\mathsf{miss}}(\mathbb{K}|\hat{\mathbb{X}}, \hat{\mathbb{L}}, \mathbb{Y}; \hat{\Sigma}),
\end{aligned} \tag{8.1}$$

which looks rather similar to (6.21) except that every term that contains the measurements is explicitly dependent on the spreading matrix. As usual, this can be grouped into a state-dependent part and an assignment prior-dependent part. The assignment prior auxiliary sub-function is identical to what we have already seen in Chap. 6, so there is no need to revisit it here. The state-dependent auxiliary sub-function is different. For the core H-PMHT, the state-dependent auxiliary sub-function is (4.64)

$$\mathcal{Q}_x^m = \sum_{t=1}^{T} ||\mathfrak{z}_t^m|| \log\left\{p\left(\mathbf{x}_t^m|\mathbf{x}_{t-1}^m\right)\right\} + \sum_{t=1}^{T}\sum_{i=1}^{I+I^U} \mathfrak{z}_t^i \frac{\int_{W^i} \log\left\{g^m\left(\mathbf{y}|\mathbf{x}_t^m\right)\right\} g^m\left(\mathbf{y}|\hat{\mathbf{x}}_t^m\right) \mathrm{d}\mathbf{y}}{G_t^{m,i}},$$

using the associated image notation introduced in Sect. 4.4. The measurement term simplifies to

$$-\frac{1}{2}\sum_{t=1}^{T}||\mathfrak{z}_t^m||\log\left\{|\Sigma_t^m|\right\}$$
$$-\frac{1}{2}\sum_{t=1}^{T}\sum_{i=1}^{I+I^U}\frac{\mathfrak{z}_t^i}{G_t^{m,i}}\int_{W^i}\left(h\left(\mathbf{x}_t^m\right)-\mathbf{y}\right)^{\mathsf{T}}\left(\Sigma_t^m\right)^{-1}\left(h\left(\mathbf{x}_t^m\right)-\mathbf{y}\right)g^m\left(\mathbf{y}|\hat{\mathbf{x}}_t^m\right)\mathrm{d}\mathbf{y}. \tag{8.2}$$

Notice that in Chap. 4, we were able to ignore terms that were independent of $\mathbf{x}_t^m$ but here we must retain anything that is a function of Σ_t^m, such as the determinant term. The difficulty with this expression is that it is a nonlinear function of two unknowns, $\mathbf{x}_t^m$ and Σ_t^m. It is not possible to simply take derivatives to optimise this expression. Instead, Streit used a generalised EM (GEM) strategy [14]. A fundamental theorem in real analysis is that any increasing sequence with an upper bound converges, for example, see [10]. In this context, it means that it is not necessary to maximise the auxiliary function, however, it is enough to increase it with each iteration. A method that increases the auxiliary function is referred to as GEM and the convergence of such methods was formally proven in the original EM work of Dempster, Laird and Rubin [6]. The GEM approach here is to optimise first for the state holding the spreading variance constant and then optimise for the spreading variance using the new state estimate. The approach of sequentially optimising over coupled variables in this way is also referred to as expectation coupled maximisation (ECM). The first phase of this coupled maximisation is the same as the core H-PMHT. Under an assumed constant Σ_t^m, the only difference is that the actual matrix used can be time varying.

We now explore the second optimisation, over Σ_t^m. This is independent of time because there is no temporal correlation in the model for the spreading matrix. For a single frame, the Σ_t^m-dependent terms are

$$-\frac{1}{2}||\mathfrak{z}_t^m||\log\left\{|\Sigma_t^m|\right\}$$
$$-\frac{1}{2}\sum_{i=1}^{I+I^U}\frac{\mathfrak{z}_t^{m,i}}{G_t^{m,i}}\int_{W^i}\left(h\left(\mathbf{x}_t^m\right)-\mathbf{y}\right)^{\mathsf{T}}\left(\Sigma_t^m\right)^{-1}\left(h\left(\mathbf{x}_t^m\right)-\mathbf{y}\right)g^m\left(\mathbf{y}|\hat{\mathbf{x}}_t^m\right)\mathrm{d}\mathbf{y}. \tag{8.3}$$

To differentiate this expression, we make use of matrix calculus identities. For an arbitrary matrix A and a constant vector v,

$$\frac{\mathrm{d}|\mathsf{A}|}{\mathrm{d}\mathsf{A}}=|\mathsf{A}|\mathsf{A}^{-1}\quad\rightarrow\quad\frac{\mathrm{d}\log\left\{|\mathsf{A}|\right\}}{\mathrm{d}\mathsf{A}}=\mathsf{A}^{-1}, \tag{8.4}$$

$$\frac{\mathrm{d}\mathbf{v}^{\mathsf{T}}\mathsf{A}^{-1}\mathbf{v}}{\mathrm{d}\mathsf{A}}=-\mathsf{A}^{-1}\mathbf{v}\mathbf{v}^{\mathsf{T}}\mathsf{A}^{-1}. \tag{8.5}$$

Applying these identities to (8.3) gives a stationary point at

$$||\mathfrak{z}_t^m|| \left(\Sigma_t^m\right)^{-1} = \sum_{i=1}^{I+I^U} \frac{\mathfrak{z}_t^{m,i}}{G_t^{m,i}} \int_{W^i} \left(\Sigma_t^m\right)^{-1} \left(h\left(\mathbf{x}_t^m\right) - \mathbf{y}\right)\left(h\left(\mathbf{x}_t^m\right) - \mathbf{y}\right)^{\mathsf{T}} g^m\left(\mathbf{y}|\hat{\mathbf{x}}_t^m\right) \left(\Sigma_t^m\right)^{-1} \mathrm{d}\mathbf{y},$$

$$\Sigma_t^m = \frac{1}{||\mathfrak{z}_t^m||} \sum_{i=1}^{I+I^U} \mathfrak{z}_t^{m,i} \mathsf{Z}_t^{m,i}, \tag{8.6}$$

where

$$\mathsf{Z}_t^{m,i} = \frac{1}{G_t^{m,i}} \int_{W^i} \left(h\left(\mathbf{x}_t^m\right) - \mathbf{y}\right)\left(h\left(\mathbf{x}_t^m\right) - \mathbf{y}\right)^{\mathsf{T}} g^m\left(\mathbf{y}|\hat{\mathbf{x}}_t^m\right) \mathrm{d}\mathbf{y}. \tag{8.7}$$

Streit referred to this matrix as the cell-level measurement covariance matrix. The estimator for Σ_t^m in (8.6) has been deliberately arranged to emphasise the structure: it is a weighted mean of these cell-level measurement covariances. Notice the similarity between this and the equivalent point measurement that uses the same weighting to form a weighted mean over cell-level centroids. It is clear that each Z_t^m is rank 1, so the estimated spreading matrix will not be full rank unless the number of pixels is greater than the dimension of the measurement space. In practice, this is highly likely to be the case: a video camera with one pixel has limited value.

The estimate in (8.6) allows for the covariance to vary with time, but in practice this estimate is too noisy [15, 16] and it is preferable to assume a constant spread. In this case, the mathematics is essentially the same except that expressions like (8.3) also contain a sum over time. The reader should be able to verify that the batch constant estimate is

$$\Sigma_t^m = \left(\sum_{t=1}^{T} ||\mathfrak{z}_t^m||\right)^{-1} \sum_{t=1}^{T} \sum_{i} \mathfrak{z}_t^{m,i} \mathsf{Z}_t^{m,i}. \tag{8.8}$$

8.2 Stochastic Gaussian Appearance

An alternative way to estimate Gaussian appearance is to treat the covariance matrix as a random variable. This means that we will require a prior distribution and an evolution function that transforms a probability of the covariance at one time to another. In the broader tracking literature, this kind of target description is referred to as a random matrix model, for example [7, 11]. The random matrix model was applied to H-PMHT by Wieneke in [2, 16, 17] and the resulting method is referred to as H-PMHT-RM. In the random matrix literature, Σ_t^m is referred to as the *extent* or *extension* matrix. Here, we continue to call it the spreading matrix.

In a Gaussian measurement system, the conjugate prior to a stochastic covariance matrix is the inverse Wishart distribution. This means that if the inverse of the covariance Σ_t^m has a Wishart prior distribution then the posterior distribution of the inverse is also Wishart. The inverse Wishart distribution for $d \times d$ symmetric positive definite matrix X is given by

$$\mathscr{IW}(\mathsf{X}; a, \mathsf{A}) \propto |\mathsf{X}|^{(-a+d+1)/2} \exp\left\{\text{trace}\left(-\tfrac{1}{2}\mathsf{A}\mathsf{X}^{-1}\right)\right\}, \tag{8.9}$$

where A is a $d \times d$ matrix that determines the mean of the distribution, and a is a scalar precision parameter. When a is high, then the distribution is compact around its mean and when a is low then the distribution is diffuse. The expected value is given by

$$\mathsf{E}\left[\mathsf{X}\right] = \frac{1}{a-d-1}\mathsf{A}, \tag{8.10}$$

where we require $a > d + 1$. The Wishart distribution is a matrix generalisation of the gamma distribution, which was used as a prior for the Poisson rate in Chap. 6. Once we have derived an estimate for the spreading matrix, we will see that it has some strong similarities with the structure of the Poisson rate estimate.

Under the random matrix model, the spreading matrix is treated as a Markov chain that is independent across targets,

$$p(\Sigma) = \prod_{m=1}^{M}\left[p\left(\Sigma_0^m\right)\prod_{t=1}^{T} p\left(\Sigma_t^m | \Sigma_{t-1}^m\right)\right]. \tag{8.11}$$

The random matrix formulation makes the unusual assumption that the state evolution is correlated with the spreading matrix, which gives

$$p(\mathbb{X}, \Sigma) = \prod_{m=1}^{M}\left[p\left(\Sigma_0^m\right)p\left(\mathbf{x}_0^m\right)\prod_{t=1}^{T} p\left(\Sigma_t^m | \Sigma_{t-1}^m\right)p\left(\mathbf{x}_t^m | \mathbf{x}_{t-1}^m, \Sigma_t^m\right)\right]. \tag{8.12}$$

Applying this leads to an auxiliary function that is slightly different from (8.1)

$$\begin{aligned}
\mathscr{Q}\left(\mathbb{X}, \mathbb{L}, \Sigma | \hat{\mathbb{X}}, \hat{\mathbb{L}}, \hat{\Sigma}\right) &= E_{\mathbb{K}|\hat{\mathbb{X}}, \hat{\mathbb{L}}, \hat{\Sigma}, \mathbb{Y}}\left[\log\left\{p_{\mathsf{comp}}(\mathbb{X}, \mathbb{L}, \mathbb{K}, \mathbb{Y}; \Sigma)\right\}\right] \\
&= \sum_{\mathbb{K}} \log\left\{p_{\mathsf{comp}}(\mathbb{X}, \mathbb{L}, \mathbb{K}, \mathbb{Y}; \Sigma)\right\} p_{\mathsf{miss}}(\mathbb{K}|\hat{\mathbb{X}}, \hat{\mathbb{L}}, \mathbb{Y}; \hat{\Sigma}), \\
&= \log\{p(\mathbb{X}, \Sigma)\} + \log\{p(\mathbb{L})\} + \sum_{\mathbb{K}} \log\{p(\mathbb{K}|\mathbb{L})\}\, p_{\mathsf{miss}}(\mathbb{K}|\hat{\mathbb{X}}, \hat{\mathbb{L}}, \mathbb{Y}; \hat{\Sigma}) \\
&\quad + \sum_{\mathbb{K}} \log\{p(\mathbb{Y}|\mathbb{X}, \mathbb{K}; \Sigma)\}\, p_{\mathsf{miss}}(\mathbb{K}|\hat{\mathbb{X}}, \hat{\mathbb{L}}, \mathbb{Y}; \hat{\Sigma}).
\end{aligned} \tag{8.13}$$

As discussed in the previous section, the Poisson rate dependent part of the auxiliary function is not influenced by the spreading matrix so the rate estimates are the same as derived in Chap. 6.

It should be no surprise that it is the treatment of the measurement-dependent term of the state auxiliary sub-function that is most difficult here. This term is the same measurement term in the previous section (8.3) and here we proceed by completing the square in much the same way as for the core H-PMHT. For now, set aside the determinant term and consider the expansion of the quadratic term from (8.3)

$$\begin{aligned}
&\sum_{i=1}^{I+I^U} \frac{\mathfrak{z}_t^{m,i}}{G_t^{m,i}} \int_{W^i} \left(h\left(\mathbf{x}_t^m\right) - \mathbf{y}\right)^{\mathsf{T}} \left(\Sigma_t^m\right)^{-1} \left(h\left(\mathbf{x}_t^m\right) - \mathbf{y}\right) g^m\left(\mathbf{y}|\hat{\mathbf{x}}_t^m\right) d\mathbf{y} \\
&= \sum_{i=1}^{I+I^U} \frac{\mathfrak{z}_t^{m,i}}{G_t^{m,i}} \int_{W^i} h\left(\mathbf{x}_t^m\right)^{\mathsf{T}} \left(\Sigma_t^m\right)^{-1} h\left(\mathbf{x}_t^m\right) g^m\left(\mathbf{y}|\hat{\mathbf{x}}_t^m\right) d\mathbf{y} \\
&\quad -2 \sum_{i=1}^{I+I^U} \frac{\mathfrak{z}_t^{m,i}}{G_t^{m,i}} \int_{W^i} \mathbf{y}^{\mathsf{T}} \left(\Sigma_t^m\right)^{-1} h\left(\mathbf{x}_t^m\right) g^m\left(\mathbf{y}|\hat{\mathbf{x}}_t^m\right) d\mathbf{y} \\
&\quad + \sum_{i=1}^{I+I^U} \frac{\mathfrak{z}_t^{m,i}}{G_t^{m,i}} \int_{W^i} \mathbf{y}^{\mathsf{T}} \left(\Sigma_t^m\right)^{-1} \mathbf{y}\, g^m\left(\mathbf{y}|\hat{\mathbf{x}}_t^m\right) d\mathbf{y}, \\
&= \sum_{i=1}^{I+I^U} \mathfrak{z}_t^{m,i} h\left(\mathbf{x}_t^m\right)^{\mathsf{T}} \left(\Sigma_t^m\right)^{-1} h\left(\mathbf{x}_t^m\right) - 2\left[\sum_{i=1}^{I+I^U} \frac{\mathfrak{z}_t^{m,i}}{G_t^{m,i}} \int_{W^i} \mathbf{y}^{\mathsf{T}} g^m\left(\mathbf{y}|\hat{\mathbf{x}}_t^m\right) d\mathbf{y}\right] \left(\Sigma_t^m\right)^{-1} h\left(\mathbf{x}_t^m\right) \\
&\quad + \sum_{i=1}^{I+I^U} \frac{\mathfrak{z}_t^{m,i}}{G_t^{m,i}} \int_{W^i} \mathbf{y}^{\mathsf{T}} \left(\Sigma_t^m\right)^{-1} \mathbf{y}\, g^m\left(\mathbf{y}|\hat{\mathbf{x}}_t^m\right) d\mathbf{y}. \qquad (8.14)
\end{aligned}$$

As with the core H-PMHT, we define the square bracket term to be the equivalent measurement

$$\tilde{\mathbf{y}}_t^m = \frac{1}{||\mathfrak{z}_t^m||} \sum_{i=1}^{I+I^U} \frac{\mathfrak{z}_t^{m,i}}{G_t^{m,i}} \tilde{\mathbf{y}}_t^{m,i}. \qquad (8.15)$$

with the cell centroid given by

$$\tilde{\mathbf{y}}_t^{m,i} = \int_{W^i} \mathbf{y} g^m\left(\mathbf{y}|\hat{\mathbf{x}}_t^m\right) d\mathbf{y}. \qquad (8.16)$$

Applying this to the quadratic expansion gives

$$\begin{aligned}
\sum_{i=1}^{I+I^U} \cdot = &\Big[||\mathfrak{z}_t^m|| h\left(\mathbf{x}_t^m\right)^{\mathsf{T}} \left(\Sigma_t^m\right)^{-1} h\left(\mathbf{x}_t^m\right) - 2||\mathfrak{z}_t^m||\tilde{\mathbf{y}}_t^m \left(\Sigma_t^m\right)^{-1} h\left(\mathbf{x}_t^m\right) \\
&+ ||\mathfrak{z}_t^m|| \left(\tilde{\mathbf{y}}_t^m\right)^{\mathsf{T}} \left(\Sigma_t^m\right)^{-1} \left(\tilde{\mathbf{y}}_t^m\right) \Big] - ||\mathfrak{z}_t^m|| \left(\tilde{\mathbf{y}}_t^m\right)^{\mathsf{T}} \left(\Sigma_t^m\right)^{-1} \left(\tilde{\mathbf{y}}_t^m\right) \\
&+ \sum_{i=1}^{I+I^U} \frac{\mathfrak{z}_t^{m,i}}{G_t^{m,i}} \int_{W^i} \mathbf{y}^{\mathsf{T}} \left(\Sigma_t^m\right)^{-1} \mathbf{y}\, g^m\left(\mathbf{y}|\hat{\mathbf{x}}_t^m\right) d\mathbf{y},
\end{aligned}$$

$$= ||\mathfrak{z}_t^m||\left(h\left(\mathbf{x}_t^m\right) - \tilde{\mathbf{y}}_t^m\right)^\top \left(\Sigma_t^m\right)^{-1}\left(h\left(\mathbf{x}_t^m\right) - \tilde{\mathbf{y}}_t^m\right) - ||\mathfrak{z}_t^m||\left(\tilde{\mathbf{y}}_t^m\right)^\top \left(\Sigma_t^m\right)^{-1}\left(\tilde{\mathbf{y}}_t^m\right)$$
$$+ \sum_{i=1}^{I+I^U} \frac{\mathfrak{z}_t^{m,i}}{G_t^{m,i}} \int_{W^i} \mathbf{y}^\top \left(\Sigma_t^m\right)^{-1} \mathbf{y}\, g^m\left(\mathbf{y}|\hat{\mathbf{x}}_t^m\right) \mathrm{d}\mathbf{y}. \tag{8.17}$$

The first term now encapsulates all of the $\mathbf{x}_t^m$ dependence and is the same as the expression we arrived at in Chap. 4. The other two terms were previously discarded because they are independent of the kinematic state but now we must retain them because they depend on Σ_t^m. Observe that

$$\sum_{i=1}^{I+I^U} \frac{\mathfrak{z}_t^{m,i}}{G_t^{m,i}} \int_{W^i} \left(\tilde{\mathbf{y}}_t^m\right)^\top \left(\Sigma_t^m\right)^{-1} \mathbf{y}\, g^m\left(\mathbf{y}|\hat{\mathbf{x}}_t^m\right) \mathrm{d}\mathbf{y}$$
$$= \left(\tilde{\mathbf{y}}_t^m\right)^\top \left(\Sigma_t^m\right)^{-1} \sum_{i=1}^{I+I^U} \frac{\mathfrak{z}_t^{m,i}}{G_t^{m,i}} \int_{W^i} \mathbf{y}\, g^m\left(\mathbf{y}|\hat{\mathbf{x}}_t^m\right) \mathrm{d}\mathbf{y}$$
$$= ||\mathfrak{z}_t^m||\left(\tilde{\mathbf{y}}_t^m\right)^\top \left(\Sigma_t^m\right)^{-1} \tilde{\mathbf{y}}_t^m, \tag{8.18}$$

because $\tilde{\mathbf{y}}_t^m$ is constant. Also we can write

$$||\mathfrak{z}_t^m|| = \sum_{i=1}^{I+I^U} \mathfrak{z}_t^{m,i} \frac{G_t^{m,i}}{G_t^{m,i}} = \sum_{i=1}^{I+I^U} \frac{\mathfrak{z}_t^{m,i}}{G_t^{m,i}} \int_{W^i} g^m\left(\mathbf{y}|\hat{\mathbf{x}}_t^m\right) \mathrm{d}\mathbf{y}, \tag{8.19}$$

which means that

$$||\mathfrak{z}_t^m||\left(\tilde{\mathbf{y}}_t^m\right)^\top \left(\Sigma_t^m\right)^{-1} \tilde{\mathbf{y}}_t^m = \sum_{i=1}^{I+I^U} \frac{\mathfrak{z}_t^{m,i}}{G_t^{m,i}} \int_{W^i} \left(\tilde{\mathbf{y}}_t^m\right)^\top \left(\Sigma_t^m\right)^{-1} \tilde{\mathbf{y}}_t^m g^m\left(\mathbf{y}|\hat{\mathbf{x}}_t^m\right) \mathrm{d}\mathbf{y}. \tag{8.20}$$

Applying this, the last two terms of (8.17) can be combined into a single quadratic.

$$-||\mathfrak{z}_t^m||\left(\tilde{\mathbf{y}}_t^m\right)^\top \left(\Sigma_t^m\right)^{-1} \tilde{\mathbf{y}}_t^m = ||\mathfrak{z}_t^m||\left(\tilde{\mathbf{y}}_t^m\right)^\top \left(\Sigma_t^m\right)^{-1} \tilde{\mathbf{y}}_t^m - 2||\mathfrak{z}_t^m||\left(\tilde{\mathbf{y}}_t^m\right)^\top \left(\Sigma_t^m\right)^{-1} \tilde{\mathbf{y}}_t^m,$$
$$= \sum_{i=1}^{I+I^U} \frac{\mathfrak{z}_t^{m,i}}{G_t^{m,i}} \int_{W^i} \left(\tilde{\mathbf{y}}_t^m\right)^\top \left(\Sigma_t^m\right)^{-1} \tilde{\mathbf{y}}_t^m g^m\left(\mathbf{y}|\hat{\mathbf{x}}_t^m\right) \mathrm{d}\mathbf{y}$$
$$- 2\sum_{i=1}^{I+I^U} \frac{\mathfrak{z}_t^{m,i}}{G_t^{m,i}} \int_{W^i} \left(\tilde{\mathbf{y}}_t^m\right)^\top \left(\Sigma_t^m\right)^{-1} \mathbf{y}\, g^m\left(\mathbf{y}|\hat{\mathbf{x}}_t^m\right) \mathrm{d}\mathbf{y}, \tag{8.21}$$

and the last two terms of (8.17) become

$$-||\mathfrak{z}_t^m||\left(\tilde{\mathbf{y}}_t^m\right)^{\mathsf{T}}\left(\Sigma_t^m\right)^{-1}\left(\tilde{\mathbf{y}}_t^m\right)^{\mathsf{T}}+\sum_{i=1}^{I+I^U}\frac{\mathfrak{z}_t^{m,i}}{G_t^{m,i}}\int_{W^i}\mathbf{y}^{\mathsf{T}}\left(\Sigma_t^m\right)^{-1}\mathbf{y}\,g^m\left(\mathbf{y}|\hat{\mathbf{x}}_t^m\right)d\mathbf{y}$$
$$=\sum_{i=1}^{I+I^U}\frac{\mathfrak{z}_t^{m,i}}{G_t^{m,i}}\int_{W^i}\left(\mathbf{y}-\tilde{\mathbf{y}}_t^m\right)^{\mathsf{T}}\left(\Sigma_t^m\right)^{-1}\left(\mathbf{y}-\tilde{\mathbf{y}}_t^m\right)g^m\left(\mathbf{y}|\hat{\mathbf{x}}_t^m\right)d\mathbf{y}. \tag{8.22}$$

The notation here is very dense but conceptually the machinations are really just collecting terms and rearranging a sum of quadratics into a different set of quadratics. This is the same trick we have done repeatedly in earlier chapters.

Substituting (8.17) and (8.22), the measurement part of the auxiliary function in (8.13) becomes

$$\sum_{\mathbb{K}}\log\{p(\mathbb{Y}|\mathbb{X},\mathbb{K};\Sigma)\}\,p_{\mathsf{miss}}(\mathbb{K}|\hat{\mathbb{X}},\hat{\mathbb{L}},\mathbb{Y};\hat{\Sigma})=$$
$$-\frac{1}{2}\sum_{m=0}^{M}\sum_{t=1}^{T}\Bigg\{||\mathfrak{z}_t^m||\log\left\{\left|\Sigma_t^m\right|\right\}+||\mathfrak{z}_t^m||\left(h\left(\mathbf{x}_t^m\right)-\tilde{\mathbf{y}}_t^m\right)^{\mathsf{T}}\left(\Sigma_t^m\right)^{-1}\left(h\left(\mathbf{x}_t^m\right)-\tilde{\mathbf{y}}_t^m\right)$$
$$+\sum_{i=1}^{I+I^U}\frac{\mathfrak{z}_t^{m,i}}{G_t^{m,i}}\int_{W^i}\left(\mathbf{y}-\tilde{\mathbf{y}}_t^m\right)^{\mathsf{T}}\left(\Sigma_t^m\right)^{-1}\left(\mathbf{y}-\tilde{\mathbf{y}}_t^m\right)g^m\left(\mathbf{y}|\hat{\mathbf{x}}_t^m\right)d\mathbf{y}\Bigg\}. \tag{8.23}$$

We now make use of the matrix identity $\mathbf{v}^{\mathsf{T}}\mathsf{A}^{-1}\mathbf{v}=\mathsf{trace}\left(\mathbf{v}\mathbf{v}^{\mathsf{T}}\mathsf{A}^{-1}\right)$ to write

$$\sum_{\mathbb{K}}\log\{p(\mathbb{Y}|\mathbb{X},\mathbb{K};\Sigma)\}\,p_{\mathsf{miss}}(\mathbb{K}|\hat{\mathbb{X}},\hat{\mathbb{L}},\mathbb{Y};\hat{\Sigma})=$$
$$-\frac{1}{2}\sum_{m=0}^{M}\sum_{t=1}^{T}\Bigg\{||\mathfrak{z}_t^m||\log\left\{\left|\Sigma_t^m\right|\right\}+||\mathfrak{z}_t^m||\left(h\left(\mathbf{x}_t^m\right)-\tilde{\mathbf{y}}_t^m\right)^{\mathsf{T}}\left(\Sigma_t^m\right)^{-1}\left(h\left(\mathbf{x}_t^m\right)-\tilde{\mathbf{y}}_t^m\right)$$
$$+\,\mathsf{trace}\left[\tilde{\mathsf{Z}}_t^m\left(\Sigma_t^m\right)^{-1}\right]\Bigg\}, \tag{8.24}$$

where the equivalent spread matrix is

$$\tilde{\mathsf{Z}}_t^m=\sum_{i=1}^{I+I^U}\mathfrak{z}_t^{m,i}\tilde{\mathsf{Z}}_t^{m,i}, \tag{8.25}$$

and the pixel spread matrix as

$$\tilde{\mathsf{Z}}_t^{m,i} = \frac{1}{G_t^{m,i}} \int_{W^i} \left(\mathbf{y} - \tilde{\mathbf{y}}_t^m\right)^{\mathsf{T}} \left(\Sigma_t^m\right)^{-1} \left(\mathbf{y} - \tilde{\mathbf{y}}_t^m\right) g^m \left(\mathbf{y}|\hat{\mathbf{x}}_t^m\right) \mathrm{d}\mathbf{y}. \tag{8.26}$$

The terms in the auxiliary function that depend on the appearance matrix Σ_t^m are the matrix prior and the measurement term above. We have assumed an inverse Wishart prior, so the logarithm of this prior is

$$\log\left\{p\left(\Sigma_t^m\right)\right\} = -\frac{1}{2}(\nu_{t|t-1}^m - d - 1)\log\left\{|\Sigma_t^m|\right\} + \mathsf{trace}\left(-\tfrac{1}{2}\mathsf{X}_{t|t-1}^m \left(\Sigma_t^m\right)^{-1}\right), \tag{8.27}$$

where the appearance hyperparameters are predicted forwards in time using the heuristic decay model of [11]

$$\begin{aligned} \nu_{t|t-1}^m &= \nu_{t-1|t-1}^m \exp\left\{-\frac{\Delta t_t}{\tau_R}\right\}, \\ \mathsf{X}_{t|t-1}^m &= \frac{\nu_{t|t-1}^m - d - 1}{\nu_{t-1|t-1}^m - d - 1}\mathsf{X}_{t-1|t-1}^m, \end{aligned} \tag{8.28}$$

where Δt_t is the elapsed time between frames $t-1$ and t and τ_R is a tunable decay time that affects how long the estimator retains data about the appearance model. This is a matrix version of the dynamic model we applied to the Poisson rate parameter back in Chap. 6.

The appearance-dependent terms in the auxiliary function can now be expressed as the logarithm of a posterior inverse Wishart distribution and we find the parameters of this posterior by matching terms. The precision parameter update is an accumulation of the associated image power [17]

$$\hat{\nu}_{t|t}^m = \hat{\nu}_{t|t-1}^m + ||\mathfrak{z}_t^m||, \tag{8.29}$$

and the shape matrix consists of two terms

$$\hat{\mathsf{X}}_{t|t}^m = \hat{\mathsf{X}}_{t|t-1}^m + \mathsf{N}_{t|t-1}^m + \tilde{\mathsf{Z}}_t^m. \tag{8.30}$$

The innovation matrix is given by

$$\mathsf{N}_{t|t-1}^m = \left[\mathsf{S}_{t|t-1}^m\right]^{-1} \left(h\left(\mathbf{x}_t^m\right) - \tilde{\mathbf{y}}_t^m\right) \left(h\left(\mathbf{x}_t^m\right) - \tilde{\mathbf{y}}_t^m\right)^{\mathsf{T}} \tag{8.31}$$

where $\mathsf{S}_{t|t-1}^m$ is the innovation covariance used in the kinematic state estimation.

The H-PMHT maximisation step forms a MAP estimate for the appearance parameter Σ_t^m by maximising the auxiliary function, which we have arranged as an inverse Wishart distribution, which leads to $\hat{\Sigma}_t^m = (\nu_t^m - d - 1)^{-1}\mathsf{X}_t^m$.

8.3 General Framework for Appearance Estimation

The random matrix version of H-PMHT[1] derived in the previous section is an example of an appearance model with stochastic parameters. In fact, we can generalise this to an arbitrary known function with unknown parameters. We will denote the general appearance function as $g\left(\mathbf{y}|\mathbf{x},\theta\right)$ where the parameters θ follow independent Markov chains for each target. In the random matrix case $\theta \equiv \Sigma_t^m$, but there are many other possible models. Some of these will be explored in the remainder of this chapter. As usual Θ_t will denote the collection of parameters across targets at a single frame and Θ denotes all of the parameters in a batch. The auxiliary function in this generalised case looks like

$$\begin{aligned}\mathcal{Q}\left(\mathbb{X},\mathbb{L},\Theta|\hat{\mathbb{X}},\hat{\mathbb{L}},\hat{\Theta}\right) &= E_{\mathbb{K}|\hat{\mathbb{X}},\hat{\mathbb{L}},\hat{\Theta},\mathbb{Z}}\Big[\log\left\{p_{\mathsf{comp}}(\mathbb{X},\mathbb{L},\Theta,\mathbb{K},\mathbb{Z})\right\}\Big],\\ &= \sum_{\mathbb{K}}\log\left\{p_{\mathsf{comp}}(\mathbb{X},\mathbb{L},\Theta,\mathbb{K},\mathbb{Z})\right\}p_{\mathsf{miss}}(\mathbb{K}|\hat{\mathbb{X}},\hat{\mathbb{L}},\hat{\Theta},\mathbb{Z}).\end{aligned} \tag{8.32}$$

The parameters follow unconditionally independent Markov chains, so the complete data likelihood is

$$\begin{aligned}p_{\mathsf{comp}}\left(\mathbb{X},\Theta,\mathbb{L},\mathbb{K},\mathbb{Z}\right) &= p(\mathbf{X}_0)p(\Theta_0)p(\mathbf{L}_0)\\ &\times\prod_{t=1}^{T}\left[\prod_{m=0}^{M}p\left(\mathbf{x}_t^m|\mathbf{x}_{t-1}^m\right)p\left(\Lambda_t^m|\Lambda_{t-1}^m\right)p\left(\theta_t^m|\theta_{t-1}^m\right)\right]\\ &\times\prod_{t=1}^{T}\left[p\left(\mathbf{K}_t|\mathbf{L}_t\right)\prod_{i=1}^{I}p\left(\mathbf{z}_t^i|\mathbf{K}_t,\mathbf{X}_t,\Theta_t\right)\right].\end{aligned} \tag{8.33}$$

[1] *MakeHPMHTParams*
The H-PMHT toolbox contains two main functions for running H-PMHT. *MakeHPMHTParams* defines tracking parameters, including which version of H-PMHT to execute, and *HPMHTTracker* actually does the tracking. The toolbox contains two versions of H-PMHT-RM. The original H-PMHT-RM was built on core H-PMHT and uses the multinomial model. This can be setup using *MakeHPMHTParams('core H-PMHT-RM')*. A version incorporating the Poisson assignment prior is initialised with *MakeHPMHTParams('Poisson H-PMHT-RM')*. For details on how to use this function, refer to the H-PMHT toolbox documentation.

The difference between this expression and the Poisson H-PMHT with known appearance is the term $p\left(\theta_t^m|\theta_{t-1}^m\right)$ and the conditioning of the measurement term on the appearance parameters, $p\left(\mathbf{z}_t^i|\mathbf{K}_t, \mathbf{X}_t, \Theta_t\right)$. The log of the complete data likelihood is once again a sum

$$\log\left\{p_{\text{comp}}\left(\mathbb{X}, \Theta, \mathbb{L}, \mathbb{K}, \mathbb{Z}\right)\right\} = \sum_{m=0}^{M}\Bigg[\log\left\{p(\mathbf{x}_0^m)\right\} + \sum_{t=1}^{T}\log\left\{p\left(\mathbf{x}_t^m|\mathbf{x}_{t-1}^m\right)\right\}$$
$$+ \log\left\{p(\theta_0^m)\right\} + \sum_{t=1}^{T}\left\{p\left(\theta_t^m|\theta_{t-1}^m\right)\right\} + \log\left\{p(\Lambda_0^m)\right\} + \sum_{t=1}^{T}\left\{p\left(\Lambda_t^m|\Lambda_{t-1}^m\right)\right\}\Bigg]$$
$$+ \sum_{t=1}^{T}\left[\log\left\{p\left(\mathbf{K}_t|\mathbf{L}_t\right)\right\} + \sum_{i=1}^{I}\log\left\{p\left(\mathbf{z}_t^i|\mathbf{K}_t, \mathbf{X}_t, \Theta_t\right)\right\}\right]. \quad (8.34)$$

The missing data likelihood is again simplified using Bayes' rule

$$p_{\text{miss}}(\mathbb{K}|\mathbb{X}, \mathbb{L}, \Theta, \mathbb{Z}) = \frac{p(\mathbb{K}, \mathbb{X}, \mathbb{L}, \Theta, \mathbb{Z})}{\sum_{\mathbb{K}} p(\mathbb{K}, \mathbb{X}, \mathbb{L}, \Theta, \mathbb{Z})},$$
$$= \frac{p(\mathbb{K}|\mathbb{L})p(\mathbb{Z}|\mathbb{X}, \Theta, \mathbb{K})}{\sum_{\mathbb{K}} p(\mathbb{K}|\mathbb{L})p(\mathbb{Z}|\mathbb{X}, \Theta, \mathbb{K})}. \quad (8.35)$$

The missing data is still a product of assignment weights, the only difference is the dependence of the weights on the appearance parameters. The missing data and the logarithm of the complete data lead to the auxiliary function as usual. By now, the reader should be able to verify, or at least believe, that this auxiliary function can be written as

$$\mathscr{Q}\left(\mathbb{X}, \mathbb{L}, \Theta|\hat{\mathbb{X}}, \hat{\mathbb{L}}, \hat{\Theta}\right) = \sum_{m=1}^{M}\mathscr{Q}_\Lambda^m + \sum_{m=1}^{M}\mathscr{Q}_{X,\Theta}^m. \quad (8.36)$$

The Poisson rate term is the same as when the appearance is known, but the kinematic state term also depends on the appearance parameters because the measurement likelihood is jointly dependent on both. This is the same situation that we have seen for both of the Gaussian appearance cases in the previous sections and we dodge it using the same trick. Under ECM, we can maximise with respect to one parameter first and then with respect to the other parameter. Here, we first optimise for the kinematic state holding the appearance fixed at its previous estimate $\hat{\Theta}$ and then fix the kinematic state and optimise with respect to the appearance. For non-Gaussian appearance models, this state optimisation has to use one of the methods described in the previous chapter. The appearance optimisation depends on the form of the appearance model. A few different kinds of appearance model are now discussed.

8.4 Gaussian Mixture Appearance

The Gaussian appearance model can be a good approximation for targets that are small compared with the pixel resolution and can be loosely treated as unimodal blobs, but when the sensor resolution is fine compared with the target size, there can be significant structure to the target appearance that cannot be captured by a single mean and covariance. One option in this situation is to use a Gaussian mixture as the appearance model, namely

$$g\left(\mathbf{y}|\mathbf{x}_t^m, \theta_t^m\right) = \sum_{j=1}^{J_t^m} \Lambda_t^{m,j} \mathcal{N}\left(\mathbf{y}; h(\mathbf{x}_t^m) + \mu_t^{m,j}, \Sigma_t^{m,j}\right) \tag{8.37}$$

with

$$\theta_t^m \equiv \left[J_t^m, \left\{\Lambda_t^{m,j}, \mu_t^{m,j}, \Sigma_t^{m,j}\right\}_{j=1}^{\eta_t^m}\right]. \tag{8.38}$$

The number of components J_t^m can be unknown and time varying and the mean $\mu_t^{m,j}$ and covariance $\Sigma_t^{m,j}$ of each are random variables. In the previous chapter in Sect. 7.4, we discussed a Gaussian mixture appearance function where the parameters of the mixture were known and fixed. The difference now is that we would like to make the number of components time varying and unknown.

At least, conceptually, this is fairly straightforward if we use the idea of associated images. The associated image for track m is determined using the previous estimates of the kinematic state and the appearance parameters and is given by (4.67)

$$\mathfrak{z}_t^{i,m} = w_t^{i,m} \bar{\mathbf{z}}_t^i,$$

where here the target-pixel association probabilities are implicitly dependent on $\hat{\mathbf{X}}_t$ and $\hat{\Theta}_t$. Using the associated image, the maximisation step is then to fit the best $\mathbf{X}_t$ and Θ_t to $\mathfrak{z}_t$. Since Θ_t are the parameters of a Gaussian mixture, we can find an estimate using H-PMHT itself!

H-PMHT with an unknown Gaussian mixture appearance is analogous to a pair of nested H-PMHT algorithms. The outer one uses a known appearance mixture to find the associated image and the inner one can use H-PMHT-RM to fit the mixture parameters. Both layers use track maintenance to create or remove new tracks and then to create or remove components of the appearance mixture.

One application of a Gaussian mixture appearance model is to form targets [3, 4]. Multiple targets moving in a correlated fashion can be treated as a single group target that is a mixture of components that combine common bulk motion (the *target* kinematic state) and independent within-group motion (the mixture component stochastic mean).

Fig. 8.1 Bounding boxes overlaid on people in a foyer

8.5 Bounding Box

In video tracking applications ,a common description for non-rigid bodies is a bounding box. This box is simply the smallest rectangle that encloses all of the object pixels. It is likely that the bounding box will also contain background pixels. Figure 8.1 shows an example of bounding boxes overlaid on a video image of people in a foyer from the CAVIAR benchmark [8]. Notice that in this example, two of the boxes overlap.

The standard bounding box is not really suitable to integrate with H-PMHT: there are two problems. First, the box does not represent a probability density function, rather it is an inclusion boundary. Pixels inside the box could be from the object and those outside cannot. We can interpret this as a uniform patch, but that is not really what the box means. Second, the box describes a hard boundary beyond which there is no target energy. This means that the appearance function should be identically zero everywhere outside the box. This is rather unfortunate for H-PMHT which is fundamentally built on the log-likelihood and the log of zero is not very pleasant. To understand the implications, consider a single time slice of the state auxiliary sub-function which consists of state and parameter transitions and a measurement term

$$\mathscr{Q}_t^{\mathbf{x}} = q(\mathbf{x}, \theta) + \sum_{i=1}^{I} \bar{\mathbf{z}}_t^i w_t^{i,m} \frac{\int_{W^i} \log \left\{ g\left(\mathbf{y}|\mathbf{x}, \theta\right) \right\} g\left(\mathbf{y}|\hat{\mathbf{x}}, \hat{\theta}\right) \mathrm{d}\mathbf{y}}{G_t^{m,i}}. \tag{8.39}$$

A uniform patch bounding box implies

$$g(\mathbf{y}|\mathbf{x},\theta) = \begin{cases} A & \mathbf{y} \in \text{box}(\mathbf{x},\theta), \\ 0 & \text{otherwise}, \end{cases} \tag{8.40}$$

and thus the integral term becomes

$$\begin{aligned} &\int_{W^i} \log\left\{g\left(\mathbf{y}|\mathbf{x},\theta\right)\right\} g\left(\mathbf{y}|\hat{\mathbf{x}},\hat{\theta}\right) \mathbf{dy} \\ &= \begin{cases} A\log\{A\}|W^i| & \mathbf{y} \in \text{box}(\mathbf{x},\theta) \text{ and } \mathbf{y} \in \text{box}(\hat{\mathbf{x}},\hat{\theta}), \\ 0 & \mathbf{y} \in \text{box}(\mathbf{x},\theta) \text{ and } \mathbf{y} \notin \text{box}(\hat{\mathbf{x}},\hat{\theta}), \\ 0 & \mathbf{y} \notin \text{box}(\mathbf{x},\theta) \text{ and } \mathbf{y} \notin \text{box}(\hat{\mathbf{x}},\hat{\theta}), \\ -\infty & \mathbf{y} \notin \text{box}(\mathbf{x},\theta) \text{ and } \mathbf{y} \in \text{box}(\hat{\mathbf{x}},\hat{\theta}), \end{cases} \end{aligned} \tag{8.41}$$

where we have asserted without proof that $0\log\{0\} = 0$. What is more of a problem is that the sub-function is $-\infty$ for all state and parameter combinations that lead to a bounding box that does not cover every pixel in the bounding box from the previous EM iteration. This means that the iterations cannot move the box.

H-PMHT cannot use a bounding box in the obvious way, but it could use something like a box if it had decaying tails instead of a sharp cut-off. In the previous chapter, we discussed ways of approximating the appearance function: one of the methods was to use a Gaussian mixture. In this case, a side effect of the approximation is that the mixture appearance has Gaussian tails, which is what we want. An intuitive approximation is to use four components with equal weighting and covariance matrices. If we arrange these components as the corners of a rectangle, then the spacing between them can be adjusted to make the interior relatively flat. Since we want a rectangle, we can constrain the covariance matrix to be diagonal. Figure 8.2 shows an example where four Gaussian components have been combined to give an approximately rectangular mixture. In this case, each component has a spread

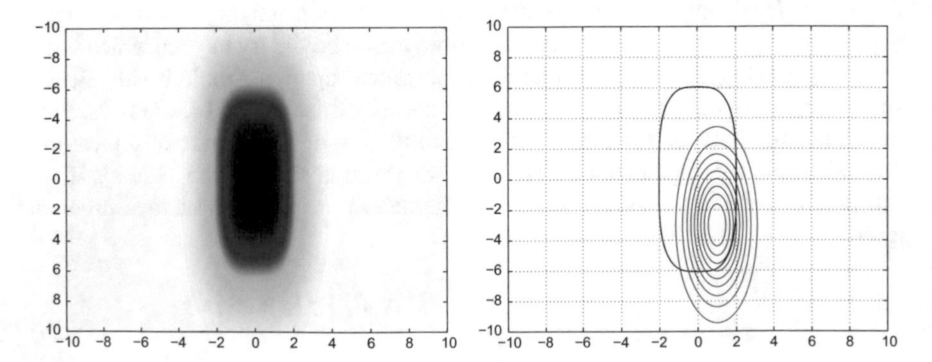

Fig. 8.2 Gaussian mixture approximate bounding box

covariance matrix of $\Sigma = \begin{bmatrix} 1 & 0 \\ 0 & 9 \end{bmatrix}$, and the components are centred on $[\pm 1, \pm 3]$. The resulting appearance function has level contours that are approximately rectangular. The left plot shows the appearance function and the right plot shows a black level contour at $g(\mathbf{y}) = 0.5 \max\{g\}$ that corresponds approximately to a rectangle of width 4 and height 12. This is four standard deviations of each individual component in each direction. The red family of curves shows the level contours for one of the four Gaussian components.

We can generalise the appearance function in Fig. 8.2 as

$$g(\mathbf{y}|\mathbf{x}, \theta_1, \theta_2) = 0.25 \sum_{j=1}^{J} \mathcal{N}\left(\mathbf{y}; \mathsf{H}\mathbf{x} + \mu^j, \Sigma\right), \tag{8.42}$$

$$\mu^j = \begin{bmatrix} -\theta_1 \\ -\theta_2 \end{bmatrix}, \begin{bmatrix} -\theta_1 \\ \theta_2 \end{bmatrix}, \begin{bmatrix} \theta_1 \\ -\theta_2 \end{bmatrix}, \begin{bmatrix} \theta_1 \\ \theta_2 \end{bmatrix}, \qquad \Sigma = \begin{bmatrix} \left(\theta^1\right)^2 & 0 \\ 0 & \left(\theta^2\right)^2 \end{bmatrix}. \tag{8.43}$$

The appearance learning task is then to estimate the parameters θ_1 and θ_2 that specify the width and height of a bevelled bounding box. An interesting feature of this model is that the parameter to be estimated appears on both the numerator and denominator of the Gaussian exponent. Optimising for these parameters will depend on the prior and dynamic model assumed for θ. We choose not to explore this model further for two main reasons. First, a bounding box is intended to be a simple description that is not generally applied to targets that are truly rectangular: proceeding once the model has become a little tricky is counterproductive. Second, experience shows that the algorithm does not need to match the assumed appearance model precisely to the true target appearance: the examples to follow later in this chapter demonstrate that a Gaussian appearance will usually suffice. In applications where video trackers might apply a bounding box, we prefer to apply H-PMHT-RM and effectively use a bounding ellipse.

8.6 Dirichlet Appearance

Using a Gaussian mixture appearance function for an arbitrary shape is a form of functional approximation. The true appearance is parametrically approximated by a known family of functions. The alternative to functional approximation is to directly model the appearance over a grid. In video tracking, this kind of appearance model is often referred to as a template model [12]. The template grid can be at the same resolution as the sensor but it does not have to be the same. Here, we describe a super-resolution model that defines a grid on a finer scale and uses the motion of the target to learn this appearance. Under our existing framework, the pixel probability is the integral of the appearance function over a pixel. For a grid appearance, this is just the sum of the grid cells that are inside the pixel. If the target state were known and changing, then each frame essentially combines a different set of grid cells and the images themselves are a set of linear equations in the appearance grid values. Under this intuition, it is possible to estimate the appearance at finer than sensor native resolution if the target motion is estimated well and there are a lot of frames. Figure 8.3 shows a very simplified example where a fine resolution appearance grid moves across a single pixel. The pixel extent W^i is shown as a red box in the figure. Figure 8.3 shows six frames and each contains a single pixel measurement that is a linear combination of the four appearance grid values. Denoting the appearance grid values as ϑ^j, $j = 1, \ldots 4$, and the relative overlap between the pixel and the jth appearance grid as $0 \leq a_t^j \leq 1$, then the cell probability can be written as a system of linear equations

$$G_t = \sum_{j=1}^{J} a_t^j \vartheta^j, \tag{8.44}$$

which can be stacked into a matrix-vector equation and solved using

$$\mathsf{G} = \mathsf{A}\theta \quad \rightarrow \quad \theta = \left(\mathsf{A}^\mathsf{T}\mathsf{A}\right)^{-1} \mathsf{A}^\mathsf{T}\mathsf{G}. \tag{8.45}$$

This strategy requires the rank of A to be at least equal to the number of appearance grid cells, so in this example $\text{rank}(\mathsf{A}) \geq 4$. The real problem we would like to address has multiple image pixels with noise and a very large number of appearance grids, we now develop this more formally.

Define the target shape over a fine grid so that $\theta_t^m = \{\vartheta_t^{m,j}\}$ with j an arbitrary index $j = 1 : N_\theta$, where N_θ is the number of grid cells in the target appearance template. The dimension of the appearance template is the same as the sensor image and again single-indexing does not limit the dimensionality of θ_t^m. The elements $\vartheta_t^{m,j}$ sum to unity and represent the spatial distribution of the target return. Effectively, these are target-pixel probabilities $G_t^{m,j}$ over a different pixel domain than the sensor image. The template gives the probability that a single shot from the target lands in grid-cell j and so they represent the parameters of a multinomial distribution for the position of data from target m. The probability that a single histogram shot from target m will fall in cell j is the sum

$$G_t^{m,i} = \sum_{j \in W_i} \vartheta_t^{m,j}, \tag{8.46}$$

where the target state influences the expression by determining which of the appearance cells j fall inside W_i, remembering that the $\vartheta_t^{m,j}$ are over a finer grid than the sensor pixels. As expected, when the continuous appearance function is replaced with a discrete approximation, the target-pixel probability integral becomes a sum.

---

[2]*MakeHPMHTParams*

The H-PMHT toolbox contains two main functions for running H-PMHT. *MakeHPMHTParams*

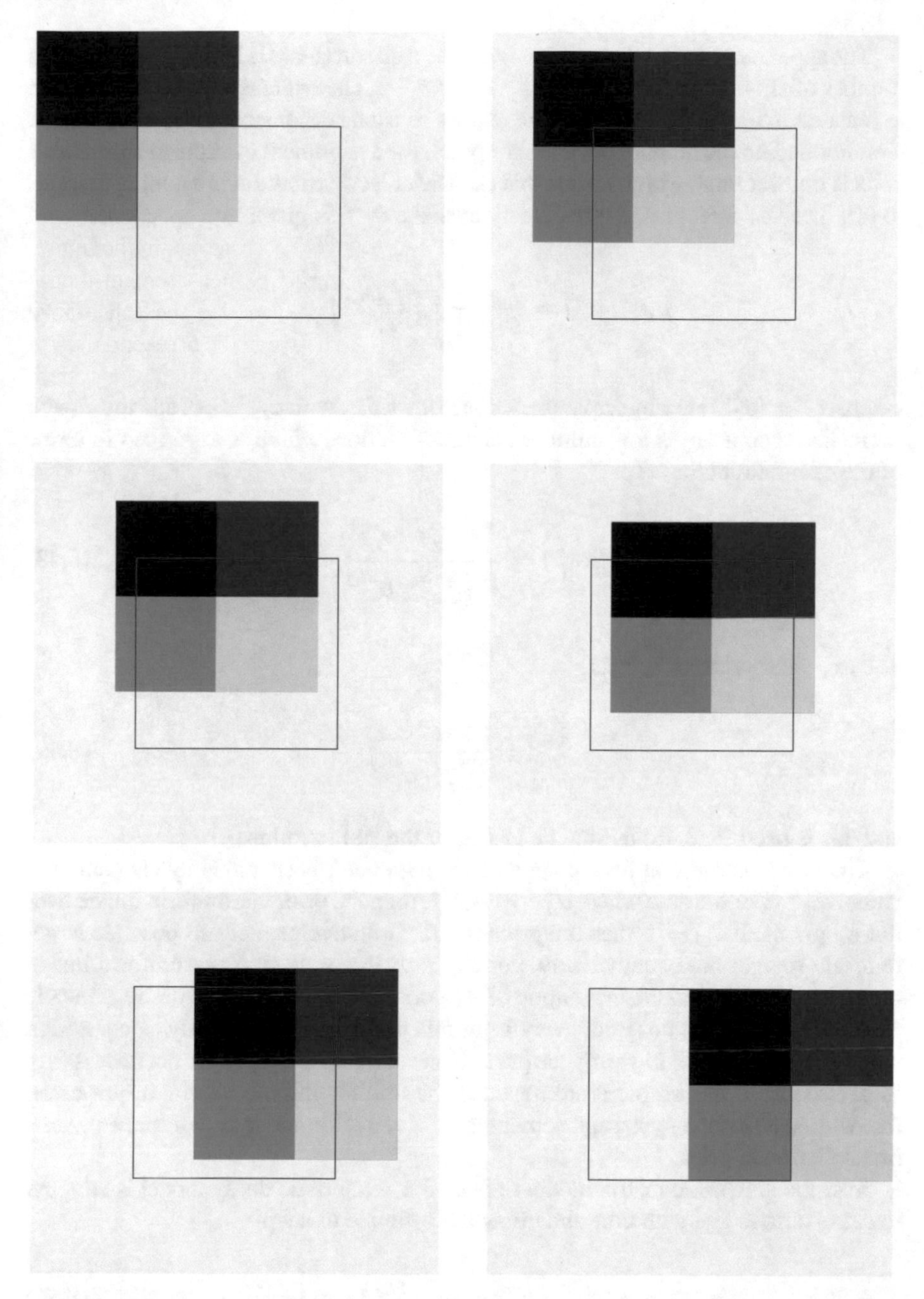

Fig. 8.3 Simplified high-resolution template: six frames of a 4-cell target moving across a single pixel

The appearance template combines with the Poisson H-PMHT[2] so that the average number of shots in the template cell j is $\Lambda_t^m \vartheta_t^{m,j}$. The expectation step defines the associated image, which gives us the energy in each sensor pixel due to this target. Conditioned on the number of shots in a pixel, the assignment of shots to appearance cells is multinomial, which means that the Dirichlet distribution is a conjugate prior to θ_t^m. The Dirichlet probability density over the $\vartheta_t^{m,j}$ is given by

$$p(\theta_t^m; \alpha_t^m) = \frac{1}{B(\alpha_t^m)} \prod_{j=1}^{N_\theta} \left(\vartheta_t^{m,j}\right)^{\alpha_t^{m,j}-1}, \tag{8.47}$$

where $\alpha_t^m = \{\alpha_t^{m,j}\}$ is a hyperparameter vector for θ_t^m, with $\alpha_t^{m,j} > 0$, and the $B(\alpha_t^m)$ normalising constant is the multinomial beta function, which is expressed in terms of the gamma function as

$$B(\alpha_t^m) = \frac{\prod_{j=1}^{N_\theta} \Gamma(\alpha_t^{m,j})}{\Gamma\left(\sum_{j=1}^{N_\theta} \alpha_t^{m,j}\right)}. \tag{8.48}$$

The expected value of $\vartheta_t^{m,j}$ is

$$\hat{\vartheta}_t^{m,j} = \frac{\alpha_t^{m,j}}{\sum_{j=1}^{N_\theta} \alpha_t^{m,j}}, \tag{8.49}$$

and this is used in (8.46) during the E-step of the EM iterations.

The prior distribution for the appearance parameter is specified by choosing the values $\alpha_0^{m,j}$. For the case where $\alpha_0^{m,j} = A_0 \forall j$, then the prior is a uniform image over the N_θ pixels. If $A_0 = 1$, then the prior is uninformative, namely all possible target template images are equally likely. For $A_0 \gg 1$, the prior prefers a uniform image over a peaked one and the magnitude of A_0 controls how much measurement data is required to move the posterior away from this preference. Effectively, using a high value of A_0 results in a low variance prior. However, a uniform prior is not necessarily ideal because it has no preferred mean: if the track is initialised with a bias error, this will be absorbed by a displacement in $\vartheta_t^{m,j}$. An alternative is to initialise with a broad Gaussian prior.

Like the stochastic matrix model in Sect. 8.2, a heuristic decay model is adopted to cause data to age with time and allow the estimate to adapt

$$\alpha_{t|t-1}^{m,j} = \alpha_0^{m,j} + \left(\alpha_{t-1|t-1}^{m,j} - \alpha_0^{m,j}\right) \exp\left\{-\frac{\Delta t_t}{\tau_\alpha}\right\}, \tag{8.50}$$

defines tracking parameters, including which version of H-PMHT to execute, and *HPMHTTracker* actually does the tracking. The default tracking parameters for the Dirichlet H-PMHT are created by *MakeHPMHTParams('Dirichlet')*. For details on how to use this function, refer to the H-PMHT toolbox documentation.

where as before τ_α is a tunable decay parameter. The Dirichlet hyperparameter vector $\alpha^m_{t|t-1}$ is updated by accumulating the associated image. For the case where the pixels of θ^m_t have the same resolution as the sensor image, the updated hyperparameter vector is given by

$$\alpha^{m,j}_{t|t} = \alpha^{m,j}_{t|t-1} + \mathfrak{z}^{m,j}_t. \tag{8.51}$$

If the resolution of θ^m_t is finer than the sensor image as implied by (8.46), then the energy in pixel i needs to be distributed amongst the pixels contained in W_i. This is achieved with a further nuisance variable. Rather than put you through that, lets take an intuitive leap and recognise that by this stage of the book, you can work out the details for yourself. The probability that a shot falls in appearance cell j contained in W^i is the relative frequency $\vartheta^{m,j}_t / \sum_{l \in W^i} \vartheta^{m,l}_t$. The resulting fine resolution update is therefore

$$\alpha^{m,j}_{t|t} = \alpha^{m,j}_{t|t-1} + \frac{\vartheta^{m,j}_t}{\sum_{l \in W^i} \vartheta^{m,l}_t} \mathfrak{z}^{m,i}_t. \tag{8.52}$$

8.7 Appearance Library

In some applications, it is possible to assume that the target appearance comes from a known library of models. These could be either fixed target templates, or each model in the library could have its own parameters. In this case, the appearance parameter is a hybrid state consisting of an index to the most likely library model and the parameter estimates for that model, if there are any. That is, $\theta^m_t \rightarrow [\kappa^m_t \theta^m_t]$. The estimate is formed by optimising the appearance auxiliary function for each model and selecting the model which gives rise to the highest auxiliary function value, namely

$$[\hat{\kappa}^m_t, \hat{\theta}^m_t] = \arg\max_{\kappa} \mathcal{Q}^{m,\kappa}_\theta, \tag{8.53}$$

where $\hat{\kappa}^m_t$ is the estimated library index and

$$\mathcal{Q}^{m,\kappa}_\theta = \log\left\{p(\theta^m_0)\right\} + \sum_{t=1}^{T} \frac{||z_t||}{H^{\Omega}_{\Theta_t}} \log\left\{p\left(\theta^m_t | \theta^m_{t-1}\right)\right\}$$
$$+ \sum_{t=1}^{T} \sum_{i} \frac{\zeta^i_t}{H^m_{ti}} \int_{B_i} h^\kappa\left(\tau | \hat{\mathbf{x}}^m_t, \hat{\theta}^m_t\right) \log\left\{h^\kappa\left(\tau | \mathbf{x}^m_t, \theta^m_t\right)\right\} d\tau. \tag{8.54}$$

For the case where the library is a collection of fixed template images, the auxiliary function simplifies significantly to

$$\mathcal{Q}_{\theta}^{m,\kappa} = \sum_{t=1}^{T} \sum_{i} \frac{\zeta_t^i}{H_{ti}^m} \int_{B_i} \vartheta_i^{m\kappa} \log \left\{ \vartheta_i^{m\kappa} \right\} d\tau. \tag{8.55}$$

This situation arises in the problem of joint tracking and classification. For example, consider a road surveillance problem where the tracker has a database of vehicle image templates (for cars, trucks and so on) and the aim is to simultaneously detect, localise and classify vehicles.

8.8 Correlated Kinematics and Appearance

Every appearance model described in this chapter assumes that the appearance is independent of the dynamics. In some cases, this is a valid assumption, but in others it is not. For example, one of the imagery applications in Chap. 12 is observing stars with an optical camera. In that application, the sensor moves while the aperture is open and the stars smear. Between frames, this same motion makes it seem that the stars are moving: the target dynamics and appearance are both functions of the camera motion. Another example is the motion of a vehicle on a road, which we might expect to be highly correlated with the orientation of the vehicle: cars usually move in the direction that the wheels point. There is no real need to modify the method for this situation. The appearance models all have access to an estimate of the kinematic state and treat this as truth, so the kinematic state can be used as a model parameter if required. When the kinematic state and the appearance are independent, then the effect of the state is to shift the target in the image plane. In the case where orientation is assumed to be the same as heading, then the kinematic state can also be used to rotate the appearance in the image plane. After this translation and rotation, the dependence on the states is removed and the estimation proceeds as described above. More complex interactions between kinematic state and appearance may arise but the details will vary with the application.

8.9 Simulated Examples

We now illustrate some of these appearance models through simulations. These examples focus on appearance estimation, so we will use simpler scenarios than the canonical multi-target scenario. The target SNR will also generally be high because appearance estimation requires better signal discrimination. We need to introduce a new performance measure to quantify the appearance estimation quality. This is the mean Kullback–Leibler divergence between the true appearance and the estimated appearance. For the case of two zero-mean Gaussian distributions with covariances Σ and $\hat{\Sigma}$, the Kullback–Leibler divergence is

$$\mathsf{div} = \frac{1}{2} \left[\log_e |\hat{\Sigma}| - \log_e |\Sigma| + \mathsf{trace} \left(\hat{\Sigma}^{-1} \Sigma \right) + 2 \right], \tag{8.56}$$

and this was averaged over Monte Carlo trials. For the case where at least one of the distributions was discrete, the divergence was calculated directly using

$$\text{div} = \sum_{j=1}^{N_\theta} \left(\vartheta^j \log_e\{\vartheta^j\} - \vartheta^j \log_e\{\hat{\vartheta}^j\} \right), \tag{8.57}$$

where ϑ^j was determined by integrating $g(\mathbf{y}|\cdot)$ over the appearance cell j for the case where the true appearance function or estimate was Gaussian. Again, this was averaged over Monte Carlo trials.

The first set of simulations considers targets that have a true appearance that is Gaussian. We compare the deterministic Gaussian approach described in Sect. 8.1 with the random matrix approach described in Sect. 8.2. These comparisons were first presented in [16, 17]. The second set of simulations looks at the performance of the Dirichlet model from Sect. 8.6 for non-Gaussian appearance targets and compares this with Gaussian-based methods. These comparisons were first presented in [1, 4].

8.9.1 Gaussian Targets

The first set of simulations considers targets with a Gaussian true appearance function. First, the appearance estimation performance is investigated using a single target scenario. Second, we illustrate how appearance estimation can improve kinematic state estimation performance when two targets move close together. The simulation parameters are shown in Table 8.1: the first scenario contained only target 1, the second contained both targets. Each frame consists of an image with 100 × 100 cells with Gaussian-shaped simulated objects in unit variance Rayleigh noise.

In the single target scenario, the minor and major axes of the appearance ellipse were aligned with the target velocity direction and perpendicular to the velocity vector respectively. The size of the appearance ellipse had a 1-sigma longitudinal extent of two cells along the velocity vector and a transverse extent of 10 cells. Figure 8.4b shows an example realisation of the first simulated data. The true object data is plotted in Fig. 8.4a. The duration of each sequence is 70 scans. A process noise of $\mathbf{v}t = 1.0$ cell was applied to the true positions as shown with a red line. The disturbed positions are shown as blue crosses. The true Gaussian object extent is drawn at three example scans. The situation is challenging, since the object is making a turn and the signal amplitude is relatively weak. 500 Monte Carlo sequences of data were simulated.

The single target sequence was processed by the following tracking algorithms:

- the H-PMHT-RM developed in Sect. 8.2;
- the H-PMHT with independent deterministic extent estimates described in Sect. 8.1, labelled H-PMHT-IE;
- core H-PMHT, which has a fixed object extent;

Table 8.1 Simulation parameters

	Target 1	Target 2
Start position	[20, 55]	[37, 10]
End position	[71, 67]	[58, 76]
Speed	1 pixel/frame	1 pixel/frame
Start heading	$-\pi/8$	0.4π
Turn start	Frame 35	N/A
Turn end	Frame 44	N/A
Acceleration	0.1 pixels/frame^2	0 pixels/frame^2
Amplitude	2.5	2.5
1-σ length	2 pixels	2 pixels
1-σ width	10 pixels	2 pixels
Extent initialisation	$\mathsf{X}_0^m = 4 \times \mathsf{I}$ and $\nu_0^m = 4$	
Kinematic initialisation	Two point difference	
π initialisation	$\hat{\pi}_0^m = 0.1$	
Detector threshold	3 × image mean	
Image size	100 × 100 pixels	
Process noise	$\mathsf{Q}t = \begin{bmatrix} 1 & 1 \\ 0 & 1 \end{bmatrix}$	
Convergence test	$\left|\hat{\mathbf{x}}t^{m(it)} - \hat{\mathbf{x}}t^{m(it-1)}\right|^2 < 0.001$	
Clutter density (PDAF)	Num. measurements/area	
Promotion threshold	1 dB	
Termination threshold	−8 dB	

- point measurement PMHT, labelled PT PMHT;
- PDAF for point objects and point measurements.

The PMHT and H-PMHT approaches were all applied with a batch length of 1 scan, so there was no smoothing step, also referred to as retrodiction. In the implementations of H-PMHT-RM and H-PMHT-IE, the initial target extent was set to a scaled identity matrix $\Sigma_0^m = 4 \times \mathsf{I}$ with an initial Wishart parameter $\nu_0^m = 4$ for each new track. All other approaches were run with a fixed extent matrix Σ. Each of the H-PMHT variants and the PMHT used a convergence test based on the track state estimates: the iterations were declared converged when the squared error between state estimates in successive iterations for all tracks was less than a threshold of 0.001.

The point measurement trackers require a single-frame detector to extract measurements from the simulated images. The target appearance in this example is relatively broad, so a simple threshold detector would produce multiple point measurements on each target. Instead, we use an image segmentation-based approach similar to [9]. The single-frame detector reports a centroid of pixels that have been associated together. For a low contrast target, the membership of each pixel fluctuates from

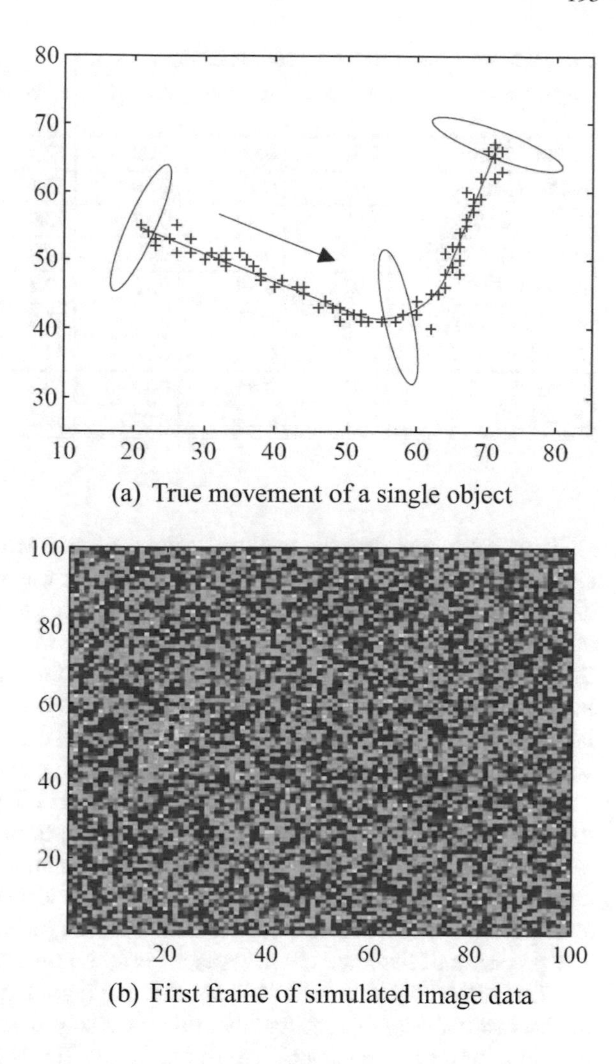

Fig. 8.4 Scenario 1: a single object in Rayleigh noise

frame to frame, which means that the location of the centroid can be quite noisy. It is also possible that the target response can be incorrectly segmented into multiple point measurements. This is relatively uncommon for the Gaussian-shaped targets in this section, but is more frequent for the irregular shapes in the next section and in the imagery applications. This is referred to as over-segmentation and breaks the single-measurement assumption used in the PDAF. Over-segmentation can lead to duplicate tracks and apparent target manoeuvres. The point measurement PMHT can assign multiple measurements to each track, so is less affected. All of the H-PMHT variants used the track management strategy described in Chap. 5. The discretised continuous time almost constant velocity kinematic model was used within each of the five tracking approaches.

Table 8.2 Tracking statistics for 500 simulations of scenario 1

Algorithm	Av. total	Av. true	Av. lost	Lifetime	Birth time	CPU
H-PMHT-RM	1.01	0.99	0.02	62.9	8.0	2.1 ± 0.2
H-PMHT-IE	1.07	0.22	0.86	55.6	10.4	2.8 ± 0.5
$\bar{\Sigma}_t \leftarrow 0.2\bar{\Sigma}_t$	1.01	0.90	0.12	63.0	8.0	
H-PMHT, $\Sigma = 2^2\mathrm{I}$	3.01	0.84	2.17	17.7	30.1	1.03 ± 0.05
$\pi_t^0 \leftarrow 10^{-8}\pi_t^0$	1.07	0.89	0.18	63.1	9.6	
PT PMHT, $\Sigma = 2^2\mathrm{I}$	2.76	0.57	2.20	24.5	27.3	1.04 ± 0.03
$\Sigma = 5.5^2\mathrm{I}$	1.04	0.84	0.20	62.7	8.6	
PDAF, $\Sigma = 2^2\mathrm{I}$	1.01	0.84	0.17	59.3	11.7	1 ± 0.03
$\Sigma = 5.5^2\mathrm{I}$	1.01	0.94	0.07	59.3	11.7	

Table 8.2 summarises the trackers' outputs for 500 Monte Carlo realisations of the single target scenario. Whereas the H-PMHT-RM was implemented as advertised, each of the other algorithms needed a little tweaking to get good performance. The table gives the performance of both the direct and tweaked implementations. It lists the total number of tracks formed, averaged over Monte Carlo trials and also identifies how many of these were valid target tracks and the number of lost tracks. A track is labelled as a lost track if it started with the target, but then any of its position estimates was more than $2.0\times$ the true extent away from the true position. A true track is a track whose estimates are all not lost. None of the trackers produced false tracks because the target extent is relatively large and the single-frame peak detection algorithm exploited this to achieve an extremely low false alarm rate. The track lifetime is the average number of scans for which the target has a track and the birth time is the average number of frames for the algorithm's track management to promote a track. The precision of the track count statistics is approximately 0.02 tracks per trial.

The H-PMHT-RM results indicate that the algorithm generated only one track on the vast majority of runs and that this track stayed within a tolerable distance of the truth. Similar performance was found for the PDAF, although it required a fairly high measurement noise covariance. The point PMHT produced too many duplicate tracks when the measurement noise was low because it tended to put multiple tracks on the target at one time. This effect was reduced by using a wider measurement noise because the PMHT uses only the measurement noise for data association. The H-PMHT-IE showed a high proportion of lost tracks. This occurred because the Kalman filter part of the algorithm treats the estimated shape matrix as the measurement error. In this example, the target is relatively large so the state estimator tended to discount the measurements in preference for the prediction, leading to an overshoot response when the target turned. This was mitigated by scaling the estimated shape matrix in the state estimation stage by a factor of 0.2. The deflated covariance caused the state estimates to follow the measurements more closely and the rate of lost tracks was greatly reduced. For the core H-PMHT, a smaller covariance matrix was used

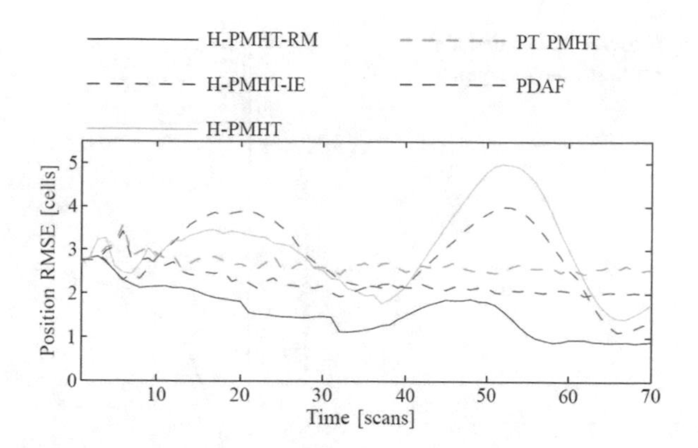

Fig. 8.5 Position RMSE over all non-lost position estimates

so that the state estimate could follow the turn but this lead to too many duplicate tracks. In this case, the remedy was to artificially deflate the clutter proportion π_t^0 so that the target energy was still assigned to a single track without influencing the state estimator's dynamic response.

For completeness, the relative CPU costs of each algorithm are listed in Table 8.2. These numbers are somewhat deceptive for two reasons: first, the highest cost process was the peak detector, which was used by all methods. This cost dominated all of the point measurement tracker costs. Second, the H-PMHT-RM and the H-PMHT-IE were implemented using the Reimann sum approximation, which was evaluated over the whole image instead of within a gate. Because there was no gate, this sum was also very expensive and so the H-PMHT-RM and H-PMHT-IE costs were approximately the peak detector plus the number of EM iterations multiplied by the Reimann sum cost. The H-PMHT-RM converged in an average of 5.2 iterations, whereas the H-PMHT-IE took an average of 8.1 iterations, which accounts for the CPU cost difference. It is anticipated that both of these bottlenecks would be substantially improved by optimised software.

Figure 8.5 shows that the H-PMHT-RM also has the best track accuracy if the root mean square error (RMSE) for all non-lost estimates is compared. Figure 8.6 refers only to the extended object approaches and plots the average length estimates of the major semi-axis and the variance of this length estimate. The angle estimate performance for the two algorithms was almost identical and is not presented. The extent is slightly underestimated by both the H-PMHT-RM and the H-PMHT-IE. Both approaches start their tracks with a small default circular extent with radius 2 that is then expanded to find the extent estimates based on the measurements. Therefore, the corresponding curve in Fig. 8.6a starts at a length of 2 cells and then increases until it reaches the final estimate. Both H-PMHT variants underestimate the target extent because the tails of the target response are associated with the clutter model and the covariance of this truncated distribution is smaller. This implies that it might be possible to correct for this bias, but we have not attempted this. The H-PMHT-RM curve is slower to converge to the biased estimate since its estimate enforces dynamics. The variance plots show that the H-PMHT-RM estimates are much more

Fig. 8.6 Estimated major semi-axis and variance

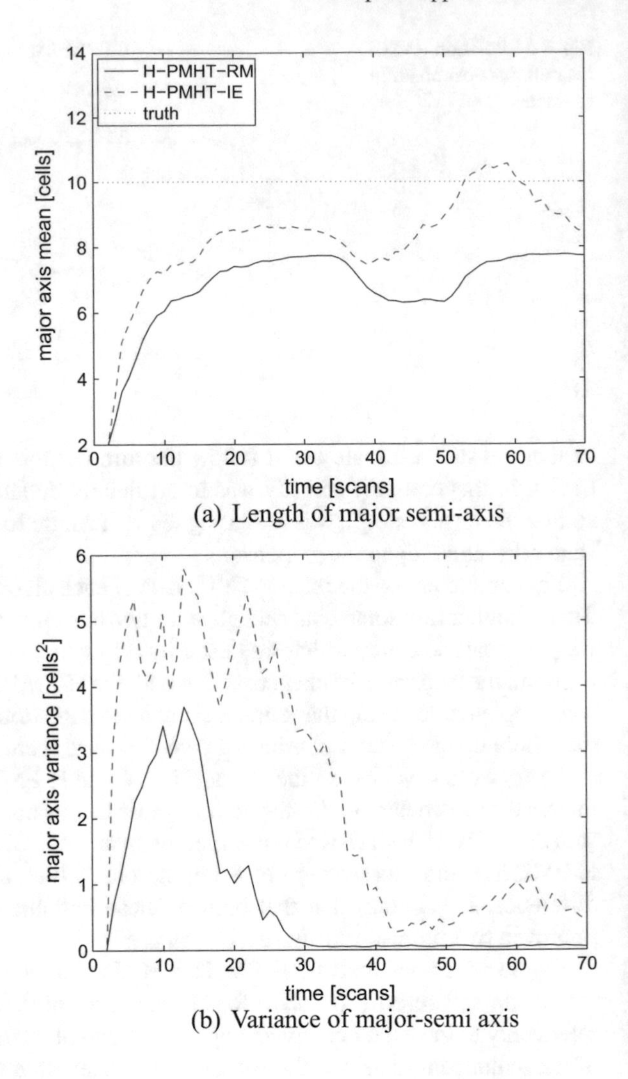

(a) Length of major semi-axis

(b) Variance of major-semi axis

consistent across the Monte Carlo trials and that the extent estimate converges to a very stable value after sufficient measurement data has been assimilated.

In another experiment, a second object of rather small extent was added to the scenario. The small object had a Gaussian appearance with a length and width of two cells. It crosses the big object and then continues moving in parallel with it. Figure 8.7 shows the true trajectories of the two targets. The paths of the target centres are marked with solid lines and dotted lines mark the 1-sigma widths of the targets. From frame 26 to frame 36, the smaller target is contained within the 1-sigma extent of the original target even though their two centres are never co-located. A count of the number of track swaps was added to performance metrics for the first scenario and the values obtained are presented in Table 8.3. A track is counted as a swapped track if

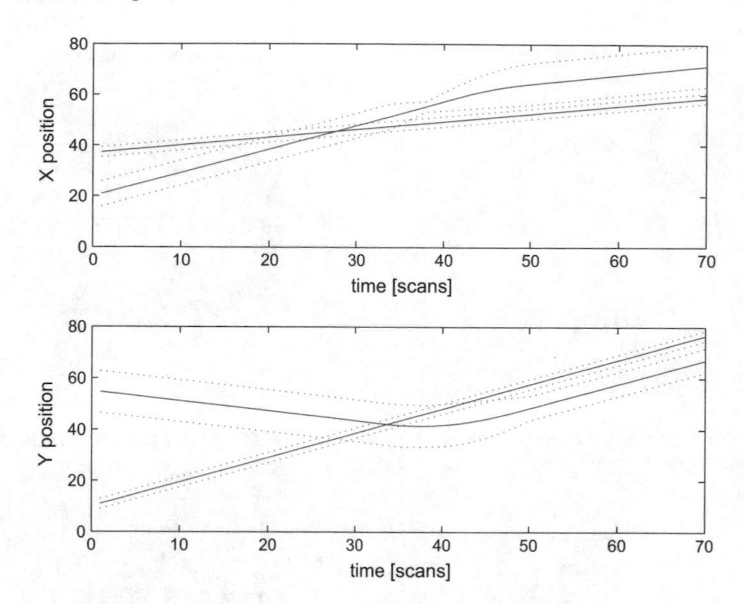

Fig. 8.7 Scenario 2: Two crossing objects in Rayleigh noise

Table 8.3 Tracking statistics for 500 simulation runs of scenario 2

Algorithm	Av. total	Av. true	Av. lost	Av. false	Av. swaps
H-PMHT-RM	2.03	1.60	0.21	0.00	0.22
H-PMHT-IE, tweaked	2.16	0.97	0.41	0.07	0.61

it changes the associated object during its lifetime. For this experiment, the factor for checking the track loss was increased from 2.0 to 2.5. Again, 500 sequences of image data were generated. The lifetime and birth measures are not reported because they are almost the same for the two algorithms. The H-PMHT-RM produced fewer track swaps than H-PMHT-IE and is able to distinguish better between the two objects. This occurs because the Wishart prior used by the H-PMHT-RM allows the algorithm to remember the shape it has previously seen and less of the energy from the broad target becomes associated with the straight line narrow one. In the H-PMHT-IE, the shape estimation is time independent, so once the two targets are unresolved, both tracks tend to equally share all of the target energy and the track on the narrow target is more easily diverted by the broader target.

8.9.2 Non-Gaussian Targets

We now illustrate the Dirichlet appearance model that uses an image template to learn an arbitrary appearance function. This model is compared with H-PMHT-RM for Gaussian and non-Gaussian appearance targets. Two different types of

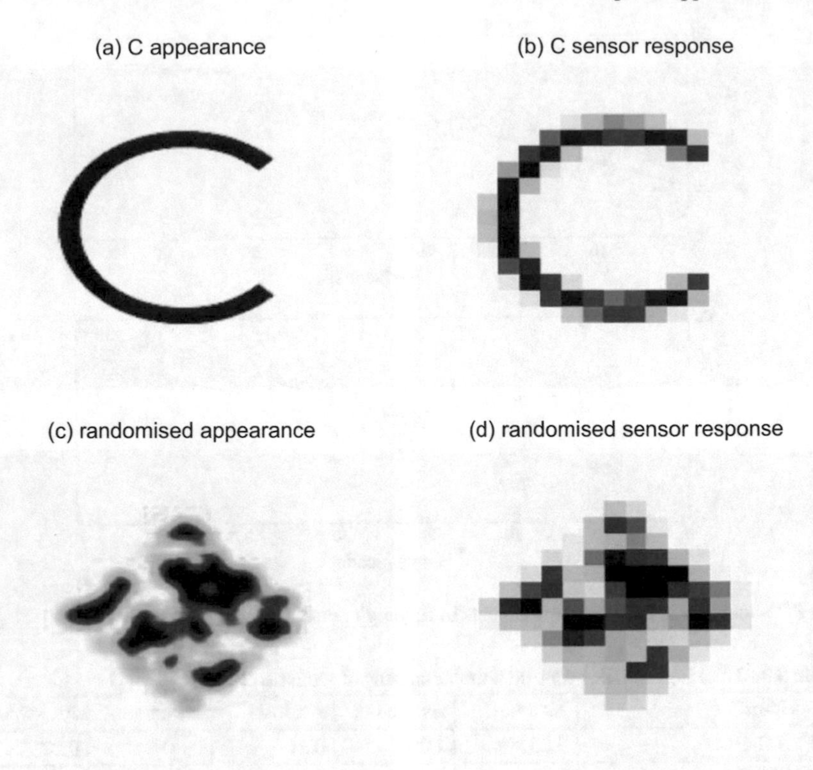

Fig. 8.8 Non-Gaussian appearance targets

non-Gaussian targets are considered: the C-target introduced in Chap. 7 and a randomised appearance function that has features over a range of different scales of granularity.

The C-target is defined by the appearance function

$$h(r, \theta) = \begin{cases} A & \text{if } 5 \geq r \geq 6 \text{ and } |\theta| > \frac{\pi}{4}, \\ 0 & \text{otherwise}, \end{cases} \tag{8.58}$$

where A is a normalising constant. This response is shown in Fig. 8.8a. The contribution of the target to each pixel is the integral of $h(r, \theta)$ over that pixel. An example of this is shown in Fig. 8.8b.

The randomised target was generated within a diamond defined by $|x| + |y| \leq 6$. A very fine grid binary image was created with equal probability of zero or unity for each pixel. This was integrated over the N_θ grid, smoothed using a rectangular kernel and truncated below 0.5. An example realisation is shown in Fig. 8.8. Again, the full resolution version is shown in Fig. 8.8c and a single frame at sensor resolution is shown in Fig. 8.8d.

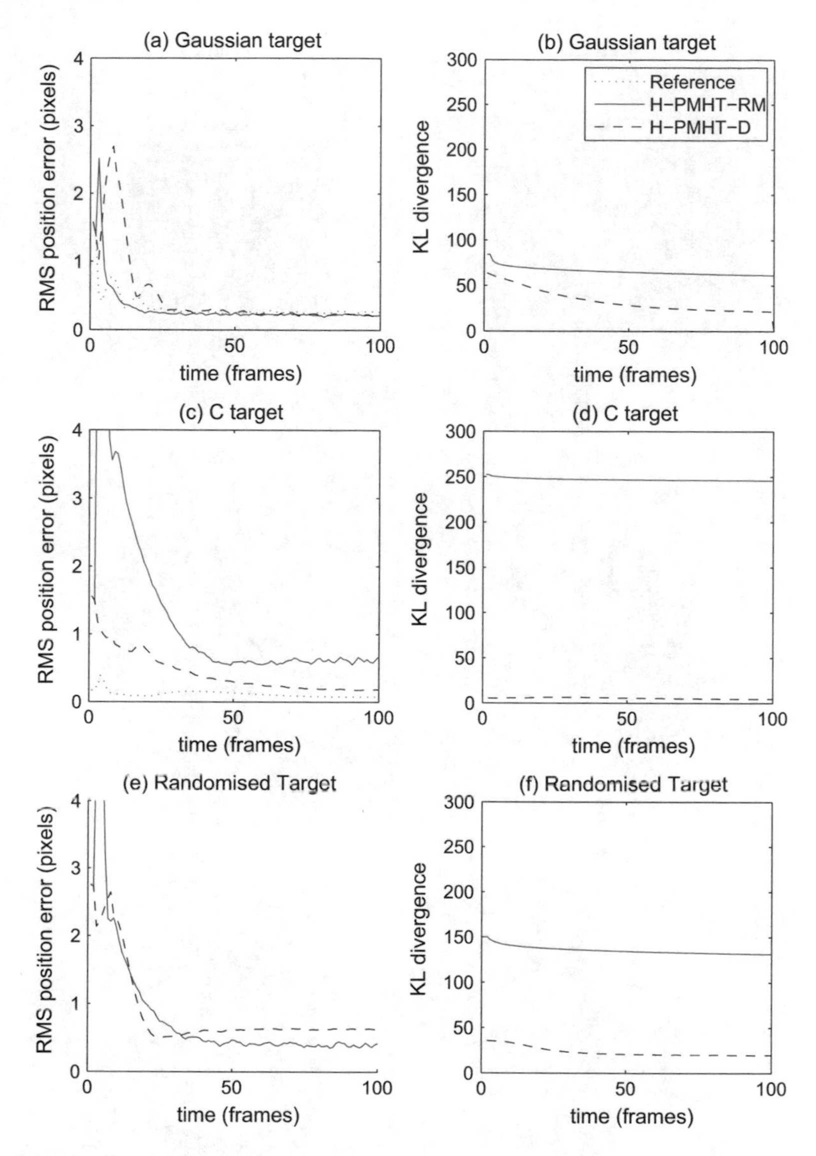

Fig. 8.9 Dirichlet H-PMHT appearance and state estimation performance

The RMS estimation error and mean Kullback–Leibler divergence for the three different appearance functions are shown in Fig. 8.9. The left column is the state estimation error and the right column is the appearance error. For the deterministic appearance functions, a reference curve is drawn in the state estimation plots that corresponds to the performance achieved when the appearance function is known. In each case, the H-PMHT-RM converges to an appearance estimate very quickly because the symmetrical matrix it uses to represent shape has only three parameters.

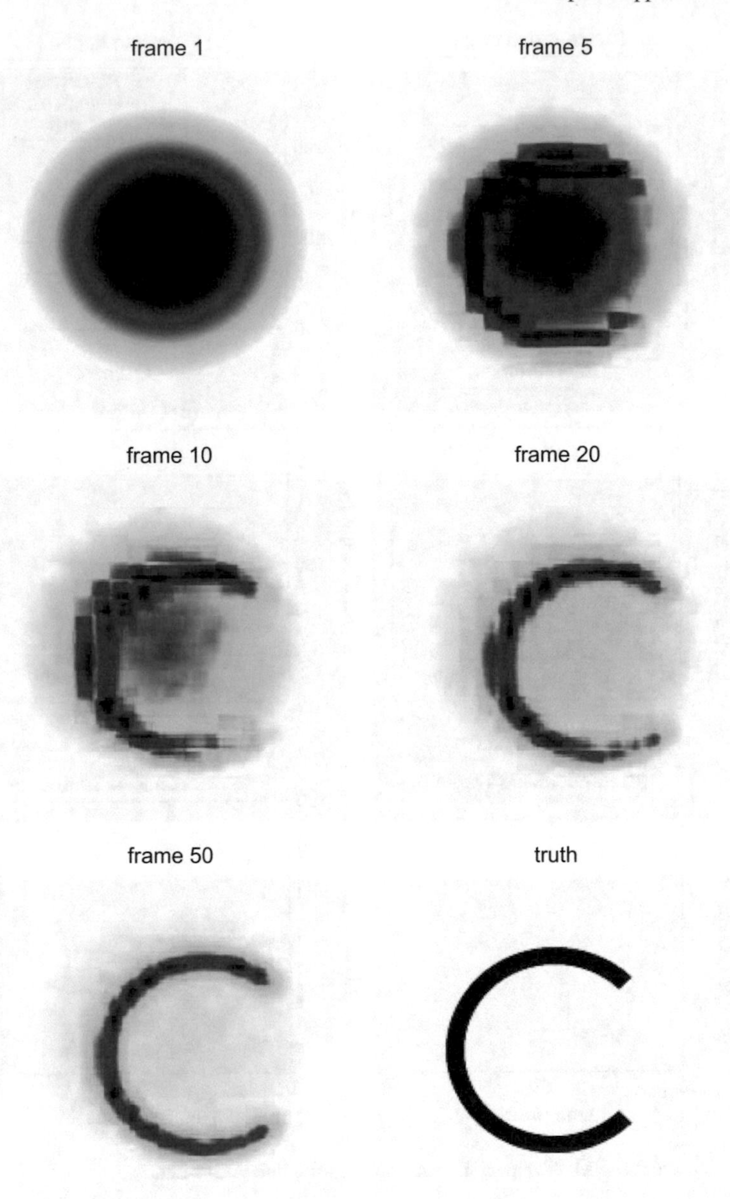

Fig. 8.10 Dirichlet H-PMHT appearance estimates for C target: selected frames and true appearance

In contrast, the Dirichlet H-PMHT uses a template that is hundreds of cells across. It has a very large number of parameters that take more frames to converge. Once it has converged, the Dirichlet H-PMHT gives very good estimates of the appearance.

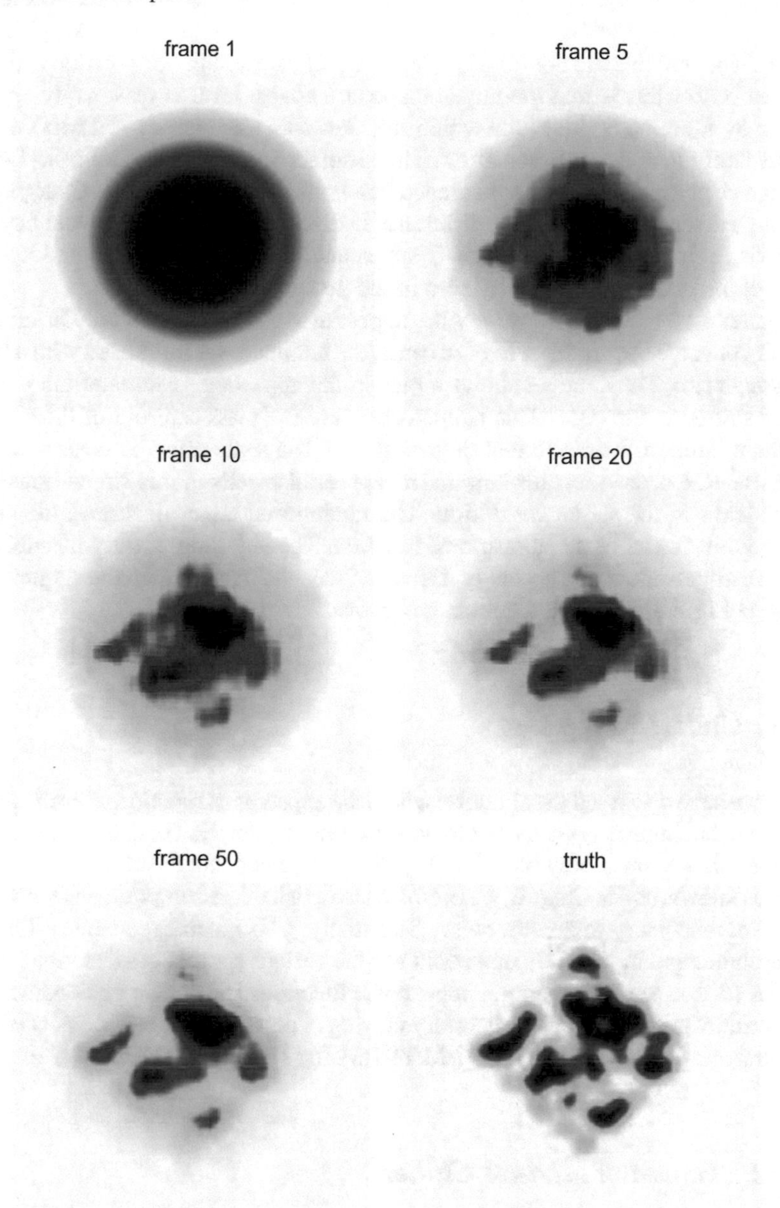

Fig. 8.11 Dirichlet H-PMHT appearance estimates for randomised target: selected frames and true appearance

An interesting feature of the randomised case is that the H-PMHT-RM and Dirichlet H-PMHT give similar position estimation performance even though the target shape is not Gaussian. The Dirichlet H-PMHT did a much better job of estimating the target shape, but this did not translate into a significantly superior position estimate. The reason for this is that the target shape at a coarse level is reasonably approximated by a unimodal blob, for which the Gaussian assumed by H-PMHT-RM is a sufficiently detailed representation. The absolute values of the Kullback–Leibler divergence between different appearance functions show that the letter-C target gave very high divergence and that the difference in divergence was much greater between methods. This is because the letter-C appearance is so non-Gaussian and because many of the pixels are identically zero in the true appearance function.

Figures 8.10 and 8.11 show how the appearance estimates from the Dirichlet H-PMHT converge over frames. In both cases, the template was initialised with a broad Gaussian prior. The C target shows a bias in the appearance estimate: the C looks a little shifted to the right. This happens because the mass centroid of the shape is not the middle of the circle that defines the C. The same effect is observed when H-PMHT-RM is used for this target. The appearance cells in this simulations were at one-tenth of the sensor resolution. The randomised target illustrates the ability to learn fine details of the appearance function. The algorithm clearly has not been able to estimate the shape perfectly. However, when compared with the single-frame image in Fig. 8.8, there is a significant improvement in detail.

8.10 Clutter Mapping

So far we have discussed the situation where the appearance function of each target is unknown, but implicitly we have retained an earlier assumption that the clutter spatial distribution is known. Way back in Chap. 4, we assumed that the clutter distribution $g^0(\mathbf{y})$ is known, and in Chap. 6, we promised to revisit this assumption once there was sufficient machinery to do otherwise. Essentially, $g^0(\mathbf{y})$ is the appearance function of the clutter and in principle one could use the various approaches described in this chapter on it as well as the target appearance function. The challenge becomes how to discriminate spatially and temporally varying clutter from targets. We now discuss two strategies that are based on H-PMHT-RM and Dirichlet H-PMHT.

8.10.1 Gaussian Mixture Clutter

The extended problem is now to also create a time-varying estimate of the pdf of clutter measurements. In the general case, the pdf at any instant in time will not conform to any convenient functional form. However, it is often not necessary to accurately model the clutter. Instead, the aim is to detect targets, and we introduce a clutter map because performance is poor without it. In this situation, the model needs to be good enough to prevent the tracker from making too many false tracks without suppressing valid tracks. Approximating the spatial distribution as a Gaussian mixture is frequently sufficient.

The principle behind a Gaussian mixture clutter model is that we use H-PMHT-RM to track everything that departs from uniformity in the video sequence. Some of the resulting tracks will be components of the clutter model and others will correspond to targets. The tracker itself does not discriminate between the different components, except that some are established tracks with higher priority access to the imagery. Tracks are presented to the user at the output interface if they conform to a collection of prior rules. This was applied in the video applications to follow [5], where the sensor geometry was approximately known and the expected size of a target could be specified: vehicle tracks are declared when the size is between upper and lower bounds, and when the estimated SNR is sufficiently high.

8.10.2 Markov Random Field Clutter

The computation cost of H-PMHT-RM is proportional to the number of tracks in the mixture, which is low when each track represents a target. However, the number of tracks retained by the algorithm can be very much larger when tracks are used to represent clutter. The use of a Gaussian mixture allows the model to be highly adaptive but it also makes it difficult to impose smoothness constraints. An alternative is to use a grid-based representation of the clutter. This is essentially approximating the true underlying clutter density with a piecewise constant function. Smoothness can be imposed on the resulting estimate by modelling the value in each cell as dependent on the neighbouring values. This kind of model is often referred to as a Markov random field.

8.10.3 Clutter Mapping Simulations

The Gaussian mixture approach to clutter mapping is now illustrated through simulation. We extended the canonical multi-target scenario by modifying the background characteristics, the target behaviour remained the same. The simulated noise was first elevated in the first 50 rows of the image. Recall that the standard noise is Rayleigh distributed: we made the noise variance in the first 50 rows 36 times higher than the rest of the image. A stationary but time-varying patch was introduced by adding a Gaussian with an evolving amplitude. This Gaussian was centred at $(100, 100)$ with a spread variance of $\Sigma = \begin{bmatrix} 2500 & 0 \\ 0 & 2500 \end{bmatrix}$; its amplitude was given by $\Lambda_t = \pi |\Sigma|^{0.5} \cos^2(\frac{2\pi t}{200})$. Two moving components were also introduced with paths defined by

$$\begin{bmatrix} X_t \\ Y_t \end{bmatrix} = \begin{bmatrix} 100 + 150 \cos\left(\frac{2\pi t}{200} \times \frac{1}{4}\right) \\ 150 + 150 \cos\left(\frac{2\pi t}{200} \times \frac{1}{4} + \frac{\pi}{20}\right) \end{bmatrix}, \qquad \Sigma = \begin{bmatrix} 3600 & 0 \\ 0 & 3600 \end{bmatrix};$$

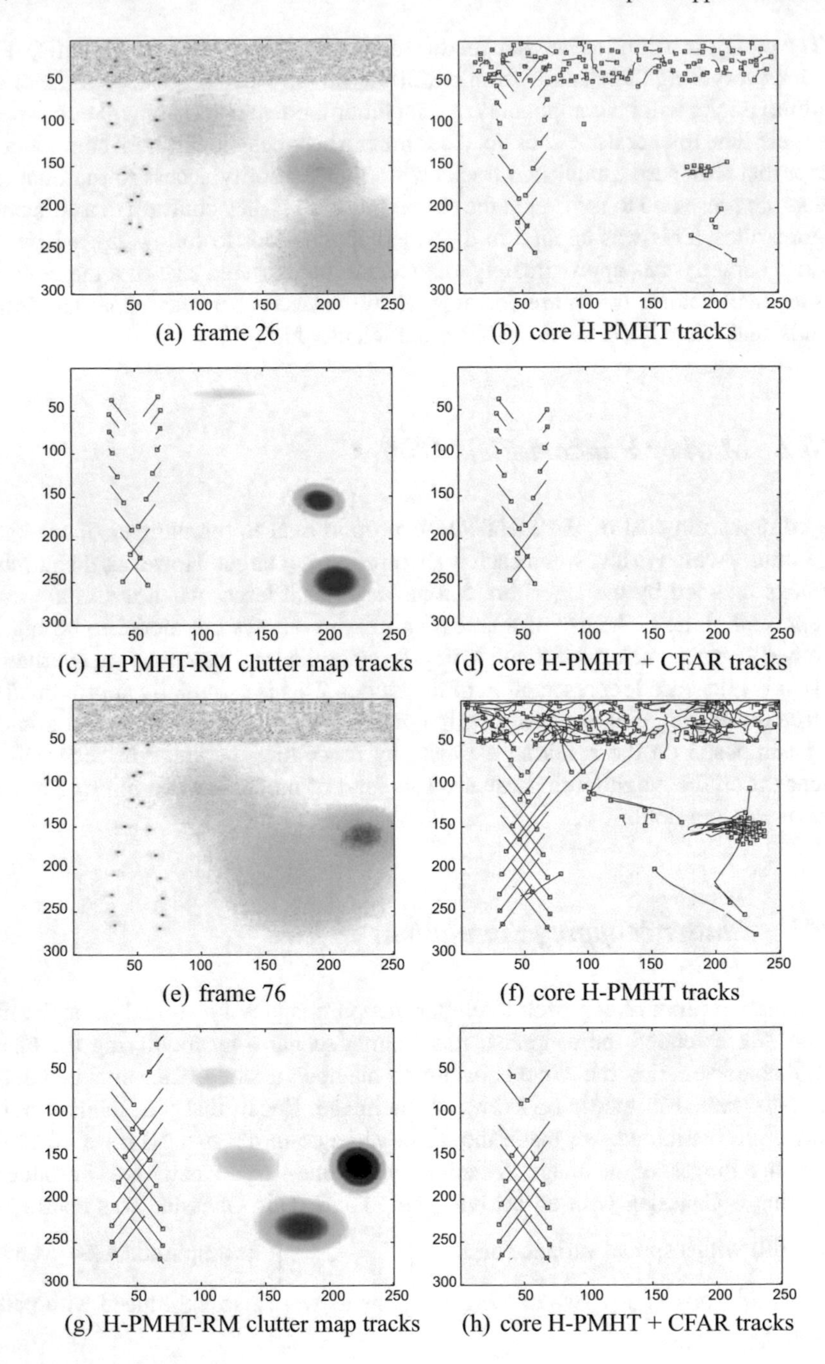

(a) frame 26
(b) core H-PMHT tracks
(c) H-PMHT-RM clutter map tracks
(d) core H-PMHT + CFAR tracks
(e) frame 76
(f) core H-PMHT tracks
(g) H-PMHT-RM clutter map tracks
(h) core H-PMHT + CFAR tracks

Fig. 8.12 Clutter mapping

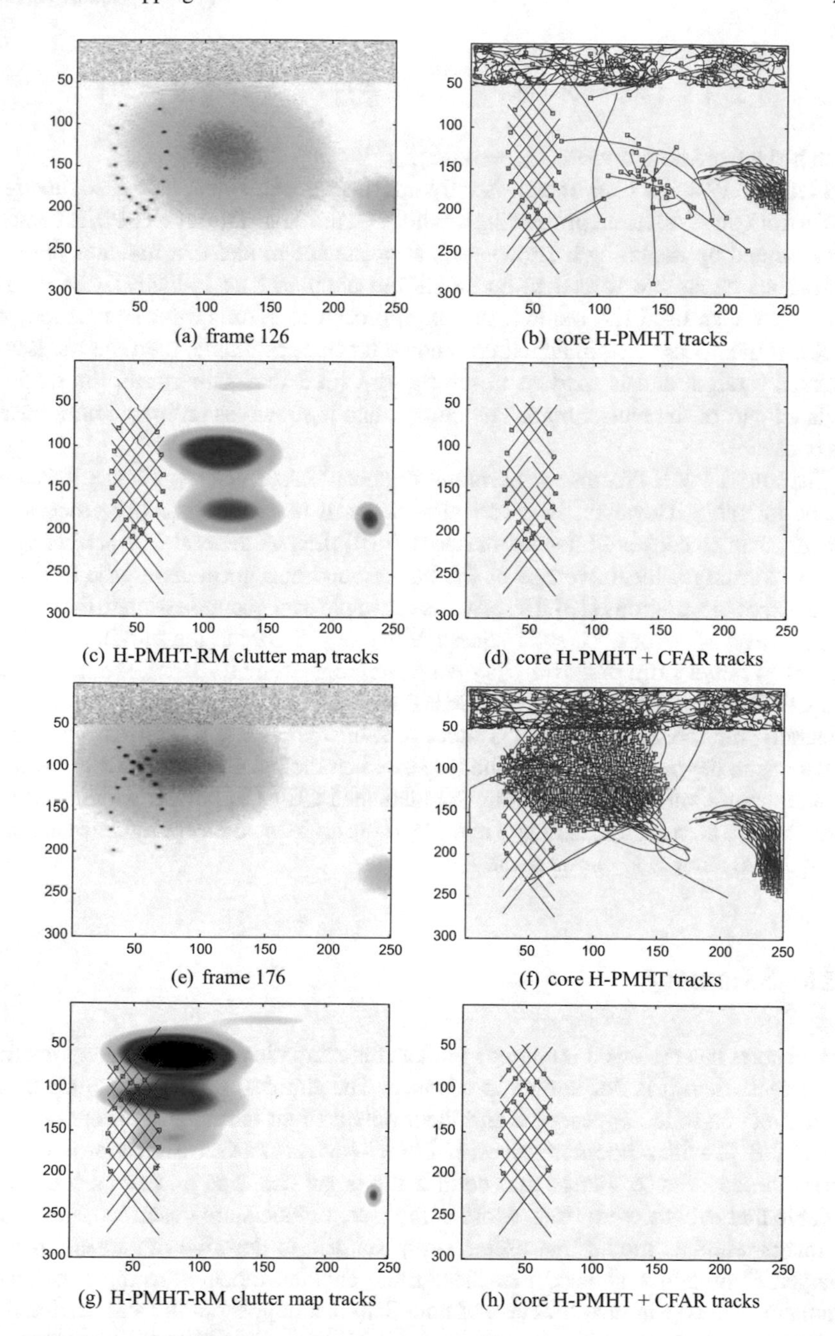

(a) frame 126 (b) core H-PMHT tracks

(c) H-PMHT-RM clutter map tracks (d) core H-PMHT + CFAR tracks

(e) frame 176 (f) core H-PMHT tracks

(g) H-PMHT-RM clutter map tracks (h) core H-PMHT + CFAR tracks

Fig. 8.13 Clutter mapping

$$\begin{bmatrix} X_t \\ Y_t \end{bmatrix} = \begin{bmatrix} 125 + 125\cos\left(\frac{2\pi t}{200} \times \frac{1}{4}\right) \\ 200 - 50\cos\left(\frac{2\pi t}{200} \times \frac{1}{2} + \frac{\pi}{20}\right) \end{bmatrix}, \qquad \Sigma = \begin{bmatrix} 400 & 0 \\ 0 & 400 \end{bmatrix}.$$

Both had a fixed amplitude of $\Lambda_t = 0.8\pi|\Sigma|^{0.5}$.

Figures 8.12 and 8.13 show selected frames from the sequence. They also illustrate tracks from three different processing methods on the data. The core H-PMHT tracks were formed by assuming a known target appearance model that matches the true targets: all tracks are forced to be small and circular. The H-PMHT-RM clutter mapping tracks used the random matrix approach to estimate the appearance of tracks. If the appearance matched the known target appearance, then the track was declared a target and is marked in the figure with a line. Otherwise, the track is declared part of the clutter map. The clutter map is shown as an image underneath the tracks.

The core H-PMHT forms an enormous number of false tracks which would clearly not be tolerable. However, in practical systems, it is common to pre-process the imagery using a constant false alarm rate (CFAR) filter. A general CFAR filter operates by forming a local average of the background and normalising the image to remove spatial variations [13]. Here, we use a simple background estimate that makes a spatial average over a 21 × 21 square with a 5 × 5 hole in the middle. The hole prevents a target from distorting its own noise floor estimate. Tracks formed using this CFAR processing and core H-PMHT are also shown in Figs. 8.12 and 8.13. Generally, the CFAR processing is successful in suppressing the false tracks but it also tends to degrade tracking on the targets when the local background is elevated around them. Quite possibly a more sophisticated CFAR approach would perform better but the point of this example is more to illustrate the concept rather than make an objective numerical comparison.

8.11 Summary

This chapter has reviewed numerous models for estimating the appearance function of a target when it is not known in advance. The simplest models use a Gaussian approximation to the appearance and incur only a slight increase in execution cost. The H-PMHT with random matrices, H-PMHT-RM, uses a Gaussian approximation where the covariance matrix that defines the target shape is treated as a random variable that evolves with time. More complicated models are described, including an image-template model that uses a sub-pixel grid to describe the appearance of a target. Simulations showed that these more detailed models are much better at estimating the appearance function of non-Gaussian targets but the state estimation performance is often not significantly degraded by a poor appearance estimate.

References

1. Davey, S.J., Wieneke, M.: H-PMHT with an unknown arbitrary target. In: Proceedings of ISSNIP 2011 (2011)
2. Davey, S.J., Wieneke, M.: Tracking groups of people in video with histogram-PMHT. In: Defence Applications of Signal Processing (2011)
3. Davey, S.J., Wieneke, M., Gordon, N.J.: H-PMHT for correlated targets. In: proceedings of SPIE Signal and Data Processing of Small Targets, vol. 8393, 83930R. Baltimore, USA (2012)
4. Davey, S.J., Wieneke, M., Vu, H.X.: Histogram-PMHT unfettered. IEEE J. Sel. Top. Signal Process. **7**(3), 435–447 (2013)
5. Davey, S.J., Vu, H.X., Arulampalam, S., Fletcher, F., Lim, C.C.: Clutter mapping for histogram PMHT. In: Statistical Signal Processing, pp. 153–156. Gold Coast, Queensland (2014)
6. Dempster, A.P., Laird, N.M., Rubin, D.B.: Maximum likelihood from incomplete data via the EM algorithm. J. Royal Stat. Soc. **140**, 1–38 (1977)
7. Feldmann, M., Franken, D., Koch, W.: Tracking of extended objects and group targets using random matrices. IEEE Trans. Signal Process. **59**(4), 1409–1420 (2011)
8. Fisher, R.: PETS04 surveillance ground truth data set. In: Proceedings of the Sixth IEEE International Workshop on Performance Evaluation of Tracking and Surveillance, pp. 1–5, 2004
9. Haralick, R.M., Shapiro, L.G.: Computer and Robot Vision. Addison–Wesley, Reading (1992)
10. Howie, J.M.: Real Analysis. Springer, Berlin (2001)
11. Koch, W.: Bayesian approach to extended object and cluster tracking using random matrices. IEEE Trans. Aerosp. Electron. Syst. **44**(3), 1042–1059 (2008)
12. Li, X., Hu, W., Shen, C., Zhang, Z., Dick, A., van den Hengel, A.: A survey of appearance models in visual object tracking. ACM Trans. Intell. Syst. Technol. **4**, 1–48 (2013)
13. Skolnik, M.I.: Introduction to Radar Systems. McGraw-Hill, NewYork (2001)
14. Streit, R.L.: Tracking on intensity-modulated data streams. Technical report 11221, NUWC, Newport, Rhode Island, USA (2000)
15. Streit, R.L., Graham, M.L., Walsh, M.J.: Multitarget tracking of distributed targets using histogram-PMHT. Digit. Signal Process. **12**(2), 394–404 (2002)
16. Wieneke, M., Davey, S.J.: Histogram PMHT with target extent estimates based on random matrices. In: Proceedings of the 14th International Conference on Information Fusion, Chicago, USA (2011)
17. Wieneke, M., Davey, S.J.: Histogram-PMHT for extended targets and target groups in images. IEEE Trans. Aerosp. Electron. Syst. **50**(3) (2014)

Chapter 9
H-PMHT with Attribute Data

The bulk of this book focuses on sensors that provide intensity maps as an observed image. The location of bright spots on the intensity map is related to the location of objects that we track, so the measurement gives an incomplete observation of the kinematic state. In some applications, the sensor can also provide non-kinematic information. For example, in passive sonar, the sensor observes the surrounding sound field that is a combination of environmental sounds and those emitted by the objects we seek to track. When each object has a unique sound signature, the sensor image can contain information about these signatures that can be used to identify the objects. In the sonar example, this identifying information is the spectral content of the energy within each beam. Another example is colour video imagery where the different objects in the scene have their own colour that can be used to identify them.

The term *attribute* data is used here as a general term to refer to non-kinematic information provided by the sensor. In the context of classification, this information can also be called *feature* data. In this book, we will not make the step of combining tracking and classification, although sometimes this is possible. In many cases, the attributes can help tracking even though they are not informative enough to lead to classification. For example, using the colour of vehicles in a road tracking application would help association but would probably provide little classification information. This chapter discusses how we can use this colour information in the H-PMHT framework. We work with the general attributes because colour is just a specific application example. Other types of attribute include the frequency spectrum, image texture or range profile. An important assumption about the behaviour of the attribute data is that it is independent of the kinematic data. For example, the colour of the vehicles in the road tracking application is independent of the locations of the cars. This will simplify the maths.

We broadly consider two types of attribute data. Additive attribute data combines together in the sensor image, which is the case in the sonar example: when the orchestra plays we hear a combination of all of the instruments. With obscuring

S. J. Davey and H. Gaetjens, *Track-Before-Detect Using Expectation Maximisation*, Signals and Communication Technology,
https://doi.org/10.1007/978-981-10-7593-3_9

attribute data, we only observe the attributes of one object in each location and it blocks the others. This is what happens with video imagery, we only get to see the object closest to the camera. The attributes of an object behave differently from its kinematic state because the attributes are usually fixed and they do not have to be localised in the attribute space. The attributes can be treated as part of the appearance state of each object since they dictate how the sensor responds to the object, but here we prefer to keep them separate. For this chapter, we will return to assuming that the appearance function is known, that is the spatial distribution of an object's energy is statistically known. In practice, we will often need to estimate the appearance and the attributes but the maths is derived in a modular fashion. The reader should be able to see how they can be combined and the applications in later chapters will do exactly this.

This chapter discusses two variations of H-PMHT that incorporate attribute data. The first of these is referred to as spectral H-PMHT and is designed to deal with additive attribute data. The second is referred to as obscuring H-PMHT and deals with obscuring attribute data. This chapter concludes with illustrations of these algorithms using simulated data.

9.1 Spectral H-PMHT

Spectral H-PMHT was developed by Streit et al. and applies to sensors that supply additive attribute data [3, 4]. Mathematically, the observed attributes are a superposition of the attributes of each object inside a pixel. The attribute data consists of a spectrum, which is a collection of bins in some non-kinematic space such as frequency. The sensor now collects energy over an augmented image whose pixels are the Cartesian product of spatial bins and spectral bins. Like the separable appearance function we used in Chap. 5, the Cartesian product reflects the idea that the spectral information is statistically independent of the spatial information.

We will work with the quantised shot measurements, each of which consists of a physical position $\mathbf{y}$ and a spectral location $\mathbf{a}$. Due to this assumed statistical independence, the measurement density can be written as

$$g\left(\mathbf{y}, \mathbf{a}|\mathbf{x}_t^m, \phi_t^m\right) \equiv g_X\left(\mathbf{y}|\mathbf{x}_t^m\right) g_A\left(\mathbf{a}|\phi_t^m\right), \tag{9.1}$$

where ϕ_t^m is the spectral state of target m, $g_X(\cdot)$ is the spatial appearance function and $g_A(\cdot)$ is the spectral appearance function. In most applications, the spectrum is fixed and we can drop the time index from ϕ^m. Again, we highlight the similarity between this expression and the two-dimensional factorised density in (5.20).

The per-target cell probability (4.9) directly factorises into a product of two integrals:

$$G^{m,i,j}\left(\mathbf{x}^m, \phi_t^m\right) = \left\{\int_{W^i} g_X(\mathbf{y}|\mathbf{x}^m)\mathrm{d}\mathbf{y}\right\}\left\{\int_{A^j} g_A(\mathbf{a}|\phi^m)\mathrm{d}\mathbf{a}\right\}$$
$$\equiv G_X^{m,i}(\mathbf{x}_t^m)G_A^{m,j}(\phi^m), \tag{9.2}$$

where A^j is the region of the attribute space spanned by spectral bin j.

Following our usual convention, $\mathbf{A}_t$ and $\mathbb{A}$ represent the collection of spectral measurements at frame t and across the batch respectively. Similarly, Φ_t and Φ are sets of spectral states. When the spectrum is fixed, these two are equivalent. The count data $n_t^{i,j}$ now defines the number of shots within spectral bin j and spatial bin i. The aim is to estimate the kinematic and spectral states, along with the Poisson rates, and the missing data are the shot assignments, the spatial shot locations and the spectral shot locations. The spectral H-PMHT auxiliary function is then

$$\mathcal{Q}\left(\mathbb{X}, \Phi, \mathbb{L}|\hat{\mathbb{X}}, \hat{\Phi}, \hat{\mathbb{L}}\right) = E_{\mathbb{K},\mathbb{Y},\mathbb{A}|\hat{\mathbb{X}},\hat{\Phi},\hat{\mathbb{L}},\mathbb{N}}\Big[\log\left\{p_{\mathsf{comp}}(\mathbb{X}, \Phi, \mathbb{L}, \mathbb{K}, \mathbb{Y}, \mathbb{A}, \mathbb{N})\right\}\Big]. \tag{9.3}$$

We will quickly sketch the outline of the algorithm derivation: the details are very much the same as we have seen repeatedly already in this book and the reader will readily be able to derive these details for herself by now. Due to independence assumptions, the complete data likelihood is

$$p_{\mathsf{comp}}(\mathbb{X}, \Phi, \mathbb{L}, \mathbb{K}, \mathbb{Y}, \mathbb{A}, \mathbb{N})$$
$$= p(\mathbb{X})p(\Phi)p(\mathbb{L})p(\mathbb{N}|\mathbb{X}, \Phi, \mathbb{L})p(\mathbb{K}|\mathbb{N})p(\mathbb{Y}|\mathbb{K}, \mathbb{X})p(\mathbb{A}|\mathbb{K}, \Phi). \tag{9.4}$$

The missing data likelihood is again simplified using Bayes' rule

$$p_{\mathsf{miss}}(\mathbb{K}, \mathbb{Y}, \mathbb{A}|\hat{\mathbb{X}}, \hat{\Phi}, \hat{\mathbb{L}}, \mathbb{N}) = \frac{p(\mathbb{K}, \mathbb{Y}, \mathbb{A}, \hat{\mathbb{X}}, \hat{\Phi}, \hat{\mathbb{L}}, \mathbb{N})}{\sum_{\mathbb{K}}\int\int p(\mathbb{K}, \mathbb{Y}, \mathbb{A}, \hat{\mathbb{X}}, \hat{\Phi}, \hat{\mathbb{L}}, \mathbb{N}))\,\mathrm{d}\mathbb{Y}\,\mathrm{d}\mathbb{A}},$$
$$= \frac{p(\mathbb{K}|\mathbb{N})p(\mathbb{Y}|\mathbb{K}, \hat{\mathbb{X}})p(\mathbb{A}|\mathbb{K}, \hat{\Phi})}{\sum_{\mathbb{K}}\int\int p(\mathbb{K}|\mathbb{N})p(\mathbb{Y}|\mathbb{K}, \hat{\mathbb{X}})p(\mathbb{A}|\mathbb{K}, \hat{\Phi})\,\mathrm{d}\mathbb{Y}\,\mathrm{d}\mathbb{A}},$$
$$= \prod_{t=1}^{T}\prod_{i=1}^{I}\prod_{j=1}^{J}\prod_{r=1}^{n_t^i} \frac{\Lambda_t^m g_X\left(\mathbf{y}_t^{i,j,r}|\hat{\mathbf{x}}_t^m\right) g_A\left(\mathbf{a}_t^{i,j,r}|\hat{\phi}_t^m\right)\Big|_{m=k_t^{i,j,r}}}{\sum_{k=0}^{M}\Lambda_t^k G_X^{k,i}\left(\hat{\mathbf{x}}_t^k\right) G_A^{k,j}\left(\hat{\phi}_t^k\right)}. \tag{9.5}$$

The difference between this and the Poisson H-PMHT in Chap. 6 is that the measurement term in the weight is now a product of a kinematic measurement likelihood and a spectral measurement likelihood. The pixel-track weight is then given by

$$w_t^{i,j,m} = \int_{W^i}\int_{A^j} \frac{\Lambda_t^m g_X\left(\mathbf{y}_t^{i,j,r}|\hat{\mathbf{x}}_t^m\right) g_A\left(\mathbf{a}_t^{i,j,r}|\hat{\phi}_t^m\right)}{\sum_{k=0}^{M} \Lambda_t^k G_X^{k,i}\left(\hat{\mathbf{x}}_t^k\right) G_A^{k,j}\left(\hat{\phi}_t^k\right)} \,\mathrm{d}\mathbb{Y}\,\mathrm{d}\mathbb{A},$$
$$= \Lambda_t^m G_X^{m,i}\left(\hat{\mathbf{x}}_t^m\right) G_A^{m,j}\left(\hat{\phi}_t^m\right) \; / \; G^{i,j}. \tag{9.6}$$

The complete data likelihood and the missing data combine to give an auxiliary function that is a sum over separated terms

$$\mathscr{Q}\left(\mathbb{X}, \Phi, \mathbb{L}|\hat{\mathbb{X}}, \hat{\Phi}, \hat{\mathbb{L}}\right) = \log\{p(\mathbb{X})\} + \log\{p(\Phi)\} + \log\{p(\mathbb{L})\}$$
$$+ E_{\mathbb{K},\mathbb{Y},\mathbb{A}|\hat{\mathbb{X}},\hat{\Phi},\hat{\mathbb{L}},\mathbf{N}}\Big[\log\{p(\mathbb{N}|\mathbb{X}, \Phi, \mathbb{L})p(\mathbb{K}|\mathbb{N})\}\Big]$$
$$+ E_{\mathbb{K},\mathbb{Y},\mathbb{A}|\hat{\mathbb{X}},\hat{\Phi},\hat{\mathbb{L}},\mathbf{N}}\Big[\log\{p(\mathbb{Y}|\mathbb{K}, \mathbb{X})\}\Big] + E_{\mathbb{K},\mathbb{Y},\mathbb{A}|\hat{\mathbb{X}},\hat{\Phi},\hat{\mathbb{L}},\mathbf{N}}\Big[\log\{p(\mathbb{A}|\mathbb{K}, \Phi)\}\Big]. \tag{9.7}$$

This can be grouped into three independent sub-functions, one each for the parameters $\mathbb{X}$, Φ and $\mathbb{L}$. Each sub-function contains a prior term and a measurement term.

The Poisson rate sub-function consists of the rate prior $p(\mathbb{L})$ and a measurement term $\log\{p(\mathbb{N}|\mathbb{X}, \Phi, \mathbb{L})p(\mathbb{K}|\mathbb{N})\}$ that depends only on the assignment likelihood. This measurement term differs from the Poisson H-PMHT only through the weights definition and the span of the measurement space. The modified number of associated measurements is given by

$$\bar{n}_t^m = \sum_{i=1}^{I}\sum_{j=1}^{J} w_t^{i,j,m} n_t^{i,j}. \tag{9.8}$$

This then drives the gamma distribution parameter update

$$p(\Lambda_t^m|n_1 \ldots n_t) \sim p_{\mathsf{gamma}}(\Lambda_t^m; \alpha_{t|t-1}^m, \beta_{t|t-1}^m), \tag{9.9}$$

where

$$\alpha_{t|t-1}^m = \exp\{-(\tau_t - \tau_{t-1})/\bar{\tau}\}\, \alpha_{t-1|t-1}^m, \tag{9.10}$$
$$\beta_{t|t-1}^m = \exp\{-(\tau_t - \tau_{t-1})/\bar{\tau}\}\, \beta_{t-1|t-1}^m, \tag{9.11}$$

and

$$\alpha_{t|t}^m = \alpha_{t|t-1}^m + \bar{n}_t^m \qquad \beta_{t|t} = \beta_{t|t-1} + 1. \tag{9.12}$$

The remaining sub-functions in $(\mathbb{X}, \mathbb{Y})$ and $(\Phi, \mathbb{A})$ are analogous to each other: in each case, the measurement term is a projection of the likelihood onto one of the orthogonal components. The spatial measurement term in the kinematic $(\mathbb{X}, \mathbb{Y})$ sub-function is

$$
\begin{aligned}
&E_{\mathbb{K},\mathbb{Y},\mathbb{A}|\hat{\mathbb{X}},\hat{\Phi},\hat{\mathbb{L}},\mathbf{N}}\Big[\log\{p(\mathbb{Y}|\mathbb{K},\mathbb{X})\}\Big] \\
&\quad= \sum_{\mathbb{K}} \int\int p_{\mathsf{miss}}(\mathbb{K},\mathbb{Y},\mathbb{A}|\hat{\mathbb{X}},\hat{\Phi},\hat{\mathbb{L}},\mathbb{N}) \log\{p(\mathbb{Y}|\mathbb{K},\mathbb{X})\}\, \mathrm{d}\mathbb{Y}\, \mathrm{d}\mathbb{A}, \\
&\quad= \sum_{\mathbb{K}} \int\int \left(\prod_{t=1}^{T}\prod_{i=1}^{I}\prod_{j=1}^{J}\prod_{r=1}^{n_t^i} \frac{\Lambda_t^m g_X\left(\mathbf{y}_t^{i,j,r}|\hat{\mathbf{x}}_t^m\right) g_A\left(\mathbf{a}_t^{i,j,r}|\hat{\phi}_t^m\right)\Big|_{m=k_t^{i,j,r}}}{\sum_{k=0}^{M} \Lambda_t^k G_X^{k,i}\left(\hat{\mathbf{x}}_t^k\right) G_A^{k,j}\left(\hat{\phi}_t^k\right)} \right) \\
&\qquad \times \left(\sum_{t=1}^{T}\sum_{i=1}^{I}\sum_{j=1}^{J}\sum_{r=1}^{n_t^i} \log\left\{ g\left(\mathbf{y}_t^{i,j,r}|\mathbf{x}_t^{k_t^{i,j,r}}\right)\right\} \right) \mathrm{d}\mathbb{Y}\, \mathrm{d}\mathbb{A}.
\end{aligned} \tag{9.13}
$$

This is simplified by recognising that each term in the log sum is only a function of the missing variables related to a single measurement, so

$$
\begin{aligned}
&E_{\mathbb{K},\mathbb{Y},\mathbb{A}|\hat{\mathbb{X}},\hat{\Phi},\hat{\mathbb{L}},\mathbf{N}}\Big[\log\{p(\mathbb{Y}|\mathbb{K},\mathbb{X})\}\Big] = \\
&\sum_{m=0}^{M}\sum_{t=1}^{T}\sum_{i=1}^{I} \Lambda_t^m \left(\sum_{j=1}^{J} \frac{n_t^{i,j}\, G_A^{k,j}\left(\hat{\phi}_t^k\right)}{G^{i,j}} \right) \int_{W^i} \log\left\{g_X\left(\mathbf{y}|\mathbf{x}_t^m\right)\right\} g_X\left(\mathbf{y}|\hat{\mathbf{x}}_t^m\right)\, \mathrm{d}\mathbf{y}.
\end{aligned} \tag{9.14}
$$

The development of this expression follows the same process as used for the core H-PMHT to define (4.30). If we compare a single time-slice for a single component, then the spectral measurement term is

$$
\begin{aligned}
&\sum_{i=1}^{I} \left(\sum_{j=1}^{J} \frac{n_t^{i,j}\, \Lambda_t^m\, G_A^{k,j}\left(\hat{\phi}_t^k\right)}{G^{i,j}} \right) \frac{G_X^{i,j}}{G_X^{i,j}} \int_{W^i} \log\left\{g_X\left(\mathbf{y}|\mathbf{x}_t^m\right)\right\} g_X\left(\mathbf{y}|\hat{\mathbf{x}}_t^m\right)\, \mathrm{d}\mathbf{y} \\
&\qquad = \sum_{i=1}^{I} \left(\sum_{j=1}^{J} n_t^{i,j} w_t^{i,j,m} \right) \frac{\int_{W^i} \log\left\{g_X\left(\mathbf{y}|\mathbf{x}_t^m\right)\right\} g_X\left(\mathbf{y}|\hat{\mathbf{x}}_t^m\right)\, \mathrm{d}\mathbf{y}}{G_X^{i,j}}.
\end{aligned} \tag{9.15}
$$

This is very similar to the equivalent term from the core H-PMHT in (4.30), which is given by

$$
\sum_{i=1}^{I} n_t^i w_t^{i,m} \frac{\int_{W^i} \log\left\{g^m\left(\mathbf{y}|\mathbf{x}_t^m\right)\right\} g^m\left(\mathbf{y}|\hat{\mathbf{x}}_t^m\right) \mathrm{d}\mathbf{y}}{G_t^{m,i}}.
$$

If we define

$$n_X^i = \sum_{j=1}^{J} n^{i,j}, \qquad w_X^{i,m} = \frac{1}{n_X^i} \sum_{j=1}^{J} n^{i,j} w_t^{i,j,m}, \tag{9.16}$$

then the spectral measurement term becomes

$$\sum_{i=1}^{I} n_X^i w_X^{i,m} \frac{\int_{W^i} \log \left\{ g_X \left(\mathbf{y} | \mathbf{x}_t^m \right) \right\} g_X \left(\mathbf{y} | \hat{\mathbf{x}}_t^m \right) \, \mathrm{d}\mathbf{y}}{G_X^{i,j}}, \tag{9.17}$$

which is now exactly the same as the core H-PMHT except that the weighting term is defined differently. In the core H-PMHT, the measurement term is a convex combination of per-pixel centroids weighted by the associated power in that pixel. The spectral H-PMHT is the same except that the associated power in spatial pixel i accounts for the correlation of the observed spectral power with the component's spectrum.

The spatial sub-function has been arranged in the same way as for the core H-PMHT, so it can be maximised in the same way. Importantly, the maximisation over $\mathbb{X}$ is not directly dependent on the attribute state or the attribute part of the appearance function. The spectral information has an indirect effect by modifying the associated spatial images. This means that we can use a Kalman-based solution if the spatial appearance function is Gaussian, even if the attribute (spectral) appearance function is not Gaussian. The examples at the end of this chapter use a Gaussian spatial appearance function and a Dirichlet spectral appearance function. Streit et al. [3, 4] used a Gaussian mixture spectral appearance function.

The spectral auxiliary sub-function is in the same form as the spatial sub-function, so its simplification proceeds in exactly the same way. As for the spatial case, the measurement term can be expressed in terms of an associated spectral image that correlates the spatial spread of energy for each spectral cell with the expected component spread, specified through the pixel probabilities $G_X^{i,j}\left(\hat{\mathbf{x}}_t^m\right)$,

$$E_{\mathbb{K},\mathbb{Y},\mathbb{A}|\hat{\mathbb{X}},\hat{\Phi},\hat{\mathbb{L}},\mathbf{N}}\Big[\log \left\{ p(\mathbb{A}|\mathbb{K}, \Phi) \right\} \Big] =$$

$$\sum_{\mathbb{K}} \int \int \left(\prod_{t=1}^{T} \prod_{i=1}^{I} \prod_{j=1}^{J} \prod_{r=1}^{n_t^i} \frac{\Lambda_t^m g_X \left(\mathbf{y}_t^{i,j,r} | \hat{\mathbf{x}}_t^m \right) g_A \left(\mathbf{a}_t^{i,j,r} | \hat{\phi}_t^m \right) \Big|_{m=k_t^{i,j,r}}}{\sum_{k=0}^{M} \Lambda_t^k G_X^{k,i} \left(\hat{\mathbf{x}}_t^k \right) G_A^{k,j} \left(\hat{\phi}_t^k \right)} \right)$$

$$\times \left(\sum_{t=1}^{T} \sum_{i=1}^{I} \sum_{j=1}^{J} \sum_{r=1}^{n_t^i} \log \left\{ g_A \left(\mathbf{a}_t^{i,j,r} | \phi_t^{k_t^{i,j,r}} \right) \right\} \right) \mathrm{d}\mathbb{Y} \, \mathrm{d}\mathbb{A},$$

$$= \sum_{j=1}^{J} n_A^j w_A^{j,m} \frac{\int_{A^j} \log \left\{ g_A \left(\mathbf{a} | \phi_t^m \right) \right\} g_A \left(\mathbf{a} | \hat{\phi}_t^m \right) \, \mathbf{da}}{G_A^{j,j}}, \tag{9.18}$$

with

$$n_A^j = \sum_{i=1}^{I} n^{i,j}, \qquad w_A^{j,m} = \frac{1}{n_A^j} \sum_{i=1}^{I} n^{i,j} w_t^{i,j,m}. \tag{9.19}$$

The spectral parameters Φ are estimated in the same way as the kinematic states. The form of the estimator depends on the particular assumed spectral appearance model. Just like the kinematic states, the estimator for the spectral parameters does not directly depend on the kinematic states or the spatial appearance model. Since the spectrum is assumed to be constant, or at least very slowly varying, the attributes are more similar to the parameters of the appearance model described in Chap. 8, rather than the kinematic states.

The simulations presented in this chapter consider colour video, where the spectrum is limited to only three bins. In this case, it is appropriate to use the Dirichlet model. An equivalent way to represent the Dirichlet attribute distribution is to model each component with independent red, green and blue Poisson rates, $\Lambda_t^{m,j}$, $j = 1, 2, 3$. Once again, these rates have gamma distributions with hyperparameters updated using

$$\alpha_{t|t-1}^{m,j} = \exp\{-(\tau_t - \tau_{t-1})/\tau\} \, \alpha_{t-1|t-1}^{m,j}, \tag{9.20}$$

$$\beta_{t|t-1}^{m,j} = \exp\{-(\tau_t - \tau_{t-1})/\bar{\tau}\} \, \beta_{t-1|t-1}^{m,j}, \tag{9.21}$$

and

$$\alpha_{t|t}^{m,j} = \alpha_{t|t-1}^{m,j} + n_A^j w_A^{j,m} \qquad \beta_{t|t} = \beta_{t|t-1} + 1. \tag{9.22}$$

9.2 Obscuring H-PMHT

The obscuring H-PMHT was first referred to as attribute H-PMHT in [2] but here we use the terminology *obscuring* because spectral H-PMHT also uses non-kinematic attributes. The spectral algorithm described in the previous section models the attribute data using superposition. As discussed earlier, superposition is a good model for some circumstances, but in others, the objects in the scene can block energy from other objects from reaching the sensor, we call this obscuration. Obscuration is the norm for video sensors where close objects prevent us from seeing behind them. Under the superposition model, the sensor would measure purple if a blue object moved in front of a red background. Under obscuration, the object stays blue.

The spectral H-PMHT has pixels that span an attribute space and a physical space. The obscuring H-PMHT treats the attribute data differently. Instead, each spatial pixel has a single attribute measurement associated with it. This means that pixel i contains two pieces of measurement information: a histogram count n_t^i and an attribute measurement $\mathbf{a}_t^i$. The precise location of each shot $\mathbf{y}_t^{i,r}$ is again missing data but the attribute of each shot is assumed to be the same as the attribute measurement for the pixel, that is all shots have the same attribute $\mathbf{a}_t^{i,r} = \mathbf{y}_t^i$. The missing data likelihood now becomes

$$p_{\mathsf{miss}}(\mathbb{K}, \mathbb{Y}|\hat{\mathbb{X}}, \hat{\Phi}, \hat{\mathbb{L}}, \mathbb{N}) = \prod_{t=1}^{T}\prod_{i=1}^{I}\prod_{r=1}^{n_t^i} \frac{\Lambda_t^m g_X\left(\mathbf{y}_t^{i,r}|\hat{\mathbf{x}}_t^m\right) g_A\left(\mathbf{a}_t^i|\hat{\phi}_t^m\right)\Big|_{m=k_t^{i,r}}}{\sum_{k=0}^{M} \Lambda_t^k G_X^{k,i}\left(\hat{\mathbf{x}}_t^k\right) g_A\left(\mathbf{a}_t^i|\hat{\phi}_t^k\right)}. \tag{9.23}$$

Notice the difference between this and the corresponding term in the spectral H-PMHT, which applies quantisation across the attribute domain because it assumes a superposition model. The spectral H-PMHT uses the quantisation to perform a soft association between the targets in each spectrum bin. In contrast, obscuring H-PMHT weights the Poisson H-PMHT with an attribute likelihood term that is the same for every shot in the spatial pixel. The obscuring H-PMHT pixel-track weight is then given by

$$\begin{aligned} w_t^{i,m} &= \int_{W^i} \frac{\Lambda_t^m g_X\left(\mathbf{y}_t^{i,r}|\hat{\mathbf{x}}_t^m\right) g_A\left(\mathbf{a}_t^i|\hat{\phi}_t^m\right)}{\sum_{k=0}^{M} \Lambda_t^k G_X^{k,i}\left(\hat{\mathbf{x}}_t^k\right) g_A\left(\mathbf{a}_t^i|\hat{\phi}_t^k\right)} \, \mathrm{d}\mathbb{Y}, \\ &= \frac{\Lambda_t^m G_X^{m,i}\left(\hat{\mathbf{x}}_t^m\right) g_A\left(\mathbf{a}_t^i|\hat{\phi}_t^m\right)}{\sum_{k=0}^{M} \Lambda_t^k G_X^{k,i}\left(\hat{\mathbf{x}}_t^k\right) g_A\left(\mathbf{a}_t^i|\hat{\phi}_t^k\right)}. \end{aligned} \tag{9.24}$$

The target-pixel weights are essentially the same as the Poisson H-PMHT with an additional scaling term based on the attribute likelihood. This form is exactly that of the PMHT with classifications [1] that applies attribute data to point measurements.

The obscuring H-PMHT auxiliary function looks very similar to the spectral auxiliary function in (9.7) because the independence assumptions are the same, but there is no spectral missing data so the expectations are different

$$\begin{aligned} \mathcal{Q}\left(\mathbb{X}, \Phi, \mathbb{L}|\hat{\mathbb{X}}, \hat{\Phi}, \hat{\mathbb{L}}\right) &= \log\{p(\mathbb{X})\} + \log\{p(\Phi)\} + \log\{p(\mathbb{L})\} \\ &+ E_{\mathbb{K},\mathbb{Y}|\hat{\mathbb{X}},\hat{\Phi},\hat{\mathbb{L}},\mathbb{N},\mathbb{A}}\Big[\log\{p(\mathbb{N}|\mathbb{X}, \Phi, \mathbb{L})p(\mathbb{K}|\mathbb{N})\}\Big] \\ + E_{\mathbb{K},\mathbb{Y}|\hat{\mathbb{X}},\hat{\Phi},\hat{\mathbb{L}},\mathbb{N},\mathbb{A}}\Big[\log\{p(\mathbb{Y}|\mathbb{K}, \mathbb{X})\}\Big] &+ E_{\mathbb{K},\mathbb{Y}|\hat{\mathbb{X}},\hat{\Phi},\hat{\mathbb{L}},\mathbb{N},\mathbb{A}}\Big[\log\{p(\mathbb{A}|\mathbb{K}, \Phi)\}\Big]. \end{aligned} \tag{9.25}$$

The same sub-functions are formed because the independence assumptions are the same. The assignment term $E_{\mathbb{K},\mathbb{Y}|\hat{\mathbb{X}},\hat{\Phi},\hat{\mathbb{L}},\mathbb{N},\mathbb{A}}\Big[\log\{p(\mathbb{N}|\mathbb{X},\Phi,\mathbb{L})p(\mathbb{K}|\mathbb{N})\}\Big]$ remains the expectation of a term that is independent of the measurement values: the log of the likelihood of observing a given number of measurements from each component. This means that the only change to the Poisson rate sub-function is the way in which the weights are defined. Once more, the key is the mean number of shots associated with the component, which can be defined in terms of an associated image

$$\bar{n}_t^m = \sum_{i=1}^{I} \mathfrak{z}_t^{i,m}, \qquad \mathfrak{z}_t^{i,m} = w_t^{i,m} n_t^i. \tag{9.26}$$

The difference here between the obscuring H-PMHT and the Poisson H-PMHT is that the former modifies the weights likelihood ratio by including a dependence on the measured attribute in each pixel.

Similarly, the kinematic state sub-function is just like the core H-PMHT except for the modified weights. We presume that the reader can do without solving it again.

The attribute expected likelihood is simplified

$$\begin{aligned}
E_{\mathbb{K},\mathbb{Y}|\hat{\mathbb{X}},\hat{\Phi},\hat{\mathbb{L}},\mathbb{N},\mathbb{A}}\Big[\log\{p(\mathbb{A}|\mathbb{K},\Phi)\}\Big] &= \sum_{t=1}^{T}\sum_{i=1}^{I}\sum_{r=1}^{n_t^i} E_{\mathbb{K},\mathbb{Y}|\hat{\mathbb{X}},\hat{\Phi},\hat{\mathbb{L}},\mathbb{N},\mathbb{A}}\Big[\log\left\{p(\mathbf{a}_t^i|\mathbb{K},\Phi)\right\}\Big], \\
&= \sum_{t=1}^{T}\sum_{i=1}^{I} n_t^i E_{k_t^{i,r}|\hat{\mathbb{X}},\hat{\Phi},\hat{\mathbb{L}},\mathbb{N},\mathbb{A}}\left[\log\left\{p\left(\mathbf{a}_t^i|\phi_t^{k_t^{i,r}}\right)\right\}\right], \\
&= \sum_{m=0}^{M}\left[\sum_{t=1}^{T}\sum_{i=1}^{I} \mathfrak{z}_t^{i,m}\log\left\{p(\mathbf{a}_t^i|\phi_t^m)\right\}\right].
\end{aligned} \tag{9.27}$$

Each component has I weighted attribute measurements at each frame. The weights depend on the proximity of the pixel to the kinematic state estimate, the measurement power in that pixel, and the match between the measured attribute and the current estimated attribute state. Notice that we have treated the attributes of each shot as independent variables even though we have simultaneously asserted that they all have the same value. This peculiarity has the same effect as the resampled kinematic prior: it ensures that the attribute measurement contribution does not vanish in the limit as the quantisation is taken to an infinitesimal step.

The attribute state estimate depends on the assumed distributions $p(\phi_t^m|\phi_{t-1}^m)$, $p(\phi_0^m)$ and $p(\mathbf{a}|\phi)$. The next section demonstrates obscuring H-PMHT applied to colour video imagery, in which case, the attribute measurement is a three-element vector. For obscuring H-PMHT, we generate an intensity difference image that gives the count in each pixel n_t^i and this is treated as independent from the colour vector $\mathbf{a}_t^i$. Although the intensity is derived from the colour image, it is a difference between the background colour and the current pixel. It is not obvious what the dependence between absolute colour and colour difference should be. Once more, we model the measurement value in each colour channel using a Dirichlet model where each target has independent red, green and blue Poisson rates, $\phi_t^m \equiv \{\Lambda_t^{m,j}\},\ j = 1, 2, 3$. The obscuring H-PMHT rate estimates are updated using

$$\alpha_{t|t-1}^{m,j} = \exp\{-(\tau_t - \tau_{t-1})/\bar{\tau}\}\, \alpha_{t-1|t-1}^{m,j}, \tag{9.28}$$

$$\beta_{t|t-1}^{m,j} = \exp\{-(\tau_t - \tau_{t-1})/\bar{\tau}\}\, \beta_{t-1|t-1}^{m,j}, \tag{9.29}$$

and

$$\alpha_{t|t}^{m,j} = \alpha_{t|t-1}^{m,j} + \sum_{i=1}^{I} \mathbf{a}_t^{i,j} w_t^{j,m} \qquad \beta_{t|t} = \beta_{t|t-1} + 1. \tag{9.30}$$

Note that the main practical difference between spectral H-PMHT and obscuring H-PMHT is in how the weights are defined. Spectral H-PMHT uses a different weight for each colour channel in a pixel and can therefore be understood to assign each colour independently to the targets. Obscuring H-PMHT has a single weight for each pixel that depends on how well the measured colour vector matches each target.

9.3 Performance Examples

In this chapter, we used a modified version of the canonical multi-target scenario to investigate the performance of the different attribute models. In the standard canonical multi-target scenario, pairs of targets cross during the highest speed part of their sinusoidal paths, that is at times where the phase approximately corresponds to 0 for a position following $\pm\sin(wt)$. In this modified version, the targets follow the same sinusoidal trajectories but we modify the horizontal offset between them so that the crossing time is close to the stationary point of the path. This means that the targets move together and then slow and turn while they are overlapping. The crossing occurs at a time where the phase is approximately $\pi/2$, which leads to prolonged times where the targets are unresolved and higher ambiguity. It is extremely difficult for the core H-PMHT to reliably resolve which target is which after the crossing time. The attribute algorithms rely on colour information to overcome this.

We present both superposition and obscuration-based simulations and compare the core H-PMHT with spectral H-PMHT and obscuring H-PMHT on each scenario. In the obscuring case, we consider two variations where the background is either dark or light.

9.3.1 Modified Canonical Multi-Target Scenario

The simulations contained 20 coloured targets interacting in a 250×300 pixel image. The target positions follow the canonical multi-target paths with $\Delta_Y = 20$ except that the even targets were shifted to the right to modify the crossing points. The result is that the states of the odd-numbered targets were defined by

$$\begin{bmatrix} X_t^m \\ Y_t^m \end{bmatrix} = \begin{bmatrix} 50 + 20\cos\left(\frac{2\pi t}{200} + \frac{(m-1)\pi}{20}\right) \\ 54 + 40\cos\left(\frac{2\pi t}{200} + \frac{(m-1)\pi}{20}\right) + 10(m-1) \end{bmatrix},$$

and even-numbered targets by

$$\begin{bmatrix} X_t^m \\ Y_t^m \end{bmatrix} = \begin{bmatrix} 85 - 20\cos\left(\frac{2\pi t}{200} + \frac{(m-2)\pi}{20}\right) \\ 50 + 40\cos\left(\frac{2\pi t}{200} + \frac{(m-2)\pi}{20}\right) + 10(m-2) \end{bmatrix}.$$

This nominal position was corrupted with noise uniform over [0, 1]. Target pairs m and $m+1$ have nominal Y positions slightly offset by 4 pixels and mirrored X paths that cross near the stationary points of the cosine functions.

Each target's spatial appearance was elliptical in shape where pixels inside the object satisfied the inequality

$$g_X(\mathbf{y}|\mathbf{x}) = \begin{cases} c, & (\mathbf{y}-\mathbf{x})^{\mathsf{T}}\mathsf{B}^{-1}(\mathbf{y}-\mathbf{x}) \leq -2\log\{0.1\} \\ 0, & \text{otherwise}, \end{cases} \tag{9.31}$$

with $\mathsf{B} = \begin{bmatrix} 2 & 0 \\ 0 & 25 \end{bmatrix}$, and c constant. That is, the object is flat within an ellipse that has a minor axis width of approximately 6 pixels in the X direction and a major axis height of approximately 21 pixels in the Y direction. Unique colours were defined for each target by selecting a uniform grid in hue and transforming this into red-green-blue. Within the target's ellipse, each pixel was filled in with its colour. Figure 9.1 shows the colours of each target positioned on a colour wheel. The colours were chosen so that targets that pass very close together are well separated on the wheel. For example, target 1 is orange in the North-West quadrant whereas target 2 is blue in the South-East quadrant. Targets 3 and 4, which are also close to target 1, are in the North-East and South-West quadrants respectively.

Two measurement modes were simulated: superposition and obscuration. The superposition measurements followed the spectral H-PMHT model where the colours add. When targets become unresolved, mixture colours are visible in the image. Most of these mixture colours are close to white because the pairs of most highly interacting targets are positioned on opposite sides of the colour wheel in Fig. 9.1. In the obscuration measurement model, targets with a higher index m occluded targets with a lower index, for example, target 3 moved in front of target 1 and 2 but behind target 4. Figure 9.2 compares frames at the same time using the two different measurement modes. Figure 9.3 shows a selection of frames from the simulated sequence using obscuration: there are frequent occlusions and periodically the targets clump together in a vertical line.

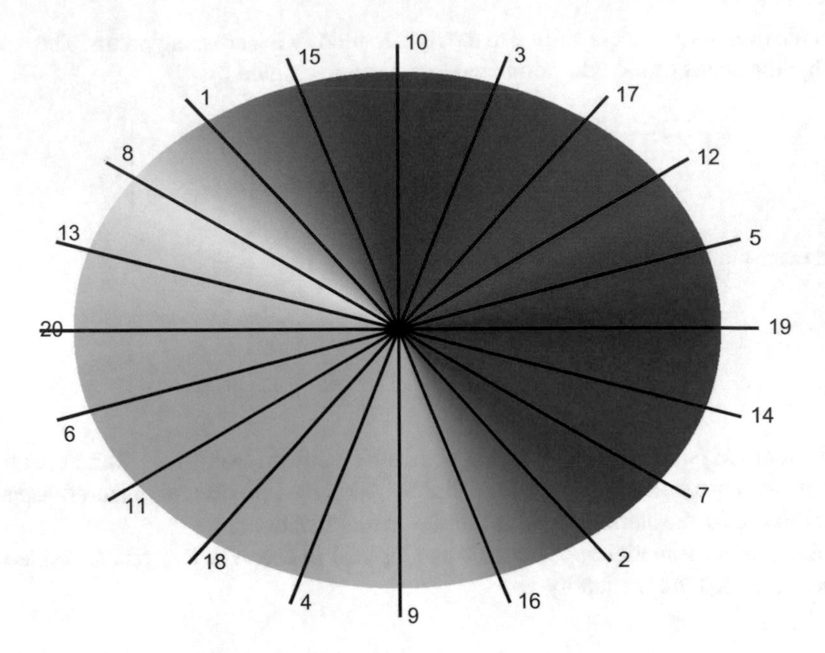

Fig. 9.1 Simulated target colours

We compare Poisson H-PMHT with spectral H-PMHT and obscuring H-PMHT. All three algorithms use a Poisson rate estimate to measure track quality and the track management strategies described in Chap. 5. The Poisson H-PMHT cannot make use of the colour information so an intensity image was constructed by taking the root mean square of the colour channels, that is we treat the colour values like a three-dimensional position vector and find the vector length.

9.3.2 *Superposition Measurements*

Monte Carlo trials were used to calculate our usual association and accuracy performance measures as discussed in Sect. 1.6; the RMS estimation error, the redundant track count, the missed target count and the number of swaps were each averaged over 100 Monte Carlo trials and are shown in Fig. 9.4 for images following the superposition model. Both of the colour H-PMHT variants follow the targets well compared with the intensity-only H-PMHT and have significantly lower estimation error. The obscuring H-PMHT is able to follow the targets when they cross so it does not produce target swaps. However, the obscuring measurement model does not have a mechanism to explain the white pixels that occur when targets overlap. It incorrectly identifies these as new targets and creates redundant tracks. When the targets separate, the white pixels vanish and the redundant tracks are terminated. In

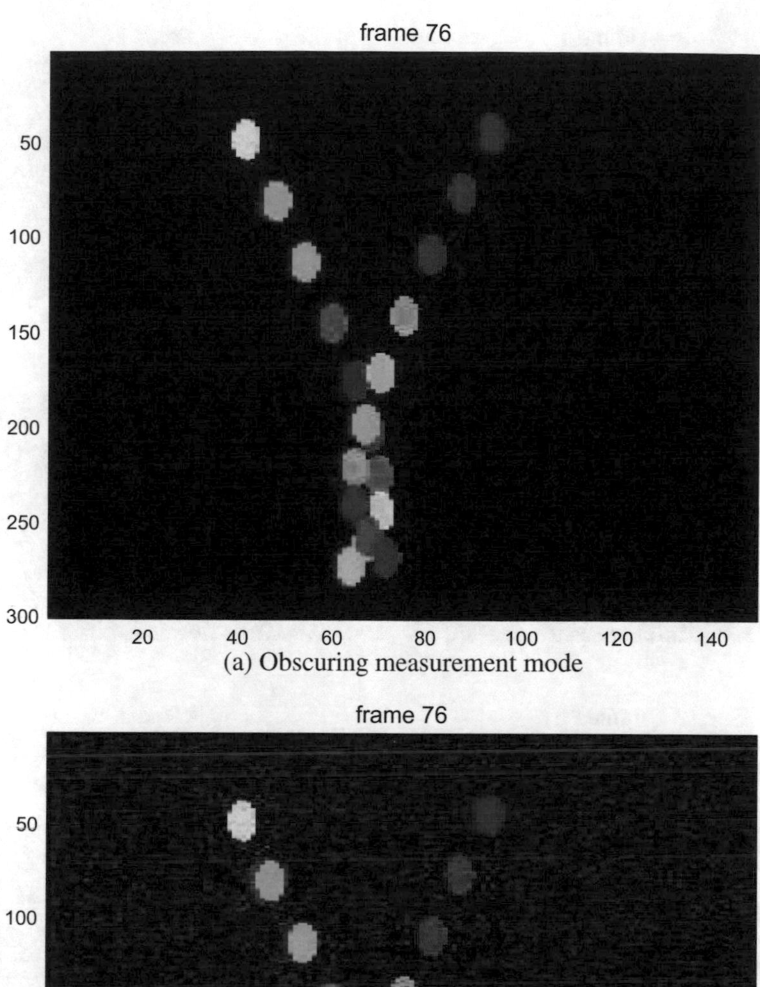

(a) Obscuring measurement mode

(b) Superposition measurement mode

Fig. 9.2 Different measurement models

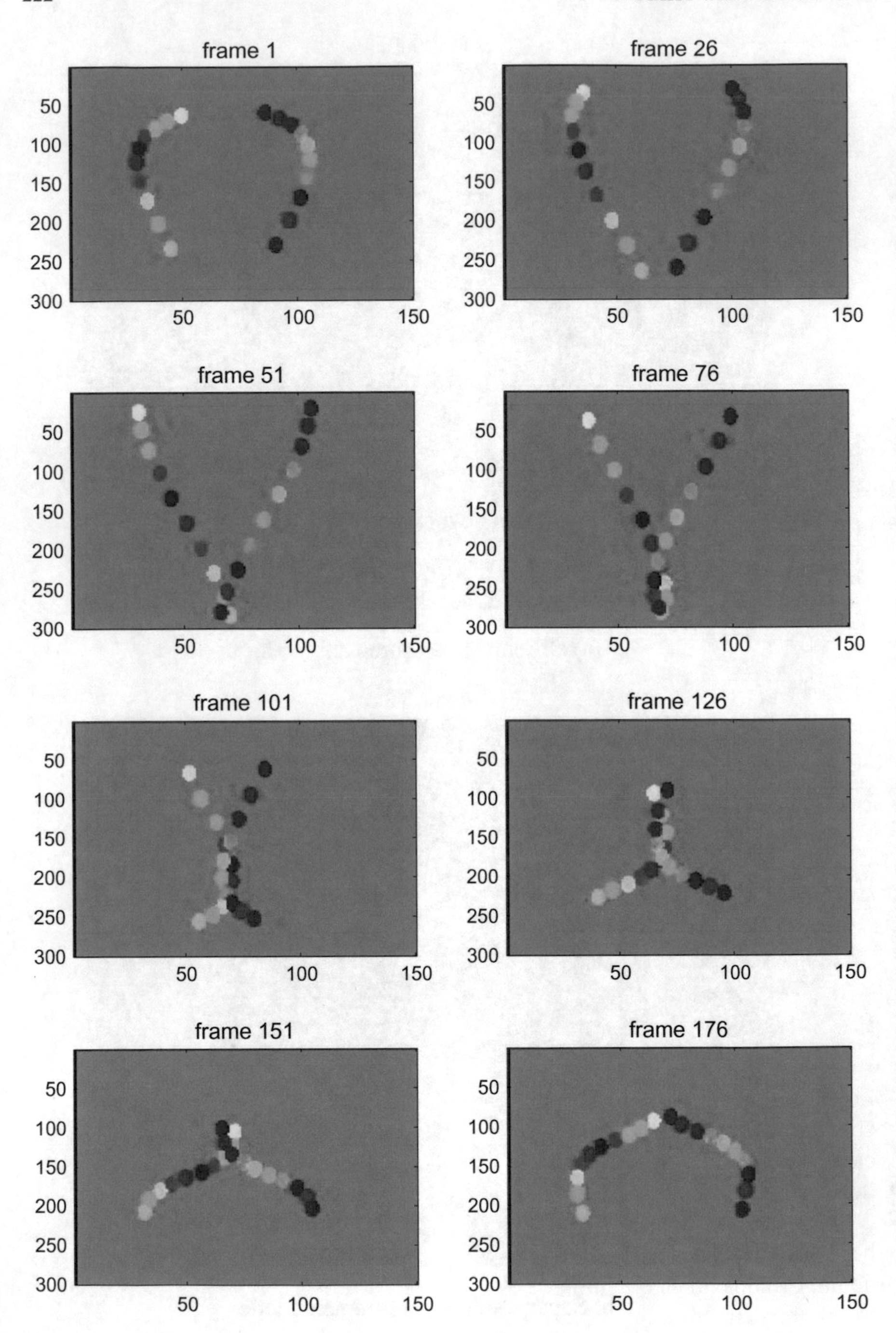

Fig. 9.3 Example simulated frames for the obscuring model

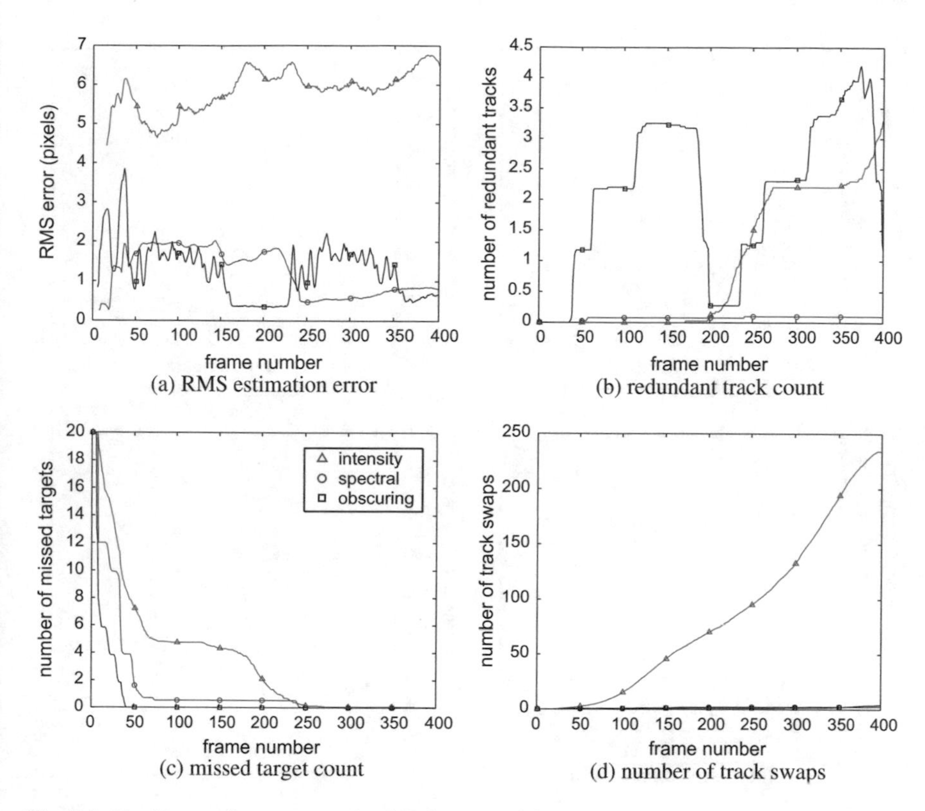

Fig. 9.4 Tracking performance on simulated superposition video

contrast, the intensity-only H-PMHT gradually builds up more and more swaps and redundant tracks as time evolves.

9.3.3 *Obscuring Measurements*

Figure. 9.5 shows the averaged performance metrics for obscuring imagery. It should come as no surprise that the obscuring H-PMHT has the best performance in this case. When the H-PMHT attribute model matches the true sensor behaviour, then the best performance is achieved.

In this case, the spectral H-PMHT erred by making track swaps. This happens because it expects the colours of overlapping targets to combine. Since the colours do not combine in an obscuring sensor, the H-PMHT adjusts its colour estimates to match the data. Both overlapping tracks effectively become the same colour and it is somewhat arbitrary which follows each target after they move apart. Again the intensity-only H-PMHT has the worst performance. The obscuring H-PMHT RMS

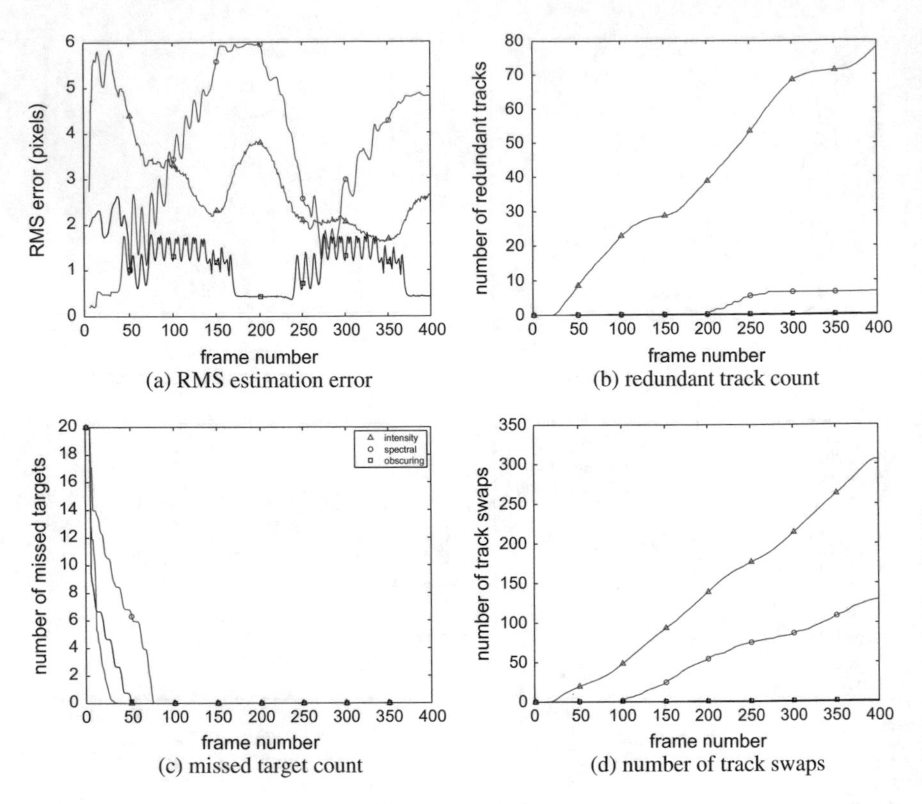

(a) RMS estimation error
(b) redundant track count
(c) missed target count
(d) number of track swaps

Fig. 9.5 Tracking performance on simulated obscuring video

curve shows a lot of structure because of the cyclic nature of the scenario; the repeated occlusions lead to bias.

9.4 Summary

This chapter developed two variants of H-PMHT that deal with attribute data. The spectral H-PMHT extends the superposition model of core H-PMHT to a non-spatial measurement dimension, referred to as the spectrum. The obscuring H-PMHT adopts an occlusion model that chooses a single attribute for each pixel. Both of the algorithms differ from the core H-PMHT by an additional attribute term that scales the measurement likelihood in the target-pixel weight.

The choice of preferred algorithm between the spectral and obscuring H-PMHTs depends on the sensor modality: the simulations clearly show that the algorithm that matches the sensor gives the best results.

References

1. Davey, S.J., Gray, D.A., Streit, R.L.: Tracking, association and classification - a combined PMHT approach. Digital Signal Processing **12**, 372–382 (2002)
2. Davey, S.J., Vu, H.X., Fletcher, F., Arulampalam, S., Lim, C.C.: Histogram probabilistic multi-hypothesis tracker with color attributes. IET Radar Sonar Navig. **9**(8), 999–1008 (2015)
3. Streit, R.L.: Tracking targets with specified spectra using the H-PMHT algorithm. Technical report 11291, NUWC, Newport, Rhode Island, USA (2001)
4. Streit, R., Graham, M., Walsh, M.: Tracking in hyper-spectral data. In: Proceedings of the Fifth International Conference on Information Fusion, vol. 2, pp. 852–859 (2002)

Chapter 10
Maximum Likelihood H-PMHT

The methods described in this book so far are all Bayesian formulations of the tracking problem. This means that we have treated the kinematic state as a random variable with a prior distribution $p(\mathbf{x}_0^t)$ and an evolution distribution $p(\mathbf{x}_t^m|\mathbf{x}_{t-1}^m)$ that is driven by noise. The tracker amounts to deriving the joint probability $p(\mathbb{X}, \mathbb{Z})$ or a point estimate based on its moments. This idea of kinematic states as a random process is just a model: in practice the movements of objects in the scene will not be a realisation of the assumed motion model. A different way to model the kinematic state is to treat the estimation as a curve-fitting problem. For example, we can fit a straight line through point measurements. This is a deterministic model of the state evolution and is not Bayesian; instead of finding the posterior probability density of the state, the curve parameters are obtained by optimising a cost function. For example, we could choose parameters to maximise the measurement likelihood or minimise the mean squared error. For Gaussian noise, these are equivalent. An advantage of using a deterministic model is that it is simple to impose constraints and the resulting estimate is forced to come from a desired family of trajectories. A disadvantage is that we are forced to work with a batch of data. Mathematically, we can define the kinematic state as a deterministic function

$$\mathbf{x}_t^m = f(t; \theta^m), \tag{10.1}$$

where θ^m is a fixed set of parameters for the mth object. In this regime, we no longer deal with expressions such as $p\left(\mathbf{x}_t^m|\mathbf{x}_{t-1}^m\right)$ because the state at every time is specified by the parameters θ^m, not a Markov chain. In the simplest case, $f(t; \theta^m)$ can be a line and θ^m would be the slope and intercept of the line. In many cases, this model is sufficient because we will likely implement the estimator as a sliding window batch, which means that limiting $\mathbf{x}_t^m$ to a line is equivalent to making a piecewise linear approximation to its real-time evolution.

The maximum likelihood probabilistic data association (ML-PDA) algorithm [7] uses this approach to solve tracking for low amplitude targets. The original ML-PDA assumes linear target motion and finds the parameters that optimise the measurement

S. J. Davey and H. Gaetjens, *Track-Before-Detect Using Expectation Maximisation*, Signals and Communication Technology,
https://doi.org/10.1007/978-981-10-7593-3_10

likelihood. That is, the goal is to solve

$$\hat{\theta} = \arg\max \prod_{t=1}^{T} p\left(\mathbf{y}_t | f(t;\theta)\right). \tag{10.2}$$

The complication is that the measurement that belongs to the target is unknown, so ML-PDA uses probabilistic data association. It assumes at most one target. Under unknown association, the optimisation problem becomes

$$\hat{\theta} = \arg\max \prod_{t=1}^{T} \sum_{\mathbf{K}_t} \left\{ p\left(\mathbf{K}_t\right) p\left(\mathbf{Y}_t | f(t;\theta), \mathbf{K}_t\right) \right\}, \tag{10.3}$$

where the assignment probability $p(\mathbf{K}_t)$ is determined using PDA, which enforces the assumption that the target can form at most one measurement in each frame. This assumption is effectively a linear constraint on $\mathbf{K}_t$.

ML-PDA has been shown to give detection performance almost as good as image-based TkBD algorithms [4] in the single target case. Important extensions from the original ML-PDA address the problems of efficient numerical optimisation [2] and statistical methods for detection [3]. The track-before-detect algorithms in this book are based on using sensor images as the algorithm input. However, sometimes this title is also applied to point measurement trackers that use a very low single-frame detector threshold. ML-PDA is an example of this kind of TkBD.

ML-PDA assumes a single target, which is too restrictive for most applications. The obvious extension is to use joint PDA (JPDA) which applies the PDA method to the multiple target case [1, 3]. Superficially, this amounts to the joint optimisation

$$\hat{\Theta} = \arg\max \prod_{t=1}^{T} \sum_{\mathbf{K}_t} \left\{ p\left(\mathbf{K}_t\right) p\left(\mathbf{Y}_t | F(t;\Theta), \mathbf{K}_t\right) \right\}, \tag{10.4}$$

where $\Theta \equiv \theta^{1:M}$ and $\mathbf{X}_t = F(t;\Theta)$ is now a vector function that defines the states of all the targets. The difficulty with this is that the assignment probability must satisfy single-measurement constraints for each target and this prevents factorisation of the sum. The number of terms in the sum is the number of permutations of measurements and tracks, which is prohibitively large for many practical problems.

One solution to the computation complexity of ML-JPDA is to use a different measurement model to define $p(\mathbf{K}_t)$ that scales well with the number of targets. The conditionally independent assignment model of PMHT was a good fit and ML-PMHT was born [11, 13]. The development of ML-PMHT is largely due to the work of Schoenecker, Willett and Bar-Shalom.

This work is actually very closely related to a much earlier classification method referred to as maximum likelihood artificial neural system (MLANS) due to Perlovsky [8, 9]. MLANS modelled the feature vector of each class as a multivariate Gaussian with a time varying mean and formed a maximum likelihood estimate of the class parameters by assigning a class probability to each measurement. MLANS is essentially the same as ML-PMHT with different terminology: target states are instead class feature distribution parameters; data association is instead fuzzy classification. The work of Perlovsky is not widely recognised in the tracking community.

The ML-PMHT is a deterministic target approach to point measurement tracking. The same target models were applied to image measurements by extending this to H-PMHT image association, leading to the ML-H-PMHT [5, 14]. This method has subsequently been given the moniker Quanta Tracking because the name ML-H-PMHT was getting laborious [6]. This is an active research area while we write and is being led by Dunham, Willett, Ogle and Balasingam. We now review the ML-PMHT and ML-H-PMHT (Quanta) methods.

10.1 Point Measurements

The ML-PMHT is a point measurement tracking method using a deterministic target model. It finds the parameters of the target model that maximise the measurement likelihood (10.4)

$$\hat{\Theta} = \arg\max \prod_{t=1}^{T} \sum_{\mathbf{K}_t} \left\{ p\left(\mathbf{K}_t\right) p\left(\mathbf{Y}_t | F(t;\Theta), \mathbf{K}_t\right) \right\}.$$

Given the product form of this likelihood, it can be convenient in implementations to deal with its logarithm. This makes no difference to the maximising parameters because the logarithm is a monotonic function.

As we know, the PMHT measurement model treats each assignment as independent with

$$p\left(\mathbf{K}_t\right) = \prod_{r=1}^{n_t} \pi_t^{k_t^r}, \tag{10.5}$$

and the point measurements are conditionally independent given the states and the assignments

$$p\left(\mathbf{Y}_t | F(t;\Theta), \mathbf{K}_t\right) = \prod_{r=1}^{n_t} g\left(\mathbf{y}_t^r | f\left(t;\theta^{k_t^r}\right)\right) \tag{10.6}$$

Following the usual point measurement development, we can rearrange the likelihood from sum-product form into a product of sums

$$\begin{aligned} \hat{\Theta} &= \arg\max \prod_{t=1}^{T} \sum_{\mathbf{K}_t} \prod_{r=1}^{n_t} \left\{ \pi_t^{k_t^r} g\left(\mathbf{y}_t^r | f\left(t;\theta^{k_t^r}\right)\right) \right\}, \\ &= \arg\max \prod_{t=1}^{T} \prod_{r=1}^{n_t} \sum_{k=0}^{M} \left\{ \pi_t^k g\left(\mathbf{y}_t^r | f\left(t;\theta^k\right)\right) \right\}. \end{aligned} \tag{10.7}$$

This likelihood has low computation complexity and is referred to as ML-PMHT. The expression has implicitly assumed that $f(t; \theta^0)$ means something, but, in practice, the clutter is usually assumed uniform and we acknowledge that θ^0 exists only to simplify notation. The priors π_t^m are generally assumed known in ML-PMHT: typically $\pi_t^m = P_D/n_t, m > 0$.

The two key differences between ML-PMHT and the standard PMHT derived in Chap. 3 are the dynamic model and the statistical objective. ML-PMHT assumes a deterministic target whereas PMHT assumes a randomly evolving target state. The second difference is that PMHT uses EM to seek an approximation of the state that optimises the joint probability of the state sequence and the assignments whereas ML-PMHT directly optimises the marginal state likelihood. ML-PMHT is able to do this because of the deterministic target model. It does not use EM so there are no iterations. ML-PDA has been shown to give detection performance almost as good as image-based TkBD algorithms [4] and we might expect that ML-PMHT would have similar performance.

10.2 Image Measurements

The Quanta, or ML-H-PMHT, algorithm applies the same maximum likelihood objective and deterministic target model to the H-PMHT image association method [5, 6, 14]. Just like the point measurement case, the key difference is the way that the target state estimates are produced. Fundamentally, the algorithm looks very much like core H-PMHT with a different state update. It assumes a Gaussian appearance function with a known circular spreading covariance Σ.

The algorithm proceeds as follows:

1. Initialise the mixing proportion estimates $\hat{\Pi}$ and target parameters $\hat{\Theta}$.
2. Determine target-pixel probabilities as usual from (4.8)

$$G_t^{m,i} = \int_{W^i} g\left(\mathbf{y}_t^r | f\left(t; \theta^k\right)\right) \mathbf{dy}$$

   except that the target state at t is derived from the parameters $\hat{\theta}^m$.
3. Calculate association weights using (4.24):

$$w_t^{i,m} = \frac{\hat{\pi}_t^m G_t^{m,i}}{\sum_{s=0}^{M} \hat{\pi}_t^s G_t^{s,i}}.$$

   and hence the associated images (4.67) $\mathfrak{z}_t^{i,m} = w_t^{i,m} \bar{\mathbf{z}}_t^i$.
4. Update the prior estimates using (4.69)

$$\hat{\pi}_t^m := \frac{||\mathfrak{z}_t^m||}{||\bar{\mathbf{z}}_t||}.$$

5. For each track m:
   - Determine the pixel centroids (4.34)

$$\bar{\mathbf{y}}_t^{m,i} = \frac{\int_{W^i} \mathbf{y}\mathcal{N}\left(\mathbf{y}; h\left(\hat{\mathbf{x}}_t^m\right), \Sigma\right) d\mathbf{y}}{G_t^{m,i}}.$$

   - Determine the synthetic measurement vector using (4.53)

$$\tilde{\mathbf{y}}_t^m = \frac{\sum_{i=1}^{I+I^U} \bar{n}_t^i w_t^{i,m} \bar{\mathbf{y}}_t^{m,i}}{\sum_{i=1}^{I+I^U} \bar{n}_t^i w_t^{i,m}},$$

   - Update the state parameter estimates $\hat{\theta}^m$
6. Repeat steps 2 to 5 until convergence.

The parameterisation in [14] uses the start and end states to describe each linear target path, namely $\theta^m = \left[\mathbf{x}_0^m, \mathbf{x}_T^m\right]$. Using this parameterisation, the state at any intermediate time is found by interpolating

$$\mathbf{x}_t^m = \frac{t}{T}\mathbf{x}_0^m + \left(1 - \frac{t}{T}\right)\mathbf{x}_T^m \equiv \mathsf{H}_t\theta^m. \tag{10.8}$$

The updated parameter estimates are then expressed in terms of a pseudo inverse

$$\hat{\theta}^m = \left(\mathsf{H}^\mathsf{T}\mathsf{H}\right)^{-1} \mathsf{H}^\mathsf{T}\tilde{\mathsf{Y}}, \tag{10.9}$$

$$\mathsf{H} = \begin{bmatrix} ||\mathfrak{z}_1^m|| \, \mathsf{H}_1 \\ \vdots \\ ||\mathfrak{z}_T^m|| \, \mathsf{H}_T \end{bmatrix}, \tag{10.10}$$

$$\tilde{\mathsf{Y}} = \begin{bmatrix} ||\mathfrak{z}_1^m|| \, \tilde{\mathbf{y}}_1^m \\ \vdots \\ ||\mathfrak{z}_T^m|| \, \tilde{\mathbf{y}}_T^m \end{bmatrix}. \tag{10.11}$$

The expression in (10.9) is a weighted curve-fitting solution where each frame t has a measurement $\tilde{\mathbf{y}}_t^m$ and a weight $||\mathfrak{z}_t^m||$. That is, the associated power in each frame determines the weight applied to the fit error for that frame. Using this intuition, we can see how the more general situation of a higher order state model could be solved.

At the time that we write, Quanta has not been applied to appearance estimation or non-Gaussian targets. However, it is quite clear that the methods in the previous few chapters could readily be adapted to the deterministic kinematic state context.

10.3 Implementation

The examples of ML-H-PMHT and ML-PMHT in the literature mostly use linear motion. The exception is [12] that uses a straight-turn-straight model for manoeuvring targets. In principle, implementation can be as simple as encoding the likelihood function and deploying one's favourite optimisation engine. In practice, this can be inefficient and may not converge. Figure 10.1 shows an illustrative ML-PDA likelihood surface. There are two problems with this surface from a general optimisation perspective. First, there are a large number of local maxima that could easily distract a hill-climbing method. Second, between the maxima, there are large expanses where the surface is completely flat. This occurs because there are no measurements at all inside the association gate and so the likelihood is independent of the state. Blanding estimated that this plateau region occupies perhaps as much as 70% of the state space [2]. Again, this will cause much angst for gradient methods. One solution is to use a tailored approach that initialises multiple local searches on each of the local maxima. This is quite feasible because the maxima are naturally associated with point measurements. This is referred to as the directed subspace search [2] and is also applicable to ML-PMHT.

Another area where the ML-PMHT and ML-H-PMHT literature differs from the methods described earlier in this book is in the area of track management. In Chap. 5, we described a method of track acceptance or rejection based on a track quality statistic that is generally Bayesian in nature, such as the estimated Poisson rate. For ML-PMHT the more standard approach is to apply a Neyman–Pearson test to the value of the likelihood surface at the optimal state. This is based on extreme

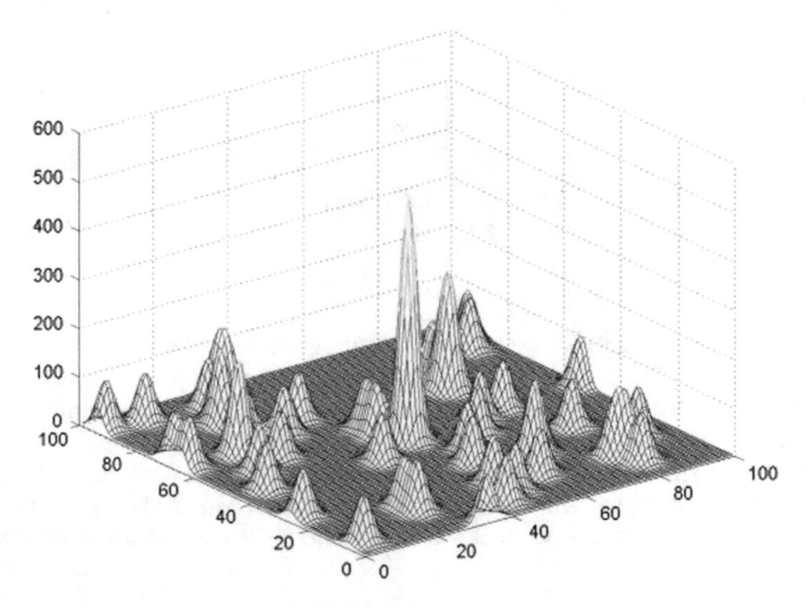

Fig. 10.1 Illustrative ML-PDA likelihood surface

value analysis applied to the statistics of false tracks [3]. In [14], the improvement in the likelihood surface by introducing a new track was tested against a Minimum Data Length criterion [10].

10.4 Summary

The ML-PMHT and ML-H-PMHT are methods that apply a deterministic target model to the tracking problem. One consequence of this deterministic model is that the resampled target prior is not required because there is no target prior. These methods are quite new and have only a relatively small body of literature. However, all of the extensions presented in the earlier chapters of this book could be applied to the deterministic target case.

References

1. Bar-Shalom, Y., Willett, P.K., Tian, X.: Tracking and Data Fusion: A Handbook of Algorithms. YBS, Storrs (2011)
2. Blanding, W.R., Willett, P.K., Bar-Shalom, Y., Lynch, R.S.: Directed subspace search ML-PDA with application to active sonar tracking. IEEE Trans. Aerosp. Electron. Syst. **44**, 201–216 (2008)
3. Blanding, W., Willett, P.K., Bar-Shalom, Y.: ML-PDA: advances and a new multitarget approach. EURASIP J. Adv. Signal Process. (2008)
4. Davey, S.J.: SNR limits on Kalman filter detect-then-track. IEEE Signal Process. Lett. **20**, 767–770 (2013)
5. Dunham, D.T., Ogle, T.L., Willett, P.K., Balasingam, B.: Advancement of an algorithm. In: proceedings of SPIE Signal and Data Processing of Small Targets, vol. 9092 (2014)
6. Dunham, D.T., Willett, P.K., Ogle, T.L., Balasingam, B.: Quanta tracking algorithm for multiple moving targets. In: proceedings of the 19th International Conference on Information Fusion, pp. 1119–1124 (2016)
7. Kirubarajan, T., Bar-Shalom, Y.: Low observable target motion analysis using amplitude information. IEEE Trans. Aerosp. Electron. Syst. **32**, 1367–1384 (1996)
8. Perlovsky, L.I.: MLANS neural network for track before detect. In: IEEE National Aerospace and Electronics Conference (1993)
9. Perlovsky, L.I., McManus, M.: Maximum likelihood neural networks for sensor fusion and adaptive classification. Neural Netw. **4**, 89–102 (1991)
10. Rissanen, J.: Modeling by shortest data description. Automatica **14**, 465–471 (1978)
11. Schoenecker, S., Willett, P., Bar-Shalom, Y.: A comparison of the ML-PDA and the ML-PMHT algorithms. In: Proceedings of the 14th International Conference on Information Fusion, Chicago, USA (2011)
12. Schoenecker, S., Willett, P.K., Bar-Shalom, Y.: The ML-PMHT multistatic tracker for sharply maneuvering targets. IEEE Trans. Aerosp. Electron. Syst. **49**, 2235–2249 (2013)
13. Schoenecker, S., Willett, P.K., Bar-Shalom, Y.: ML-PDA and ML-PMHT: comparing multistatic sonar trackers for VLO targets using a new multitarget implementation. IEEE J. Ocean. Eng. **39**, 303–317 (2014)
14. Willett, P.K., Balasingam, B., Dunham, D.T., Ogle, T.L.: Multiple target tracking from images using the maximum likelihood HPMHT. In: proceedings of SPIE Signal and Data Processing of Small Targets, vol. 8857 (2013)

Chapter 11
Radar and Sonar Track-Before-Detect

The previous set of chapters introduced a collection of different extensions to the core H-PMHT that expand its range of applications and improve the quality of its output. These chapters have been rather theoretical and highly mathematical. We now shift the focus to applications and provide a series of examples of how the various versions of H-PMHT described in this book can be used in practice. This chapter presents applications of H-PMHT to surveillance problems using radar and sonar. Five out of the six applications use archived sensor data from experimental systems, the other is simulation-based. A common feature of these applications is that the time between measurements is relatively high compared with the target dynamics. In most of the applications, the targets are small compared with the sensor resolution, although this is not always the case.

The first example uses data collected by Defence Research and Development Canada (DRDC) and was collected by an airborne maritime surveillance radar observing a highly manoeuvrable boat on a rough sea surface. The challenge in this application is the rate with which the boat's signature moves through the range cells and the variation in its speed. The second example is Defence Science and Technology Group's experimental passive bistatic radar system. In this application, the illumination is due to a terrestrial digital video transmission. The receiver collects the direct signal and scattered returns from objects in the field of view. The final radar application is a simulation of skywave Over-the-Horizon Radar (OTHR). Skywave OTHR uses the propagation properties of signals in the HF band to observe objects at very long ranges, over 1000 km. The propagation medium is a region of the atmosphere called the *ionosphere* and in some conditions it supports multiple signal paths. The treatment here compares alternative approaches for dealing with this multipath environment. The OTHR case uses a three-dimensional measurement image.

The remaining examples in this chapter deal with sonar applications. The first of these uses data from a Defence Science and Technology Group experimental towed active sonar system referred to as CASSTASS. The data sets contain observations

S. J. Davey and H. Gaetjens, *Track-Before-Detect Using Expectation Maximisation*, Signals and Communication Technology,
https://doi.org/10.1007/978-981-10-7593-3_11

of an echo repeater collected off the coast of Western Australia. The second sonar example is a passive sonar application where the acoustic sensor listens to the sounds emitted by other vessels. In this application, the sensor image is a frequency spectrum rather than spatial bins. The final sonar example is a Defence Science and Technology Group demonstration system referred to as Aura that uses acoustic signals in the air to illustrate signal processing concepts. It is implemented as an application for IOS devices, such as an iPAD, and is intended to be used by STEM students.

11.1 Airborne Maritime Surveillance

The first application is an experiment to observe a small manoeuvrable boat using an airborne radar. The data used was collected by Defence Research and Development Canada (DRDC) in Halifax harbour, Canada, using the DRDC Ottawa X-band Wideband Experimental Airborne Radar. The sensor observed a small speedboat, as shown in Fig. 11.1. A detailed description of the experiment and the data characteristics can be found in [21–23]. The results in this section were first presented in [9].

We first discuss some salient features of the experimental data and then describe alternative methods to process it. We consider a conventional detect-then-track approach based on probabilistic data association and two alternatives based on

Fig. 11.1 Small speedboat target

H-PMHT. Due to the nature of the sensing platform, the radar moves during image collection. The first H-PMHT approach ignores this motion and tracks directly in the sensor range-bearing measurement frame using core H-PMHT. The second H-PMHT uses the cell-varying Gaussian H-PMHT described in Chap. 7.

11.1.1 Data Characteristics

There are three features of the DRDC data that make it challenging: the range resolution, the sea clutter, and the platform motion. The range resolution of a radar system is a function of the waveform bandwidth and is given by $\Delta r = c/(2B)$, where c is the speed of light and B is the bandwidth. In this case, the sensor bandwidth was relatively high and so the range resolution of the sensor was quite fine compared with the physical extent of the target. The sensor swept through azimuth and collected range profiles. Again, the spacing of these azimuth beams was relatively fine compared with the beamwidth of the sensor. There are two consequences of this high resolution. First, the appearance function of the target spreads over many bins in range and azimuth. This is an advantage because it makes it easier to discriminate the spatially correlated target response from more localised clutter. The second consequence is that the target moves through a very high number of range bins from one scan to the next. If the target speed were constant, this would not be a problem, but in this case, the target speed was highly variable, so it is difficult to predict where its response will be from scan to scan.

The second feature is the clutter characteristics. In the theoretical development of the algorithms in this book, we have assumed spatially uncorrelated noise. We have shown how to deal with the case where the noise density is spatially non-uniform, but even in this case, we assume that each shot measurement is an independent draw from that non-uniform distribution. The clutter in a real system is different from that. The term clutter really refers to unwanted energy in the sensor image that arises from physical sources, not of interest to the user. In this application, the clutter comes from scattering off of waves on the sea surface. These waves are obvious spatially correlated and the responses they generate have similar correlation structure to the target response: we actually get bright blobs in the image due to clutter. Even worse, these clutter peaks are often stronger in energy than the target response. Coupled with the manoeuvrability of the target, this makes things very awkward. McDonald [21–23] has studied the structure of this noise in detail.

The third data feature is the motion of the platform. In all of the discussion leading to this point we have described sensors as though they simultaneously capture a snapshot across an array of pixels at one time instant. This is true for some sensors, but not all. Many radars operate by steering a single narrow beam and scan this beam spatially to gather data over the scene. This means that the platform and the target are not fixed during the data collection. Since the target motion is unknown, there is little we can do about it short of a very computationally expensive feedback loop. However, the platform motion is known. The location of the platform is a parameter

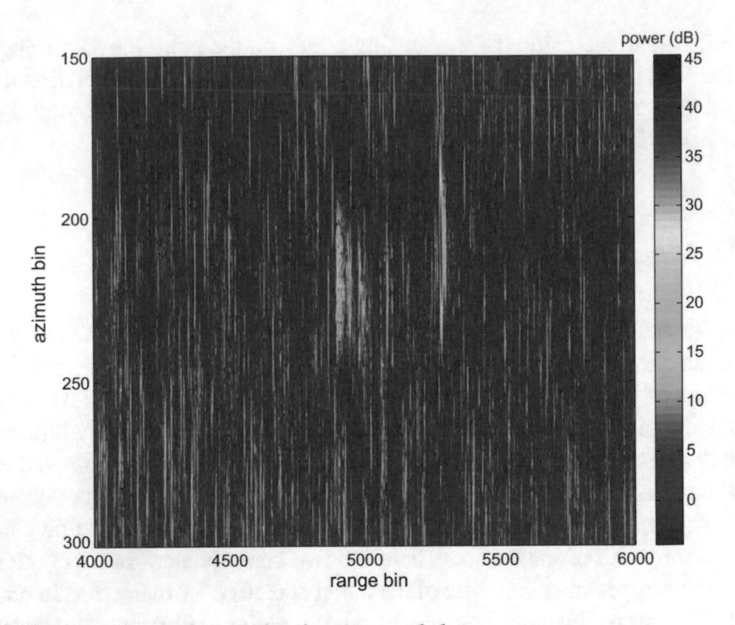

Fig. 11.2 Segment of a single scan showing target and clutter

that determines the relative range and bearing to a target, so it is natural to build this knowledge into the processing We use the cell-dependent spread function described in Chap. 7 because it is more efficient than particles [8].

The volume of data from the sensor is relatively large: each scan consists of around 13000 range bins and over 400 azimuth bins. This makes it challenging to process the data in a timely manner and provides an excellent measure of the scalability of the H-PMHT approach to realistic surveillance volumes.

Figure 11.2 shows a segment of one scan. Two strong returns are evident near the centre of the image. The rightmost return is from a buoy that happened to be near the target, and the leftmost one is the speedboat. Both buoy and boat have a response that is spread over tens of range bins and tens of azimuths. Numerous other vertical streaks are evident. These are due to sea clutter and are unfortunately target-like. Dividing the environment into background noise and interference spikes, it is clear that the signal to noise ratio is high. However, the signal to interference ratio is not. In fact, the brightest spots on each image are usually sea clutter spikes.

McDonald has characterised the clutter distribution [21–23] and found that the clutter was well described by a KA-distribution. The common and simpler Rayleigh distribution comes from assuming complex Gaussian noise with zero mean and a constant variance. The K-distribution belongs to the more general class of compound Gaussian distributions and extends Rayleigh by allowing the noise variance to be random with a known mean. The KA-distribution further extends this by including an outlier distribution; refer to [22] for more details. The tails of the distribution are highest for the KA-distribution and lowest for the Rayleigh distribution. This is of

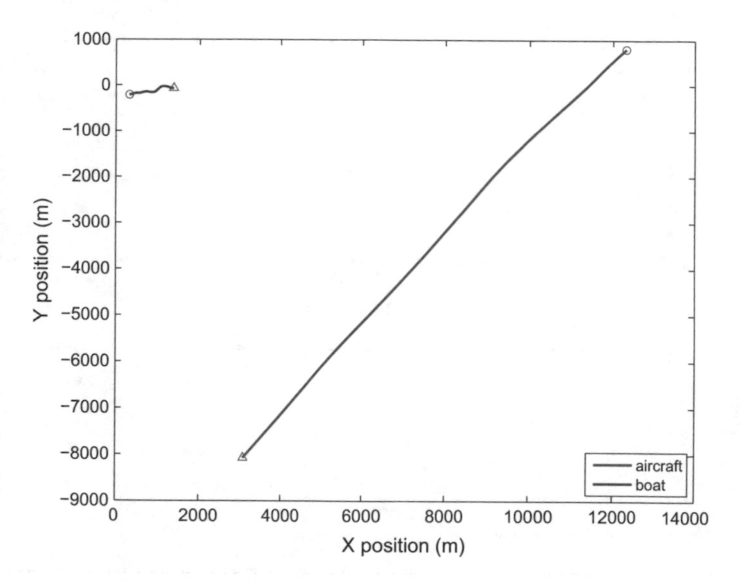

Fig. 11.3 Target-sensor geometry

no direct consequence for H-PMHT since it does not use the data likelihood ratio, but it may be expected to lead to higher false track rates.

Figure 11.2 also highlights that the target response in range is not simply a broad point spread function, but rather shows a much more complicated structure due to scattering off various parts of the boat. The target has the potential to be perceived by the sensor as several targets.

The speedboat was fitted with a GPS logger. However, registration of the GPS data with the radar data has proven difficult [23]. Thus the GPS data is useful as an indication of where the target is, but it is not useful from the perspective of determining target localization error.

The data of interest is a sequence of 40 frames collected while the speedboat was manoeuvring. The boat was initially almost stationary for around 10 frames and then it accelerated, following a snaking trajectory for the remaining frames. The signal to noise ratio during the first part of the trajectory was relatively poor, but increased when the speedboat moves more quickly.

Figure 11.3 shows the overall scenario geometry in Cartesian coordinates. Circles mark the starting position of the aircraft and the boat and triangles mark their final position. The aircraft maintained an approximately constant altitude of 300 m and moved in an approximately straight path.

Figure 11.4 shows the speedboat trajectory in local Cartesian coordinates and Fig. 11.5 shows the trajectory in the measurement frame, that is range and azimuth cells. These figures highlight the volatility of the target motion and also demonstrate that the target was moving very rapidly through the sensor field of view. Between scans 10 and 25, the target moves through roughly 1500 range cells.

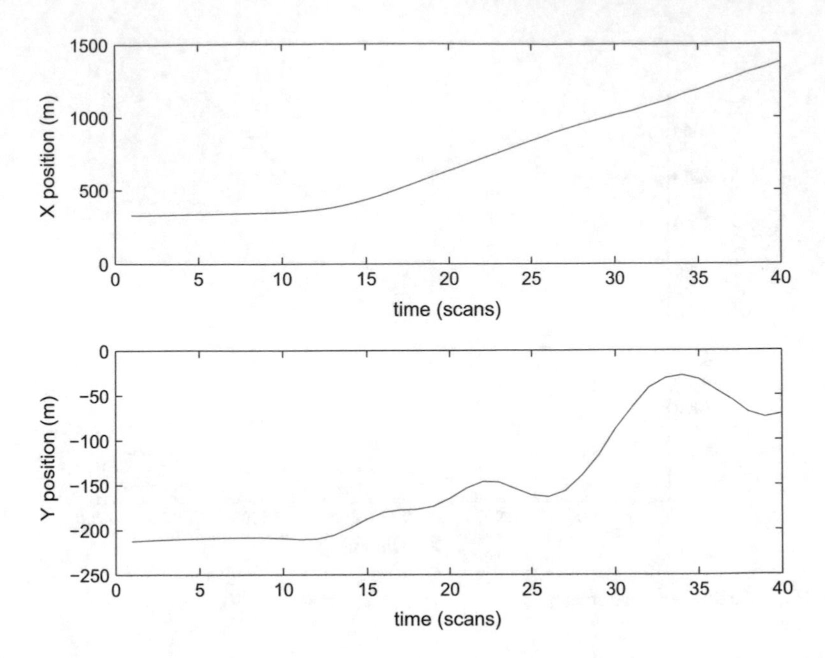

Fig. 11.4 Target trajectory (local Cartesian frame)

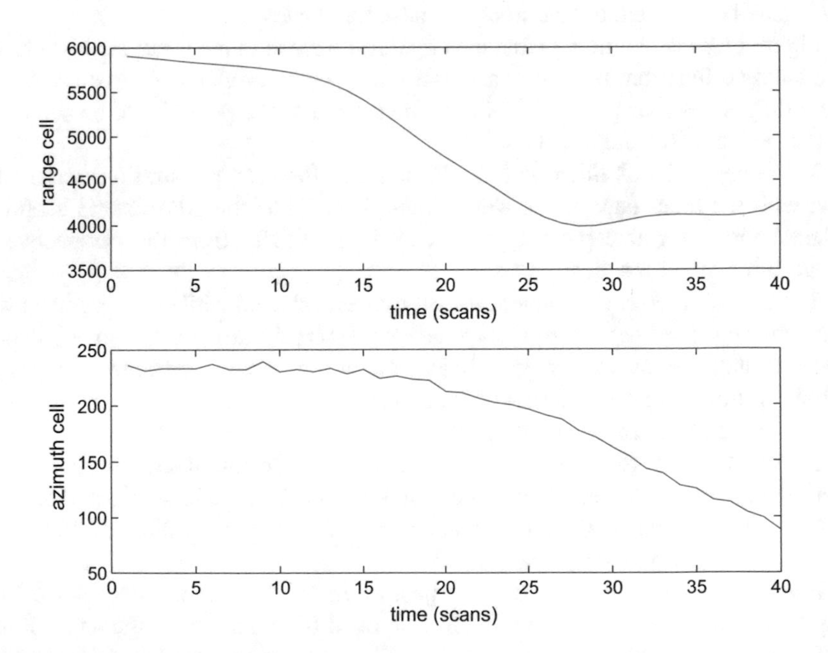

Fig. 11.5 Target trajectory (sensor frame)

11.1.2 Point Measurement Detection and Tracking

The conventional approach is to apply a single-frame detector followed by a point measurement tracking algorithm. In this case, the size of the target response is relatively large compared with the sensor resolution, so it is appropriate to use an image segmentation-based detector. The general strategy is to find the centroids of extended objects and track them using a Probabilistic Data Association Filter (PDAF), similar to [25].

A significant effort was spent in tuning the preprocessing before applying the PDAF. First, it was observed that the target signal was spread over a greater number of image cells than the impulsive noise returns. This motivated the use of spatial averaging, which was implemented by convolving the sensor image with a 10 beam by 5 range uniform rectangular kernel. A threshold was then applied to the smoothed image and adjacent pixels that exceed the threshold using a two-pass labelling algorithm [18]. This can be efficiently implemented in Matlab via the *bwlabel* function. The centroid of each connected set of pixels was determined and this centroid was mapped to a Cartesian reference space. Measurements were only retained if the number of foreground pixels (i.e. threshold crossings) associated with a particular object was within an upper and lower tolerance band. This single-frame detection scheme was very CPU-intensive to perform, but yielded a high probability of detection for very few false alarms. From this point onwards, it was straightforward to form tracks, so only a very simple point-tracker was considered.

Figure 11.6 illustrates the preprocessing used to feed the conventional tracker. Figure 11.6a shows the raw radar image with a green circle at the GPS location, which is close to the target but not co-located with it. Figure 11.6b shows the result of the smoothing filter and (c) shows the output of the segmentation algorithm. The larger target near the middle of the image is the buoy and the smaller one to the left of the buoy is the boat.

The point measurement tracker used a converted-measurement approach to deal with the nonlinearity [3], which means that each measurement was transformed into a local Cartesian reference frame on the ground using the known sensor location and then the estimation problem was treated as linear. The tracker assumed a Cartesian measurement model with a fixed sensor variance, R, as given in Table 11.1. The X-direction in the local frame was approximately aligned with the radial between the target and sensor (although this varied with time) and so the measurement noise was smaller in the X-direction than the Y-direction.

The PDAF used was a simplified version of the algorithm described in [6] and differed from a textbook algorithm (e.g. [3]) only in that it used target *visibility* to determine the merit of tracks for track maintenance decisions. Under the visibility approach, the PDAF considers $n_t + 2$ association events corresponding to the assignment of each measurement, a valid but undetected target, and an *invisible* target. The invisible target case essentially corresponds to a false track and its probability is used to make maintenance decisions. Uniform clutter was assumed, the target state was

Fig. 11.6 Single-frame detection

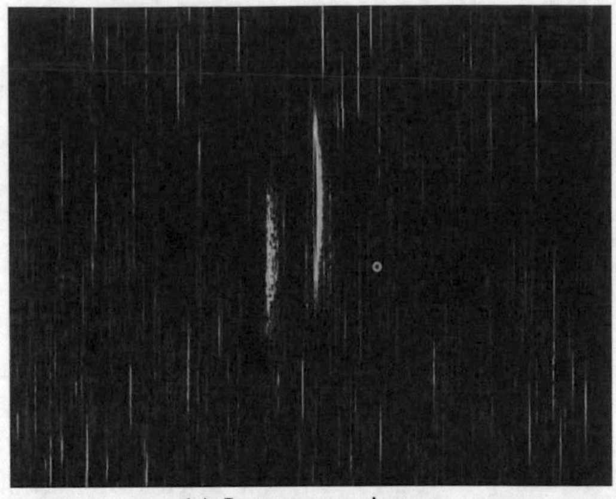

(a) Input sensor image.

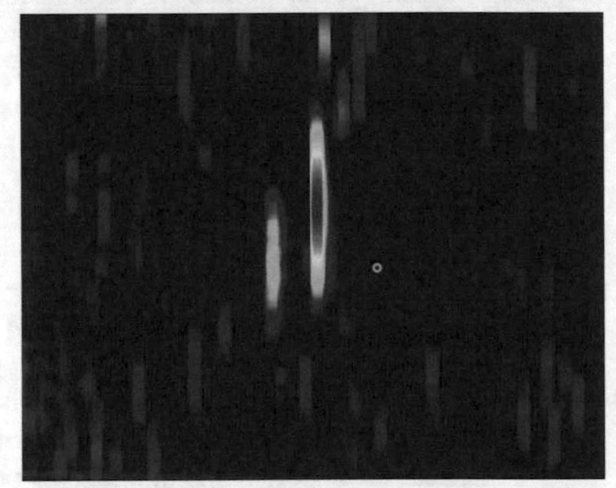

(b) Smoothed sensor image.

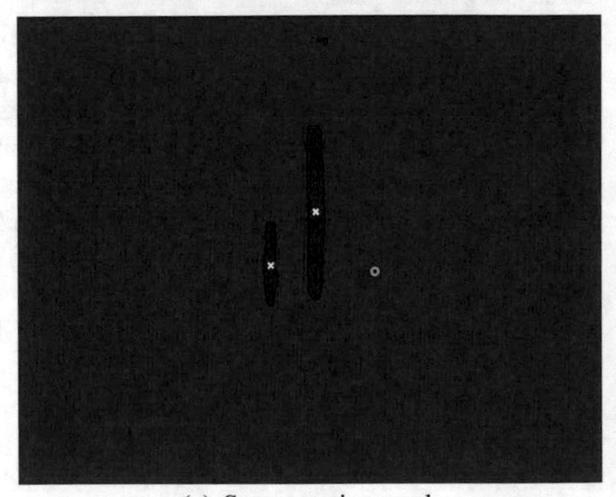

(c) Segmentation mask.

Cartesian position and velocity in two-dimensions and an almost constant velocity model was used. The process noise for this model, Q, is given in Table 11.1.

Thus the measurement association weights were given by

$$\beta^r = \frac{b^r}{\sum_{r=-1}^{n_t} b^r}, \tag{11.1}$$

where n_t is the number of measurements in the scan and

$$b^r = \begin{cases} \dfrac{1 - Pv_{t|t-1}}{Pv_{t|t-1}} & r = -1\,, \\ 1 - Pd & r = 0\,, \\ \frac{PdV}{|2\pi \mathsf{S}|} \exp\{-\frac{1}{2} d^r\} & 0 < r \le n_t\,, \end{cases} \tag{11.2}$$

with Pd the probability of detection, V the surveillance volume, S the innovation covariance, $d^r = (\mathsf{H}\mathbf{x} - \mathbf{y}^r)^{\mathsf{T}} \mathsf{S}^{-1} (\mathsf{H}\mathbf{x} - \mathbf{y}^r)$ the statistical distance between the track and measurement r, and $Pv_{t|t-1}$ the predicted target visibility, given by

$$Pv_{t|t-1} = (1 - P_{\mathsf{death}}) Pv_{t-1|t-1} + P_{\mathsf{rebirth}} (1 - Pv_{t-1|t-1})\,, \tag{11.3}$$

where the probability of target death and rebirth are tuning parameters given in Table 11.1. The updated target visibility is given by $Pv_{t|t} = 1 - \beta_{-1}$ and is used as the basis of track management decisions. Similar to the H-PMHT track manager, candidate tracks are formed using two-point differencing. Table 11.1 lists the parameters of the conventional detection and tracking approach.

11.1.3 H-PMHT Models and Parameters

The H-PMHT was implemented as a time-recursive filter for this analysis, that is, there was no batch; each scan was processed sequentially and only one scan was available to the algorithm at a time. The state estimates were not smoothed.

Two processing strategies for H-PMHT were considered. In the first, the movement of the sensor was ignored, and the target state was modelled in the measurement frame. That is, the state was pixels and pixels per frame. This makes life much easier for implementation, and the separable point spread function expressions derived in Chap. 5 can be used with core H-PMHT. This version is referred to as H-PMHT(rb) since the target state is in range-beam space.

The second strategy was to model the target state in Cartesian coordinates on the ground and use the cell-dependent nonlinear method from Chap. 7 to relate the target state to the sensor image. This approach has much higher implementation complexity but should more accurately model the true system. This version is referred to as H-PMHT(xy).

Table 11.1 Conventional detection and tracking parameters

Detector Parameters	
Minimum segment size	75 pixels
Maximum segment size	None
Segmentation threshold	2 × image mean
Tracker parameters	
Track state vector	$[x, \dot{x}, y, \dot{y}]^{\mathrm{T}}$ x and y in metres $\dot{x}$ and $\dot{y}$ in metres per frame
Measurement vector	$[x, y]^{\mathrm{T}}$ in metres
R	$\begin{bmatrix} 400 & 0 \\ 0 & 900 \end{bmatrix}$
Q	$100 \begin{bmatrix} \mathrm{Q2} & 0 \\ 0 & \mathrm{Q2} \end{bmatrix}$
Q2	$\begin{bmatrix} \frac{1}{3}\Delta_t^3 & \frac{1}{2}\Delta_t^2 \\ \frac{1}{2}\Delta_t^2 & \Delta_t \end{bmatrix}$
P_{death}	0.012
P_{rebirth}	0
Initial visibility	$Pv_0 = 0.5$
Promotion threshold	$Pv_{t\|t} > 0.6$ for any $t > 4$ frames

The background was assumed to be uniform and the target appearance function was assumed to be Gaussian with a diagonal covariance matrix as given in Table 11.2. The target was more spread in azimuth than range, so the azimuth variance was higher.

The target model was an almost constant velocity model in the plane. That is, the target motion was approximated with a constant rate of movement through range and azimuth bins. The target state was in units of bins for position and bins-per-scan for velocity. This was transformed into ground units as a post-processing stage for comparison with the other trackers and the ground truth. The process noise variance used is given in Table 11.2, also using bins and scans as units.

Candidates were promoted if they had an estimated SNR of greater than 20dB for four frames and terminated if they had an estimated SNR of less than −5dB for two consecutive frames.

The H-PMHT(xy) software was a Matlab implementation of the cell-varying psf in Chap. 7. This algorithm used the same measurement covariance matrix as the measurement-space implementation and also used an almost constant velocity target model. However, the target state was in different units, namely metres, so a different process noise covariance matrix was required, as given in Table 11.2. The same track management rules were used as for the measurement-space implementation.

For the H-PMHT(xy), the position part of the state vector was in metres, but the measurement frame was in range and azimuth cells. The estimator was, therefore,

Table 11.2 H-PMHT parameters

Detector Parameters	(used to form candidate tracks)
Minimum segment size	75 pixels
Maximum segment size	None
Segmentation threshold	3 × image mean
H-PMHT(rb) Parameters	
Track state vector	$[r, \dot{r}, b, \dot{b}]^{\mathsf{T}}$ r and b in cells $\dot{r}$ and $\dot{b}$ in cells per frame
Q	$\begin{bmatrix} 100\mathrm{Q2} & 0 \\ 0 & 10^4\mathrm{Q2} \end{bmatrix}$
H-PMHT(xy) Parameters	
Track state vector	$[x, \dot{x}, y, \dot{y}]^{\mathsf{T}}$ x and y in metres $\dot{x}$ and $\dot{y}$ in metres per frame
Q	$1.2 \times 10^5 \begin{bmatrix} \mathrm{Q2} & 0 \\ 0 & \mathrm{Q2} \end{bmatrix}$
Common Parameters	
Measurement vector	$[r, b]^{\mathsf{T}}$ in cells
R	$\begin{bmatrix} 400 & 0 \\ 0 & 900 \end{bmatrix}$
Confirmation threshold	4 frames greater than 20dB

implemented as an EKF. The measurement model included the scaling and offset from range in metres to range in cells and similarly for azimuth.

11.1.4 Results

The tracking outputs of the PDAF-based approach and the two H-PMHT approaches are now presented. Each is compared with GPS data collected from the target of interest.

Figure 11.7 shows all of the output tracks from each of the three trackers overlaid with the GPS measurements. The target tracks are roughly centred in the plot and move from left to right. As mentioned earlier, there was a buoy that was coincidentally in the region, near to the target. The buoy location is slightly north of the target starting position. There are no other known targets in the area, and the other tracks are assumed to be false.

All three trackers were able to detect the speedboat and the buoy. The PDAF tracker shows a number of false tracks to the south of the area, which is closer in

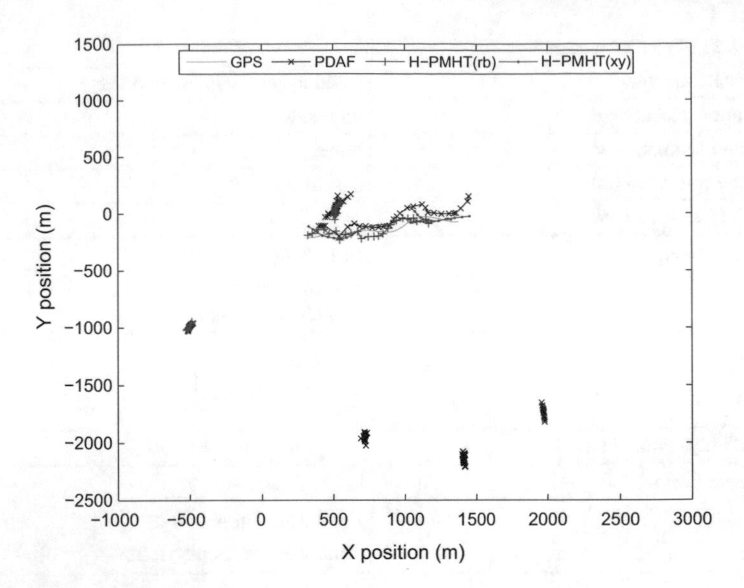

Fig. 11.7 Small speedboat target

range to the sensor. At close range, the clutter spikes are more prominent and so the tracker forms false tracks.

By manual inspection, the tracks that have followed the target were determined and are shown in Fig. 11.8. It is clear that none of the trackers were able to detect the target during the initial period where it was stationary. During this time it is not possible to detect the target by eye in the data either: the received SNR is very low. Once it starts moving and the SNR improves, the PDAF tracker took a little longer to establish track.

At approximately scan 32, the target performed a manoeuvre that caused the PDAF tracker to diverge. Unfortunately, soon after this, the target left the sensor coverage area in scan 39 and there is insufficient data to determine whether the PDAF would have recovered from the error. The measurement-space H-PMHT(rb) tracker lost the target at the same time. The ground-space H-PMHT(xy) followed the target until the second to last scan. However, during the last ten scans it produced a duplicate track on the target. The original track started out following the centre of mass of the target response, but later it drifted slightly to focus on the short-range component. Recall that the target response is a complicated superposition of multiple partially resolved scatterers. This caused the tracker to form a second track on the longer range component of the target response.

Figure 11.9 shows tracks that were manually determined to be following the buoy. The buoy has an extremely high SNR and is easily tracked by all of the algorithms. At times, the peak level of the buoy is 40 dB higher than the boat. However, the H-PMHT(xy) track is considerably smoother than the others, especially the PDAF which shows several large excursions. This is partly because the PDAF was tuned with a relatively high process noise to cope with the high manoeuvrability of the

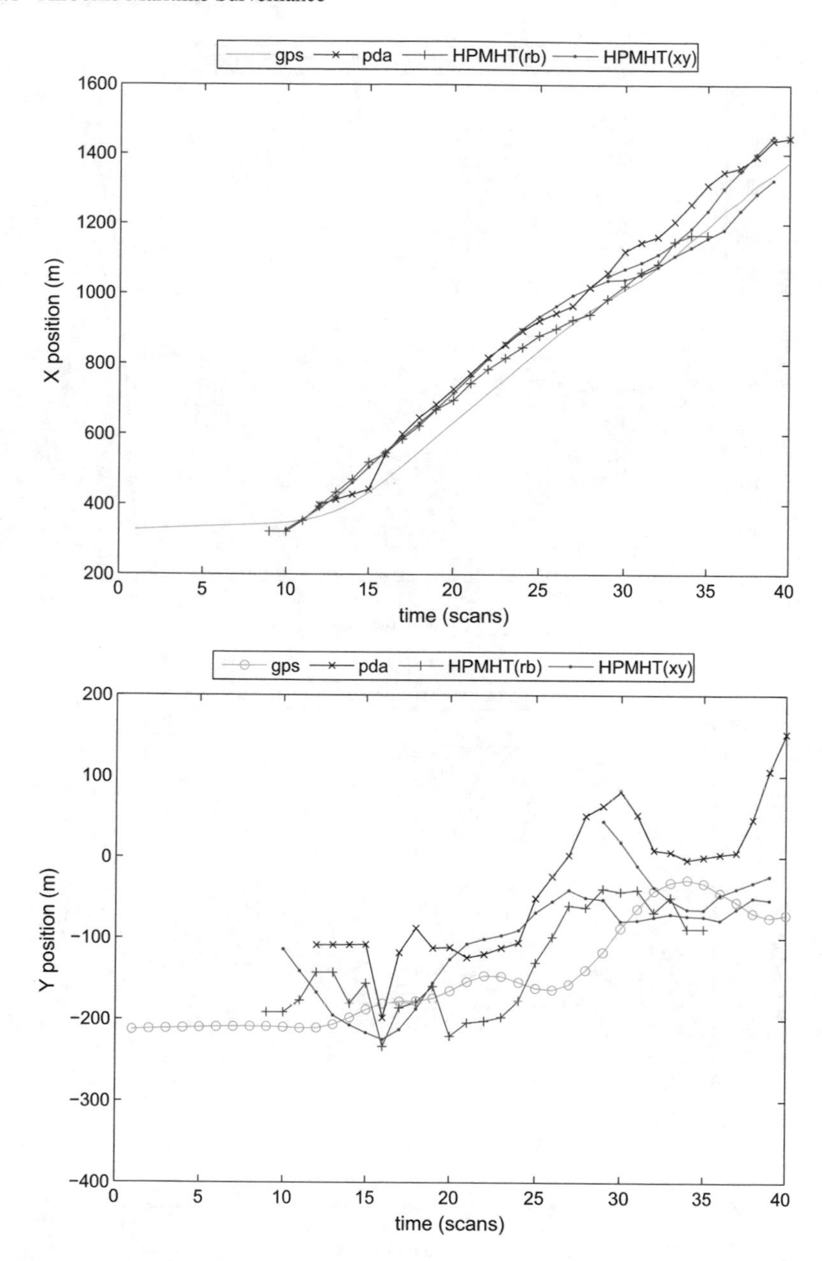

Fig. 11.8 Small speedboat target

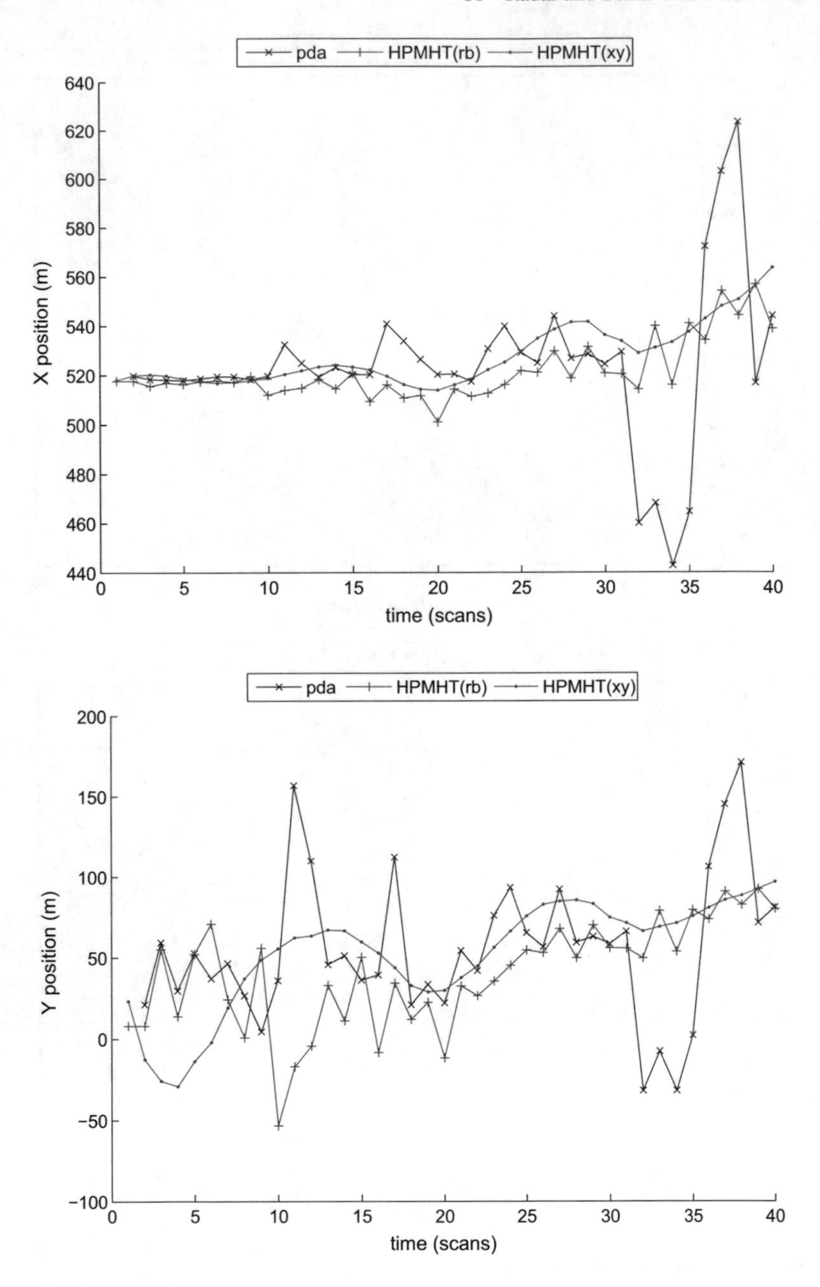

Fig. 11.9 Strong buoy target of opportunity

speedboat and partly because the detection process reports the centroid of a group of connected pixels, which can be highly variable.

The implementation of the algorithms was not optimised, so the CPU time spent by each is not a reliable measure of performance. Nevertheless, figures are given as a qualitative comparison. The H-PMHT(rb) was the fastest algorithm, requiring 280 CPU seconds to run under Matlab on a quad-core PC. For the PDAF approach, the single-frame detector was orders of magnitude more expensive than the PDAF tracker. This is because of the very large number of pixels in the sensor image and the double-pass clustering required to extract detections. In fact, the single-frame detector took around twice the computation effort of the measurement-space H-PMHT. Substantial experimentation showed that simpler detectors for the PDAF lead to an abundance of false tracks. The PDAF total cost was 478 CPU seconds.

The ground-space H-PMHT(xy) was the slowest at 575 CPU seconds. Since the sensor image is so large, the available memory limited the number of intermediate variables that could be stored. In particular, for candidate tracks, it was not feasible to store the P_l^m values, which meant that they had to be calculated twice for each EM iteration. These values alone can amount to around half a gigabyte of storage for a modest number of tracks. Thus the trade-off between memory and computation resulted in a longer execution time for the ground-space H-PMHT. For the sensor-space H-PMHT, these quantities can be stored as their marginal vectors, which are a fraction of a percent of the size. An optimised implementation of the ground-space algorithm could alleviate much of this overhead.

11.2 Passive Bistatic Radar

The second radar application is tracking on data from the Defence Science and Technology Group's passive bistatic radar [26]. The radar system exploits radio frequency transmissions from third-party sources, such as television and radio broadcasts. With the advent of digital broadcasts, these signals have reasonably good waveform properties. We provide tracking results for an example bistatic radar data set that was collected using a terrestrial television broadcast. The radar system collects the direct signal from the broadcaster and simultaneously collects reflected signals from objects in the field of view. The system coherently correlates to produce range-Doppler maps that are essentially input images for the tracker. Due to the bistatic geometry, the absolute target location can be ambiguous unless angle information is also used.

The test data set contained a car moving along a nearby road and an aircraft. The tracks produced are shown in Fig. 11.10. The aircraft is at longer range and moves much more quickly than the car. A third track was formed on what appears to be a short-lived target-like return. Example frames from the image sequence are shown in Fig. 11.11.

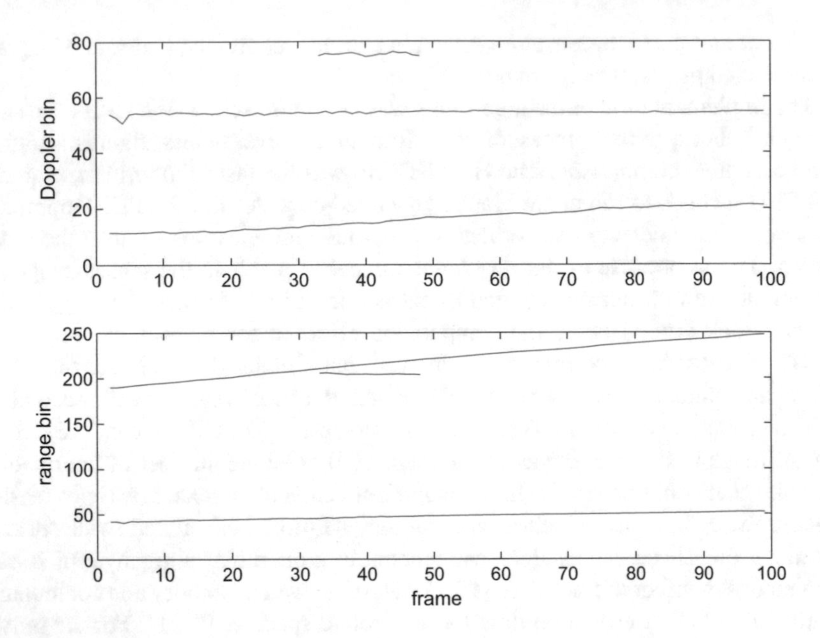

Fig. 11.10 Bistatic radar tracks

11.3 Over-the-Horizon Radar

The ionosphere is a region of the Earth's atmosphere that is ionised by solar radiation. Due to this ionisation it refracts electromagnetic waves and that provides a mechanism for propagation to ranges well beyond the line of sight. [14] Skywave over-the-horizon radar (OTHR) is a sensing modality that exploits refraction in the high-frequency band to illuminate targets at ranges up to 3,000 km. The simplest model of the ionosphere is a spherical mirror at a fixed altitude but in reality propagation is much more complex. In particular, there are multiple *layers* within the ionosphere that refract radiated signals. The layers commonly involved with skywave OTHR are referred to as the E layer, the F1 layer and the F2 layer. These different layers act as multiple propagation paths between the sensor and the targets and the range illuminated by each path is a function of the radar carrier frequency. At some combinations of range and frequency, a single target can appear as a cluster of returns at the receiver [15].

A skywave OTHR typically operates by transmitting a number of sweeps of a frequency modulated continuous wave signal and coherently integrating over time. A set of sweeps with the sensor steered to a particular location is referred to as a *dwell*. A return path from transmitter through the ionosphere to the target and back to the receiver through the ionosphere is referred to as a *mode*. Some modes propagate to the target and back along the same path whereas others go out and back along different paths. Figure 11.12 shows a two-layer example resulting in four two-way

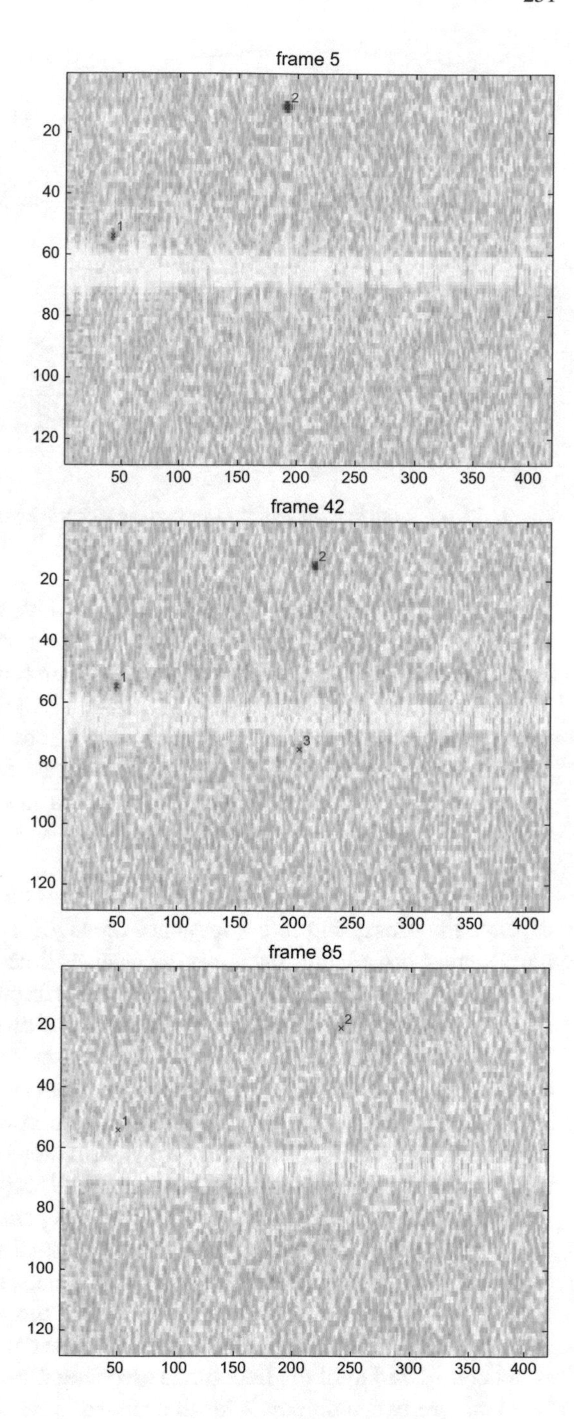

Fig. 11.11 Bistatic radar example frames

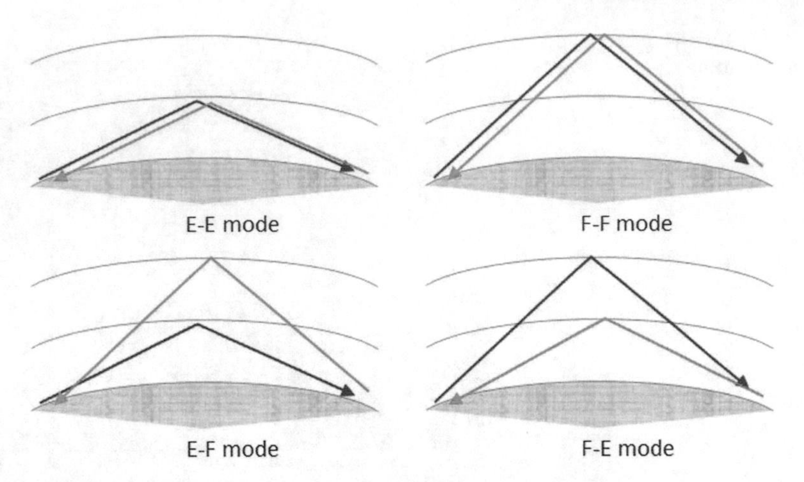

Fig. 11.12 Four mode propagation over two ionospheric layers

paths. Geometry means that paths over higher layers are longer in delay and paths returning from higher layers have a higher elevation. Assuming a linear receiver array and an off-boresight target, higher elevation translates to a change in measured azimuth due to the coning effect [15]. Figure 11.13 shows an example of how these four modes can appear to the sensor. The figure uses the sensor parameters described later for our simulations: the grid shows the range and azimuth bins; the start of each mode is marked with a circle and the end with a square. Higher elevation paths also reduce the component of the target horizontal velocity that is radial, leading to a lower magnitude Doppler shift.

The conventional way to process the data from this cluster of target returns is to separate the processing into a sequence of stages: range processing, beam-forming and Doppler processing transform the receiver time series into a three-dimensional Azimuth-Range-Doppler (ARD) data cube; interference mitigation and constant false alarm rate algorithms clean up the data; a single-dwell detector finds local maxima (peaks) in the data cube [7]; a measurement-space tracker performs multi-target tracking, for example [10]; an ionospheric model estimates the transformation from measurement space to a geographic coordinate system; and a multiple hypothesis data association engine is used to determine which tracks correspond to a family of returns from a common target, for example [29, 35]. This conventional processing chain implicitly assumes that the modes are independent at the single-dwell detector by modelling the ARD data cube as the superposition of point targets. The tracker explicitly models the modes as independent targets, so it assumes that the motion of each is not correlated with the others and that the presence of one mode does not influence the likelihood of observing others. The coupled nature of the target returns is not considered until the final mode association stage.

There are two problems with the conventional sequence of algorithms. Firstly, the detector and the tracker are both data reduction stages. While this is beneficial

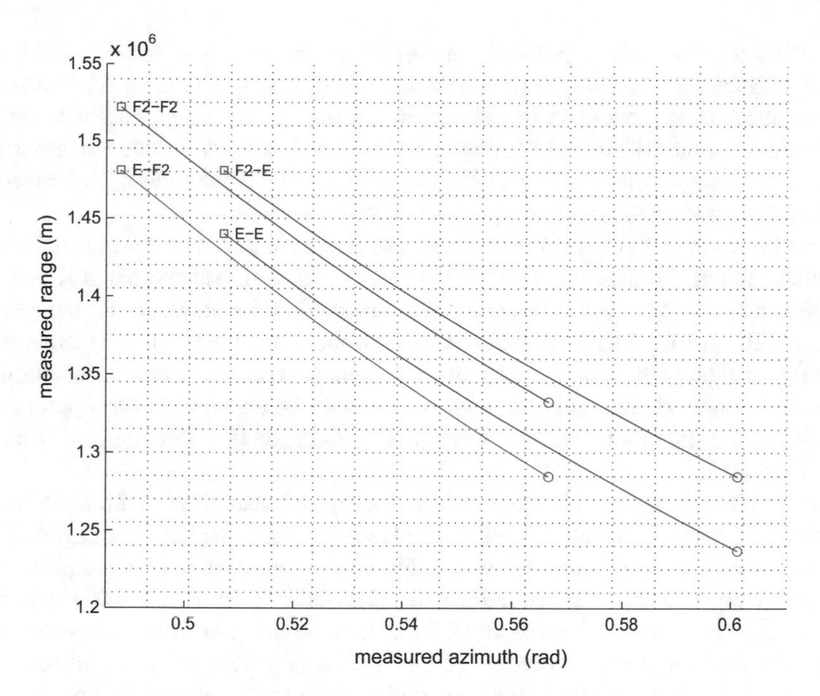

Fig. 11.13 Range-azimuth plot for two ionospheric layers

from the perspective of network communications overhead, it has the potential to discard useful data. In particular, the collection of tracks might not match the expected modes predicted by the ionospheric model. In some cases, this leads to ambiguity in the assignment of modes to tracks. For example, consider the mode pattern in Fig. 11.13, if two modes are detected at the same azimuth then the modes could be E-E and F2-E or E-F2 and F2-F2. If there is a high degree of ambiguity, the solution can be dominated by the prior, which is a function of the mode strengths estimated by the ionospheric model. The second problem is that there are two layers of one-to-one assignment required. Both of these require approximate algorithms because enumerating the association hypotheses is not feasible. The complexity is a consequence of the constraint that each measurement can be caused by only one mode and one target, and that each target can produce at most one measurement per mode.

Several alternative processing chains have been considered. Generally, these have focused on the idea of combining the mode association stage with the multi-target tracking stage. Pulford and Evans developed a Probabilistic Data Association (PDA) based tracking algorithm that associated peaks to modes first using a known ionosphere and later under fitting an assumed ionospheric model [27, 28].

A problem with PDA-based data association is that it explicitly enumerates over the joint assignment of all measurements to modes. The number of these joint assignments grows combinatorially with the number of modes and measurements. In fact, the PDA solution only requires the marginals of the joint assignments and these marginals can be efficiently estimated using belief propagation [11]. Alternatively, the measurement model can be relaxed in the manner of the probabilistic multi-hypothesis tracker (PMHT) [32] that models the assignments as independent realisations of a multinomial process. Liang et al. [19] extended the PMHT approach to marginalise over both target and mode association variables.

A different way to couple the sequential processes in the conventional chain is to combine the mode association and the ionospheric model. Anderson and Krolik [1, 2] used a maximum a posteriori association method in the presence of ionospheric uncertainty. A fully joint estimation scheme would combine all of the elements of the conventional chain, performing measurement to track and mode association as well as ionospheric modelling in a single method. An example of this approach is the Signal inversion for target extraction and registration (SIFTER) of Fridman and Nickisch [16].

A problem with methods that combine ionospheric estimation with mode association and tracking is that the models used to describe ionospheric propagation are usually nonlinear and highly detailed, using numerical estimation techniques, such as ray tracing, to estimate propagation paths [5]. Under the joint estimation approach the ionospheric states influence all of the measurements and so the measurements from different targets are correlated. Given the large number of ionospheric states this amounts to a very high dimension nonlinear state estimation problem.

Here we consider two ways of applying H-PMHT to the multipath environment. These results were first presented in [11].

11.3.1 Measurement Model

We revise notation here because the OTHR application leads to some extra concepts. As usual we have M targets with geographic state vectors $\mathbf{x}_t^m$. We use *geographic* to describe states in a fixed-Earth reference, such as latitude and longitude, whereas the sensor measures a projection of this into group delay and cone angle. We typically refer to these as range and measured azimuth.

The sensor observes the targets through the medium of the ionosphere that supports propagation over M^μ round-trip modes between the sensor and the geographic location of the target. The particular modes that support propagation vary as a function of the target location. For example at longer ranges propagation is not possible through the E-layer, which has a low altitude. Each target can be treated as a point because of the sensor resolution and the appearance function is, therefore, determined by the complex-valued blurring resulting from finite aperture and time integration along with the signal processing tapers used for beam, range and Doppler formation. The ARD data cube is the superposition of the individual per mode responses across the targets and a noise process,

$$\zeta_t^i = \sum_{m=1}^{M} \sum_{\mu=1}^{M^\mu} \exp\{-j\psi_t^{m,\mu}\} A_t^{m,\mu} G_t^{m,i,\mu} + G_t^{0,i}, \tag{11.4}$$

where we use ζ to denote a complex valued pixel measurement; $\exp\{-j\psi_t^{m,\mu}\}$ and $A_t^{m,\mu}$ are the phase and amplitude of mode μ for target m; G_t^0 is the noise, in the simplest case complex Gaussian. The analysis here does not model the ground reflected clutter, which is assumed to be removed by preprocessing algorithms. As usual, $G_t^{m,\mu}$ is the target response from mode μ in ARD cell i given by

$$G_t^{m,i,\mu} = \int_{W^i} g_t^\mu \left(\mathbf{y}|\mathbf{x}_t^m\right) \mathbf{dy}, \tag{11.5}$$

where the mode dependence of the spreading function is through the nonlinear mapping for that mode that defines the appearance mean. The amplitude of each mode is assumed to fluctuate, so $A_t^{m,\mu}$ is a random variable; the propagation loss for each path is potentially different, so the statistics of $A_t^{m,\mu}$ vary with the mode. Note that it is common for some of the modes to be unresolved in the measurement space, that is multiple modes are likely to arrive within the same ARD cell. The measurement ζ_t^i is complex valued, but the algorithms described in this book operate only on the magnitude of the cell values, $\mathbf{z}_t^i = |\zeta_t^i|$.

The sensor also builds an ionospheric model from auxiliary measurements collected by ionospheric *sounders*. It uses these along with a synoptic model of the seasonal average ionosphere and constructs a set of mappings from the geographic coordinate system into the measurement system by drawing sample paths through a finite element electron density profile. This process is referred to as numerical ray tracing [5]. The result is a family of modes supported at a particular geographic location with associated estimate of the propagation loss and mode covariances. The mean of the measured signal due to mode μ from geographic location $\mathbf{x}$ is denoted $h_t^\mu(\mathbf{x})$ and the spatial variance of the measured signal is denoted R_t^μ. Approximating the main lobe of the sensor spectral response as a Gaussian with covariance Σ, the spreading for mode μ is then

$$g_t^\mu \left(\mathbf{y}|\mathbf{x}_t^m\right) \approx \mathcal{N}\left(\mathbf{y}; h_t^\mu(\mathbf{x}_t^m), \mathsf{R}_t^\mu + \Sigma\right). \tag{11.6}$$

In the case of spherical mirrors, the family of functions $g_t^\mu(\cdot)$ differ only in the height of the mirror for each mode μ.

11.3.2 Sequential Processing

We consider two sequential strategies for detection and estimation. Under conventional processing, a single-dwell detector forms point measurements that are fed into a tracker-based on probabilistic data association. Under sequential track-before-detect, H-PMHT is applied to the amplitude images $\mathbf{z}_t$ without knowledge of modes. In both cases, the tracks are created assuming that the modes are independent targets and then associated together using the multipath track fusion algorithm [29, 35].

11.3.2.1 Conventional Tracking

The conventional processing stream begins with generating peaks by finding local maxima in $\mathbf{z}_t$ and interpolating them [7]. This produces the collection of point measurements $\mathbf{Y}_t = \{\mathbf{y}_t^r\}$ where $r = 1 \ldots n_t$ is an index into the n_t point measurements at time t. This detection is performed in ignorance of the ionospheric model. Each point measurement is a four-element vector consisting of azimuth, range, Doppler and power. The probability that a particular mode is detected is the probability that $\mathbf{z}_t^i > T$ where i here is the closest cell to $h_t^\mu(\mathbf{x})$ and T is the detection threshold. This probability is mode dependent because the modes have different path losses and is state dependent because the modes are spatially varying. It is denoted $P_\text{D}^\mu(\mathbf{x})$. In contrast, we assume that the noise is white and so the probability that a false alarm is created in a particular noise-only cell is uniform across measurement space and is denoted P_FA. The false alarms are not generated by physical objects reflecting power over ionospheric layers (because we assumed that we don't have to worry about this) and so do not have associated modes.

The peaks are then used as input into a multi-target tracking algorithm, for example, Joint Probabilistic Data Association (JPDA) [6, 10]. This tracker assumes that each mode of every target is an independent entity and tracks them accordingly. Ideally the tracker would produce $\sum_{m=1}^{M} M^\mu$ tracks corresponding to each of the modes of the targets, but of course, in practice, this does not always occur. In particular, it is common that some modes are very much stronger than others and the weaker modes can be undetected. It is also common for multiple modes to be unresolved by the sensor and to give rise to merged measurements and hence a merged track. This is more likely close to boresight because we have assumed a linear receive array and this geometry cannot discriminate elevation at boresight. In that case, mixed modes such as the E-F2 and F2-E in the bottom of Fig. 11.12 are unresolved.

The tracker assumes that the number of false alarm measurements follows a Poisson distribution with a known intensity and that the spatial distribution of these measurements is uniform. In this comparison, we use a simplified version of the Unified-PDAF in [10] described in the appendix.

11.3.2.2 Sequential H-PMHT

For sequential H-PMHT we can apply the Poisson H-PMHT directly on the OTHR dwell images. The sequential approach ignores the mode structure until after tracking, so there is no need to modify the H-PMHT here. The input image is three-dimensional, which is an implementation detail that is different to the other examples in this book, but this is really a software issue.

11.3.2.3 Multipath Track Fusion

The measurement-space tracks are then associated together using multi-hypothesis association [29, 30, 35] referred to as multipath track fusion (MPTF). The MPTF aims to identify which tracks were created from each target by associating tracks to targets and modes using a track-orientated framework. A complete hypothesis is a collection of assignments for every track that identifies the mode and target that caused that track. The number of complete hypotheses grows combinatorially with the number of tracks and the number of modes, so it is not feasible to enumerate the complete hypotheses explicitly. Instead, the track-orientated framework recognises that each complete hypothesis is built from a number of target hypotheses that contain only the tracks associated with a single target. A complete hypothesis containing multiple targets is the concatenation of the individual target hypotheses.

MPTF works by evaluating the target hypotheses and their likelihoods and then it finds the best combination of target hypotheses to implicitly locate the most likely complete hypothesis. This is a constrained linear optimisation because each track must be accounted for and no two target hypotheses can be combined if they contain a common track. Pruning methods are still required when the number of ionospheric layers is relatively high. These amount to discarding the low probability target hypotheses before the optimisation stage. More details about this algorithm can be found in [29, 30, 35].

11.3.3 Multimode H-PMHT

The alternative to sequential processing is to perform target tracking and mode association jointly. In the context of H-PMHT this amounts to defining a Gaussian mixture appearance function where the mean of each mixture component is defined by the path transform for a particular mode.

$$g_t\left(\mathbf{y}|\mathbf{x}_t^m, \mathbf{L}_t^m\right) = \sum_{\mu=1}^{M^\mu} \Lambda_t^{m,\mu} g_t^\mu\left(\mathbf{y}|\mathbf{x}_t^m\right), \tag{11.7}$$

where $\mathbf{L}_t^m = \{\Lambda_t^{m,\mu}\}$ is the collection of average mode powers for this target. Gaussian mixture H-PMHT was introduced in Sect. 7.4. The solution is a bank of decoupled rate filters for $\Lambda_t^{m,\mu}$ and a multi-measurement state estimation filter. The complication here is that the Gaussian components of the appearance mixture are nonlinearly dependent on the state. For the simulations here we address the nonlinearity through an extended Kalman filter implementation using numerical Jacobian matrices because the transform is specified numerically.

An intuitive interpretation of the multimode H-PMHT is that it treats the OTHR modes as members of a group target with a known formation. The kinematic state $\mathbf{x}_t^m$ follows the geographic target state and the measurement modes are derived from

Table 11.3 Simulation parameters

Carrier frequency	15 MHz
Bandwidth	10 kHz
Waveform repetition frequency	50 Hz
Coherent integration time	2 s
Receive array aperture	3 km
Number of beams	20
Azimuth resolution	0.38 deg
Number of range bins	60
Range resolution	15 km
Ambiguous range	3000 km
Number of velocity bins	100
Velocity resolution	5 ms^{-1}
Ambiguous velocity	500 ms^{-1}
Clutter width	10 bins
Revisit period	30 s

it. In group tracking, the geographic state is analogous to the group centroid and the relative position of each group member is determined by the mode transforms.

The multimode H-PMHT results in a separate Poisson rate estimate for each mode. This means that at each frame we now have a vector of track quality numbers rather than a scalar. An overall target cross-section estimate can be derived by scaling each of the mode rates by the ionospheric model path loss. Or we can just build a composite rule. We found it more successful to apply two layers of rules: if any single mode satisfies the regular Poisson H-PMHT promotion rules, then the whole multimode track was promoted; if two or more modes satisfied a lower set of thresholds, then the whole multimode track was promoted. Tracks were terminated when all modes failed quality tests.

11.3.4 Simulations

The different methods for forming geographic coordinate tracks from multimode data are now compared using simulations. We consider a linear target scenario using realistic radar and ionospheric parameters. The radar parameters were chosen to follow the nominal air target detection parameters described in Table 1.3 of [15]. These are presented in Table 11.3.

The quality of the track picture is quantified using accuracy and cardinality measures, as usual. Target estimation accuracy is defined as the root mean square miss distance between the true target position and the estimated track position. Cardinality is whether the tracker correctly identifies the number of targets present. We measure two types of cardinality error: a missed target occurs when there is no track within a 50 km gate of a target; and a redundant track occurs when there is a track that is not associated with any target. A redundant track can be caused by an error resolving modes that leads to more than one geographic track on a single target, or it can simply be a false track due to random noise. The number of false tracks is low because of the idealised nature of simulated noise and we make no effort to differentiate between mode errors and false tracks. The geographic tracks are associated with targets using a global nearest neighbour approach.

11.3.4.1 Simulated Ionosphere

The purpose of this study is not to estimate the parameters of an ionospheric model so we use a relatively simple description of the ionosphere for the simulations. Figure 11.14 shows an experimentally collected oblique ionogram, which is a measurement of the ionosphere that uses separate transmit and receive sites and is analogous to a one-way path from transmitter to target or target to receiver. The image shows the received power as a function of frequency and group delay [14, 15]. A number of modes are present and these manifest themselves in the figure as curves of high received signal power. Black rings are overlaid on the ionogram image: these are fitted traces that were used to derive equivalent (*virtual*) layer heights, as shown in Fig. 11.15. These range-dependent heights are used for the ionospheric truth in our simulations. The lowest ionospheric layer is sporadic E and causes masking of the F1 layer. The second lowest path is referred to as the F2 low ray and represents paths refracted from the bottom of the F2 layer. The upper two paths are referred to as the ordinary and extraordinary F2 high ray paths and result from propagation through the upper part of the F2 region. There are two F2 high rays due to magnetic field effects that are not resolvable in the low ray.

Our simulation uses the sporadic E and F2 low height profiles derived from this ionogram and allocates equal power to each. This maximises the effect of using multiple modes. In practice, there is usually a substantial difference in power between modes, in which case performance approaches that of a single mode. We compare the F2-low only case with sporadic E and F2 low.

11.3.4.2 Simulation Results

We simulate the complex ARD data cube directly. The noise was a unit variance complex Gaussian. The target position in geographic coordinates was transformed through every layer to give a family of 4 mode returns in the measurement space. Each mode sampled an amplitude and phase by generating a complex Gaussian with variance given by the mode strength and the spread of each mode is determined by the signal processing tapers. We use the spectral response of a Hanning window. The mode spread functions superpose in the complex measurement space. The result of this method is that the amplitude of each mode fluctuates following a Rayleigh

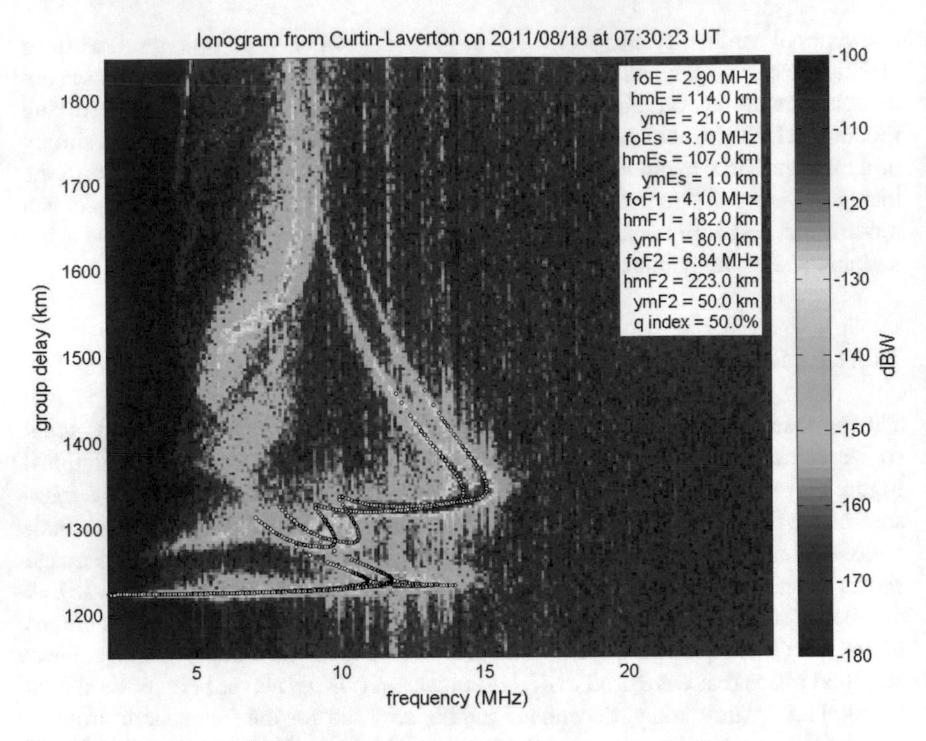

Fig. 11.14 Four modes fitted to an archived ionogram

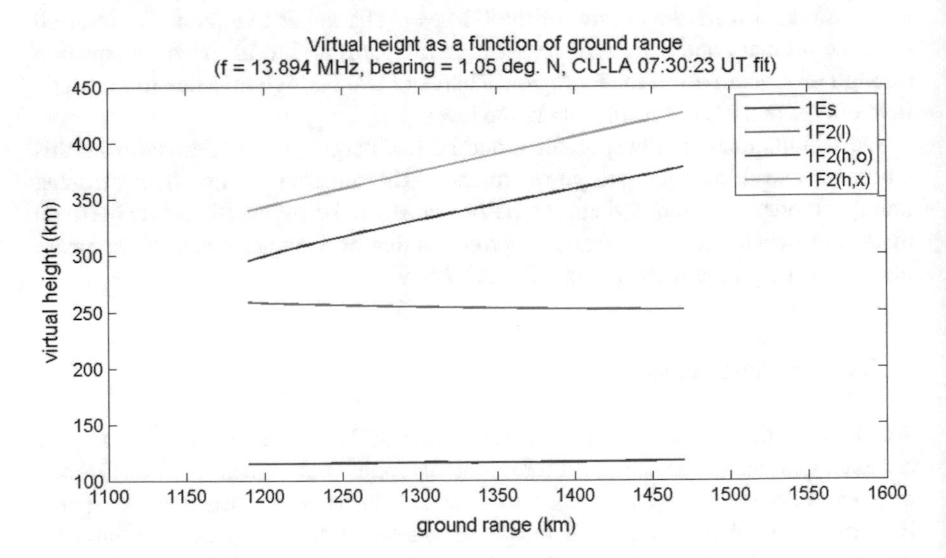

Fig. 11.15 Equivalent virtual heights of each layer

distribution and where modes are unresolved, they can interfere with each other. The target geometry is presented in the measurement frame in Fig. 11.13. Although the target motion is constant velocity, the nonlinear measurement function and the sensing geometry results in a gradual change in radial speed and so a slightly nonlinear range progression. We consider first a single target scenario and then two parallel targets.

The first scenario contains a single outbound target that starts at a geographic range of 1,200 km and a bearing of 35 degrees. It travels North at 400 knots. The paths shown in Fig. 11.13 are the Es and F2 multimode returns for this geometry.

The scene was simulated for 100 Monte Carlo trials with varying target SNR and for three ionospheric conditions. The reference case is one-mode propagation where only the F2 low is present. To illustrate the maximum potential benefit of multipath, we next simulate two equal power layers by introducing the Es. The third condition mimics the real propagation in Fig. 11.14, namely four layers with a dominant F2 low.

Figure 11.16 shows the track RMS estimation accuracy and the number of redundant tracks as a function of SNR. For clarity, the four-mode curves are not shown in these figures. A hundred Monte Carlo trials were performed at each SNR value. There is essentially no significant difference between the algorithms in these plots. There is an increase in error as the number of modes increases. One might expect the opposite behaviour: that including more modes would offer extra data that gives fusion gain. This is true but it is overwhelmed by the cases where a single mode is tracked well but incorrectly geo-registered. If there is only one track then the MPTF will label the track as the most likely mode for the sequential approaches, which is the only sensible action. However, the simulation has four equally likely modes, so there is a high probability of choosing the wrong mode and, therefore, incurring high estimation error. The same kinds of error occur with multimode H-PMHT.

Figure 11.17 shows the number of missed targets as a function of SNR. Here there is a very substantial improvement for the two H-PMHT methods compared with the point measurement tracker. There is also an improvement in sensitivity when more modes are available. The curves were correlated against the single mode point measurement tracker, which was the worst performing algorithm. The sensitivity improvement is shown in Table 11.4. The most important result is that the H-PMHT approaches both give more than 6 dB sensitivity improvement. They also make better use of the multimode data: for point measurement tracking the increased sensitivity is less than 1 dB but for H-PMHT the improvement is more. The multimode H-PMHT offers better performance than the sequential H-PMHT: based on the more detailed results in Fig. 11.17, this is mostly at very low SNR. Note that the four-mode performance is almost identical to the single mode performance because there is a single dominant mode and the other weaker modes do not provide very much additional information.

Table 11.5 gives the CPU costs for the various algorithm-scenario combinations. This time was measured running Matlab on a single desktop PC. Clearly, the multi-mode aware H-PMHT is the most costly; although the MPTF cost grows nonlinearly, it is not the bottleneck here.

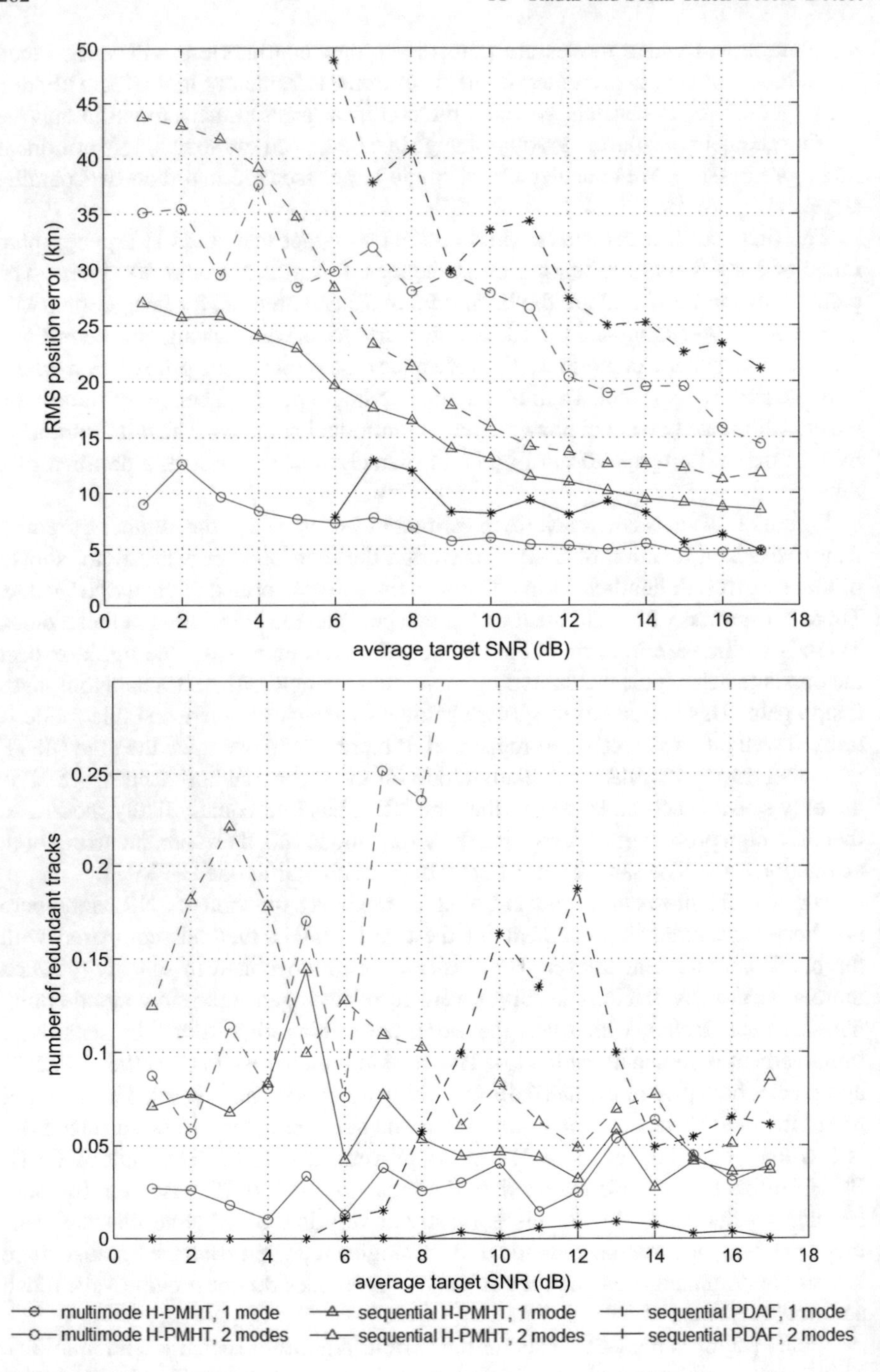

Fig. 11.16 One target, RMS estimation accuracy and number of redundant tracks

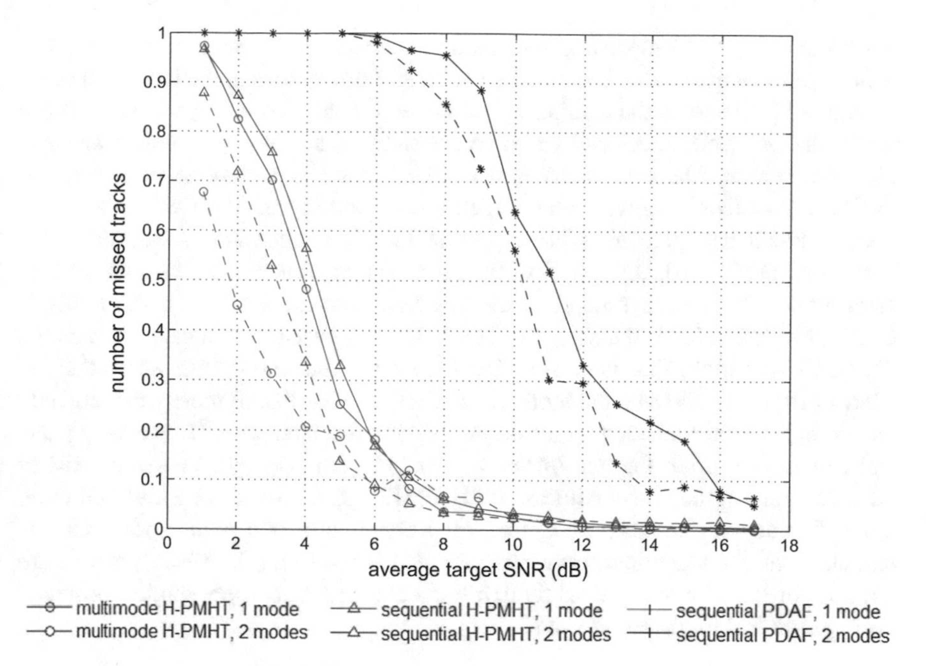

Fig. 11.17 One target, number of missed tracks

Table 11.4 Sensitivity improvement compared with 1-mode point measurement tracking

Method	1-layer (dB)	2-layers (dB)	4-layers (dB)
Conventional	0	0.7	−0.4
Sequential H-PMHT	6.5	7.7	7.0
Multimode H-PMHT	6.9	8.9	7.6

Table 11.5 Average CPU cost per 2 s dwell (seconds)

Method	1-layer	2-layers	4-layers
Conventional	0.28	0.56	0.31
+ MPTF	0.81 ms	4 ms	6.0 ms
Sequential H-PMHT	0.21	0.41	0.26
+ MPTF	1.5 ms	15 ms	38.9 ms
Multimode H-PMHT	0.41	0.70	3.55

The next scenario introduces a second target that follows the same path as the first but is ahead by 500 s. This emulates two aircraft following a flight lane. The spacing means that the E-E mode of the leading target is in the same range cell as the F2-F2 mode of the trailing target and the azimuth spacing between the two is a little over a beam. Figure 11.18 illustrates the four modes of each target in beam number and range bin for the two-mode and four-mode simulated propagation conditions. The relative SNR of the modes is shown by the weight of the lines in the four-mode case

Figure 11.19 shows the cardinality measures for the two-target scenario. These curves are not well described by simple translations, so we do not derive an SNR gain in this case. The point measurement PDAF has a slower slope so that the H-PMHT approaches reach very low missed track count faster. A missed count of 0.2 is equivalent to a 90 percent probability of tracking in this scenario. At this detection level, the PDAF is 10dB worse than the H-PMHT approaches. In the single target case, more modes always improve the available information, but in the multiple target case, it can lead to ambiguity. The PDAF performance plateaus even as the SNR gets very high. This is most noticeable for the four-mode case where there is a large difference between the mode strengths. Here the PDAF mostly detects only the strongest mode on each target but MPTF is presented with 16 two-way paths and chooses to group the two tracks as a single ground-target. The same sort of effect happens in the four-mode case for H-PMHT: the existence of additional weak modes that do not give much energy leads to a degradation of detection performance. Although the ionospheric information provided is the truth, the weaker modes are not seen in the data and it is as though there is a mismatch between the provided propagation and the observed data.

11.4 Active Underwater Sonar

The next application uses archived data from a DST Group experimental active sonar array referred to as CASSTASS. The data was collected during a series of sonar trials at two different locations in the Western Australian eXercise Area (WAXA) where the CASSTASS array was towed behind a moving surface ship. Results on this data were first reported in [33, 34]. The first dataset was collected in a shallow water environment with depths between 150 and 250 m; the second was collected in an intermediate environment with water depths between 800 and 1400 m. Each has individual characteristics and a challenging clutter environment. The sonar transmissions were designed to detect possible targets with a maximum range of 60 km with the majority of transmissions being in the aft direction. The datasets feature a fluctuating target and persistent clutter detections that are the result of reflections from canyon walls along the continental shelf. Both datasets consist of approximately 20 frames with a time period between transmissions of approximately 90 s. On some of the frames the transmission failed, in which case the measured frame is essentially a passive observation and the target cannot be detected in these.

The target is an echo repeater (ER) made to emulate the returns from a large scale ship in the ocean environment. During the trial, the average speed of the ER was 0–4 knots and the target strength was controllable: for the datasets we use the nominal SNR was 25 dB, but the actual observed SNR fluctuated and was at times as low as 13 dB. This is easily enough to form a track. However, the bigger challenge in this data is the clutter, which is typically much stronger than the target

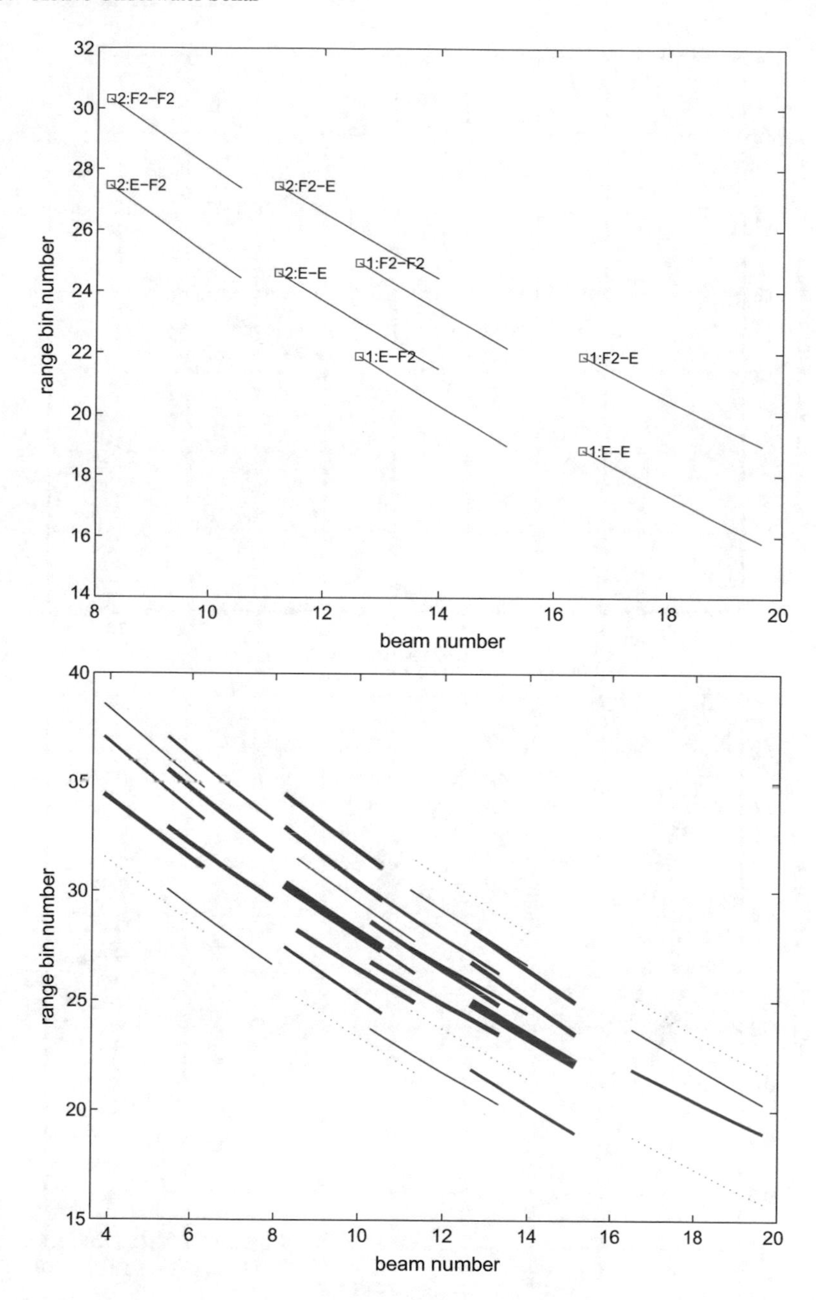

Fig. 11.18 Two target trajectory in measurement frame: 2 layers (top) and 4 layers (bottom)

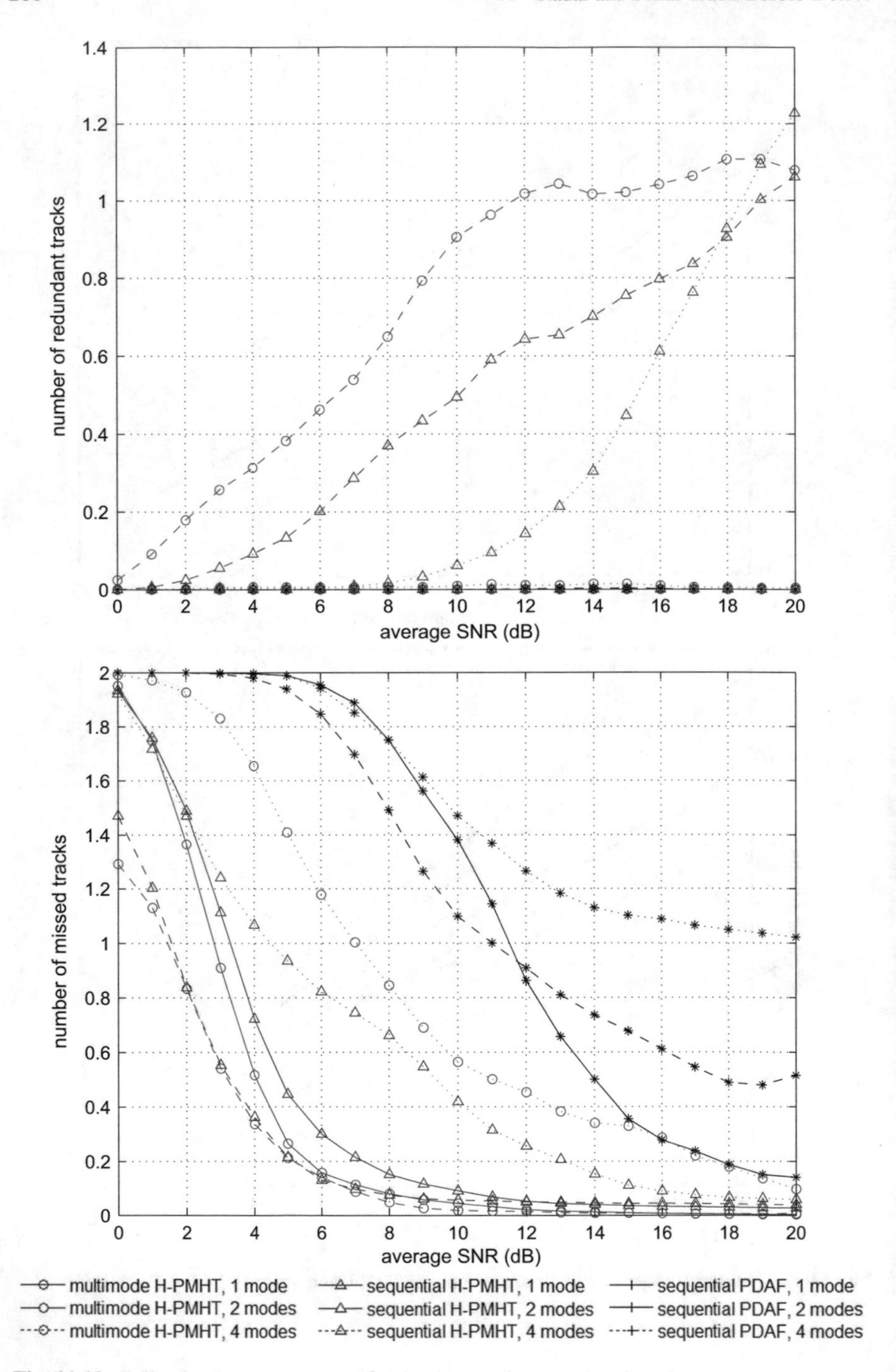

Fig. 11.19 Following targets, number of redundant tracks and missed tracks

and can lead to false tracks if not adequately handled. The echo repeater captures the transmitted sonar ping, waits, and then transmits a copy of the ping back to the sensor. The delay means that the apparent transit time of the artificial echo is longer than the actual propagation time from sensor to the repeater and back. The result is that the measured range to the repeater is longer than the physical range. Figure 11.20 illustrates the geometry of the two datasets. The vessel towing the sonar array is referred to as the ownship and is North of the repeater in both cases. The true physical position of the repeater is known from instrumentation but the apparent position of the repeater measurement was derived from manual inspection of the actual sonar data. As discussed, the apparent position is farther away from the sensor than the true position.

The array data was processed by forming two degree beams through a full 360 degree circle, although these beams are not independent. In particular, there is an ambiguity because the array can only measure angles relative to the line of the array. Much like the OTHR case previously, this means that the measured bearing is ambiguous around a cone of constant angle to the array. Assuming a surface ship target, this ambiguity means that the left and right halves of the surveillance region are aliased onto each other. In practice, this can be resolved by ownship manoeuvres but the sequences we use do not contain these so we process only one side of the beam pattern. Also, the beamwidth of the array varies as a function of steer angle. The array perpendicular is referred to as boresight, and beams close to this are relatively narrow; beams that are nearly co-linear with the array are referred to as end-fire and are very much wider. The array aliasing means that beams that are close to end-fire coalesce with their ambiguous mirrors to form a wide response with a double peak. The array follows the ownship, so this end-fire direction is along a line behind the ship and can also be referred to as aft. Figure 11.21 illustrates the spreading of the beams as the steer direction moves from boresight to end-fire. The half-height beamwidth for the beampattern is approximately 8 degrees at broadside, 50 degrees across both peaks in the near aft direction, and 42 degrees at aft. This is important for the tracker because it means that the appearance function is spatially varying. The spatial variation in the beamwidth was accounted for by adjusting the covariance of the appearance function Σ.

Figure 11.22 shows two example frames. In the first, the instantaneous SNR is 24 dB and in the second it is 13 dB. In both of these cases, the repeater is close to end-fire where the image is quite smeared in azimuth due to the spread of the beamwidth.

We now compare the performance of three approaches to tracking the echo repeater. The first method is a point measurement PDA tracker that uses an existence model for track decisions [3, 6]. The other two are core H-PMHT and Poisson H-PMHT, both using the bearing dependent appearance function. The difference between these is that the core H-PMHT uses mixing proportion estimates π_t^m that are independent over time whereas the Poisson H-PMHT uses a Poisson rate Λ_t^m with a dynamic model, as discussed in Chap. 6. In this application, the strength of the target signal is fairly good but there is a significant amount of high amplitude clutter. In order to adequately suppress this clutter, the PDA tracker required a fairly

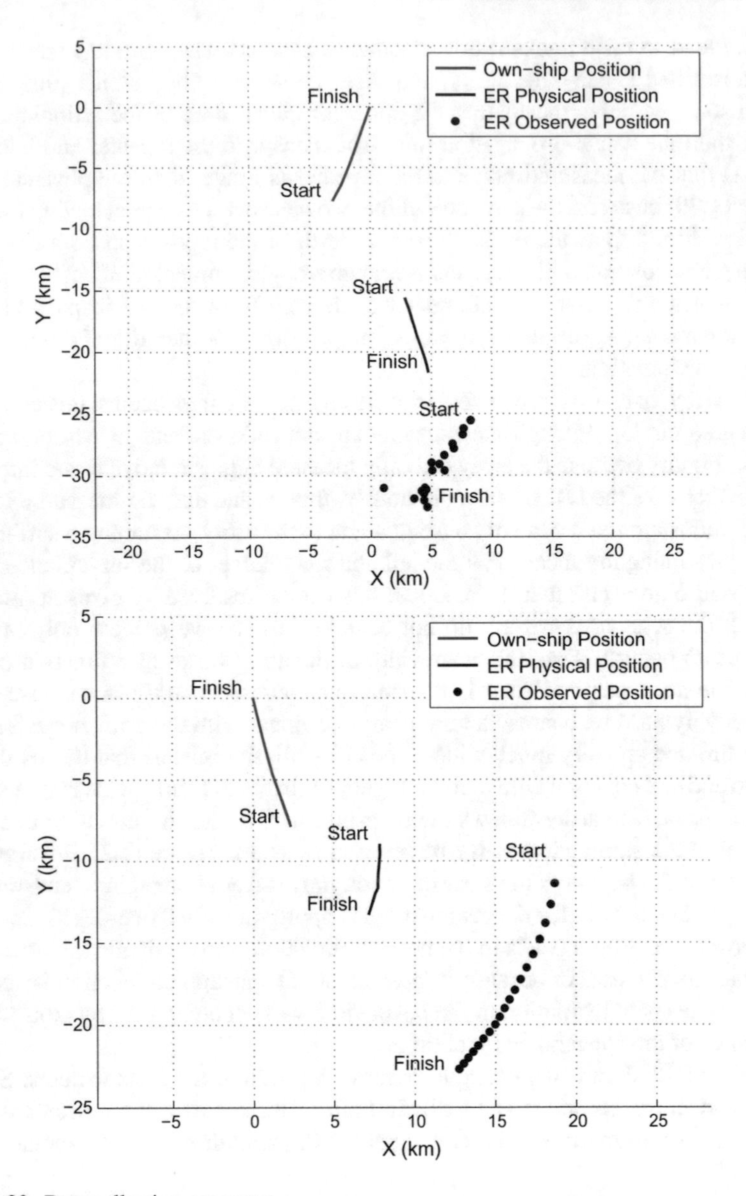

Fig. 11.20 Data collection geometry

high amplitude threshold to screen the point measurements before tracking. In an effort to keep things fair, we applied this same threshold to the candidate formation stage in the H-PMHT track manager: refer to Chap. 5 for details. We considered thresholds at 11, 13 and 15 dB. The resulting tracks for the shallow water scenario are shown in Fig. 11.23 and for the intermediate scenario in Fig. 11.24.

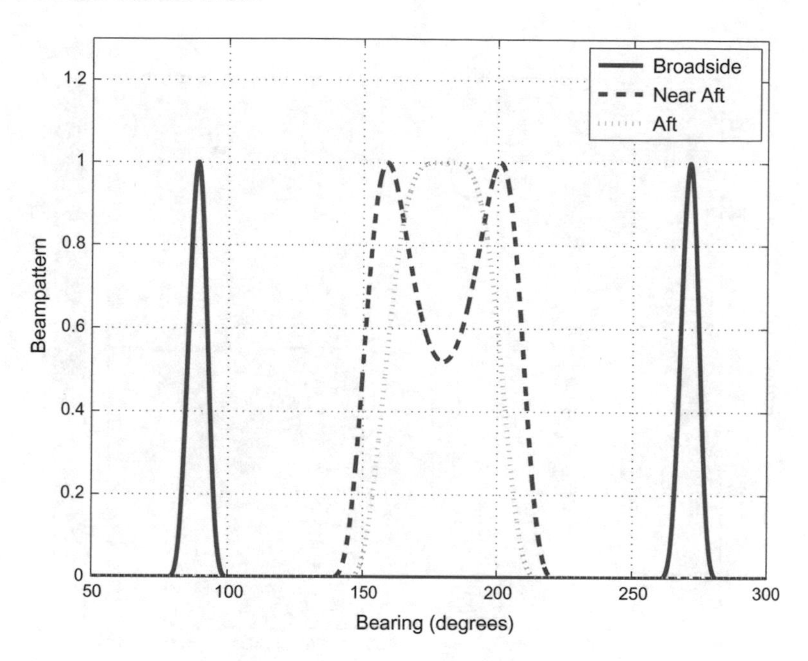

Fig. 11.21 Spreading of the beamwidth close to end-fire (aft)

Table 11.6 Number of false and divergent PDA tracks for the shallow dataset

Threshold (dB)	Average point-measurement count	Num divergent tracks	Num false tracks
11	115.4	0	8
13	4.4	0	1
15	0.8	0	0

For both cases the H-PMHT tracks were essentially not effected by the initialisation threshold: the repeater is strong enough to start a track even with the threshold and the tracker doesn't make false tracks. The PDA was unable to adequately track at the 11 dB threshold. There were numerous false tracks and the repeater track diverged. For the 13 dB threshold, the false tracks were suppressed but the repeater track still diverged in the intermediate dataset. At 15 dB both scenarios were tracked well. Tables 11.6 and 11.7 summarise the PDA false track rate and divergent track rate for the two scenarios. A divergent track was declared if a track started close to the target but then move off, such as in the 13 dB case for the intermediate water in Fig. 11.24.

The tracking results illustrate that PDA is only able to function well with a high SNR threshold on the input data, which clearly would have prevented it from detecting a weaker target. The H-PMHT variants both formed a track without such a threshold.

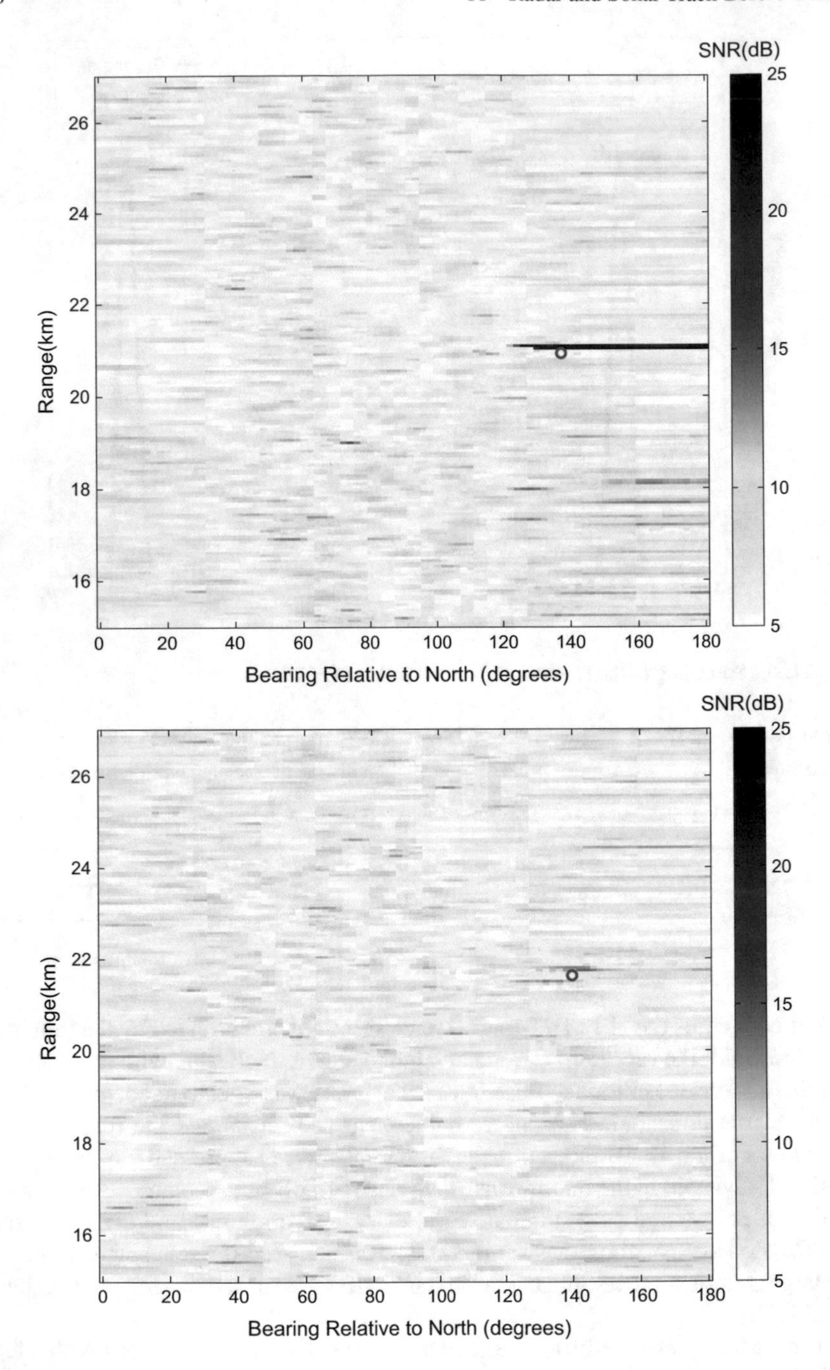

Fig. 11.22 Example sonar frames: 24 dB target (top) and 13 dB target (bottom)

Fig. 11.23 Shallow water tracks for thresholds of 11 dB (top), 13 dB (middle) and 15 dB (bottom)

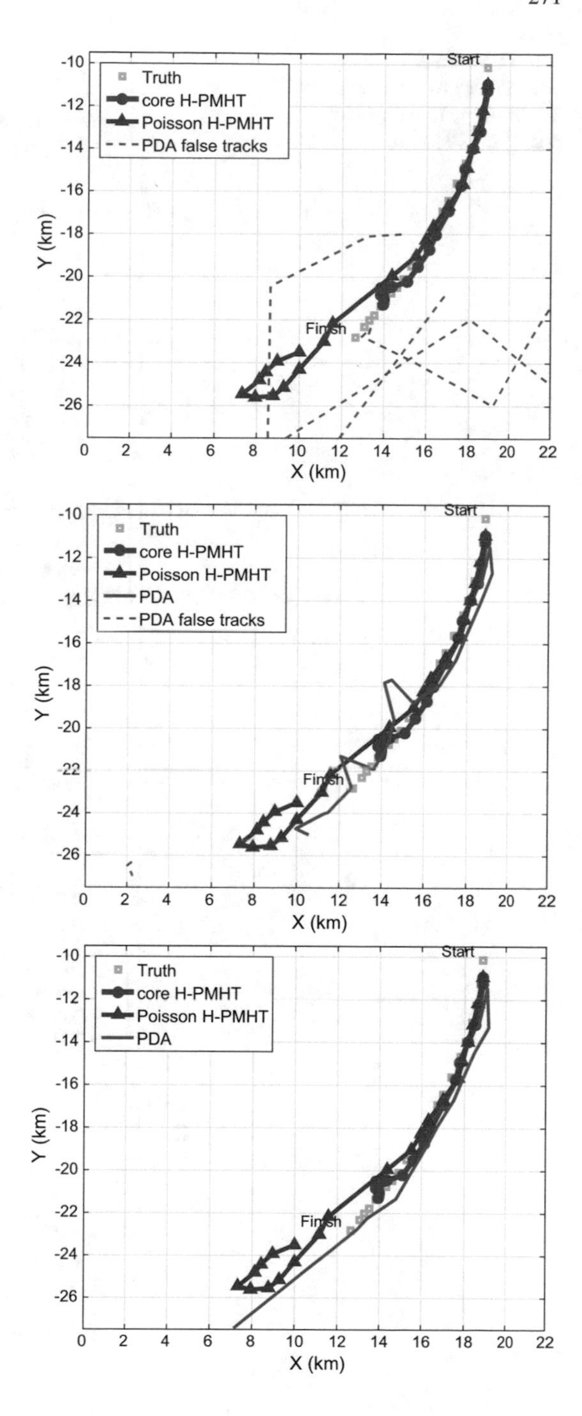

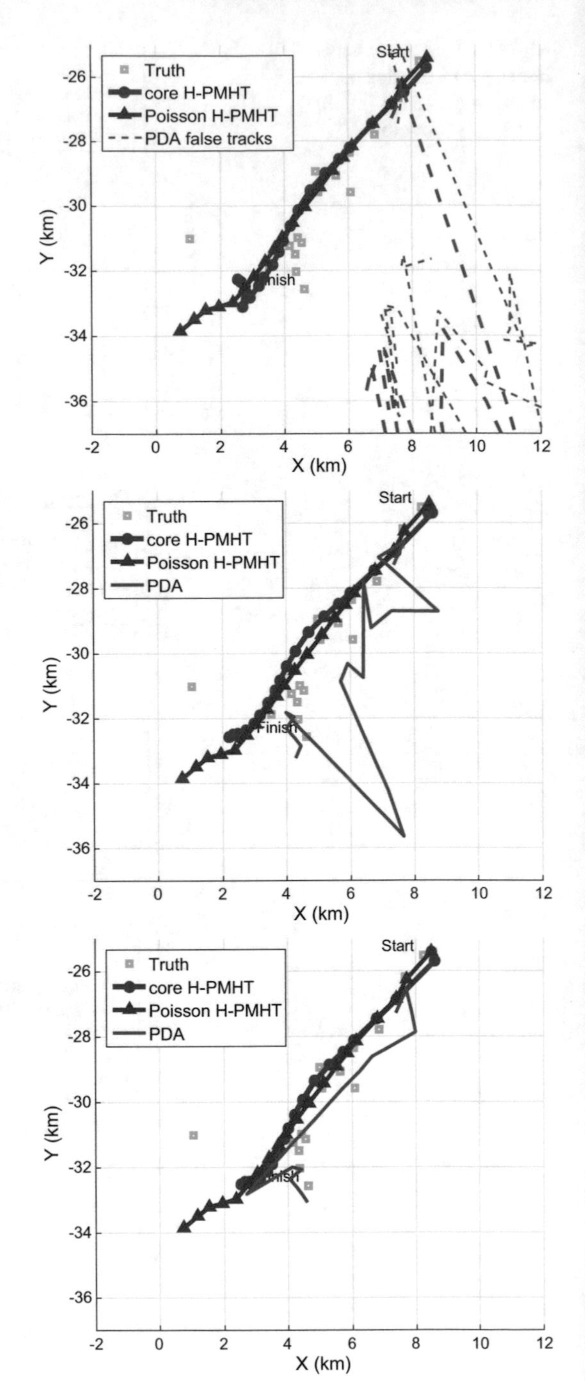

Fig. 11.24 Intermediate water tracks for thresholds of 11 dB (top), 13 dB (middle) and 15 dB (bottom)

Table 11.7 Number of false and divergent PDA tracks for the intermediate dataset

Threshold (dB)	Average point-measurement count	Num divergent tracks	Num false tracks
11	118.5	1	12
13	5.1	1	0
15	0.7	0	0

11.5 Passive Underwater Sonar

The second sonar example is a passive application. In this case, the sensor listens to the sounds generated by vessels in the ocean. The frequency spectrum of a marine vessel (such as ship, boat or submarine) is typically composed of broadband and narrowband frequency components. Broadband components are generated by many sources including propeller cavitation and impulsive events in the engine while narrowband components, or tonals, are due to harmonics of the engine speed and shaft/propeller rotation. The spectral components are characteristic for individual vessels and can be used for vessel identification. However, noise radiated by a marine vessel is normally contaminated by ambient sea noise that is produced by causes such as oceanic turbulence, different wave effects and ship traffic.

Passive sonar spectrum measurements contain components that usually fluctuate in both amplitude and frequency. The spectrum can also contain clutter artefacts and transient signals. In order to produce stable frequency tracks in the presence of clutter the track-before-detect algorithm should estimate the frequency extent of each component as well as its mean, and deal with data contaminated by outliers. The H-PMHT-RM algorithm is well suited to this problem and can be applied to both narrowband spectra and broadband spectra. The example considered here consists only of a time varying spectrum, so the sensor 'image' is one-dimensional. In the context of this application, we refer to the sensor images as *spectrograms*. This application is very closely aligned with the work of Luginbuhl and Willett [20] that was a precursor to H-PMHT. The results here were first presented in [4].

We now present the results of applying H-PMHT-RM to simulated spectrograms and spectrograms collected from an experimental array at sea. The implementation used Poisson PMHT [17] as described in Chap. 6, track management [9], and single target chip factorisation [13] as described in Chap. 5.

11.5.1 Simulated Spectrograms

The simulated spectrograms were generated from a mixture of ten target components and a uniform background. The spread covariance of each component Σ was uniformly sampled from [0.8, 16] Hz^2 and fixed within a Monte Carlo trial. The sim-

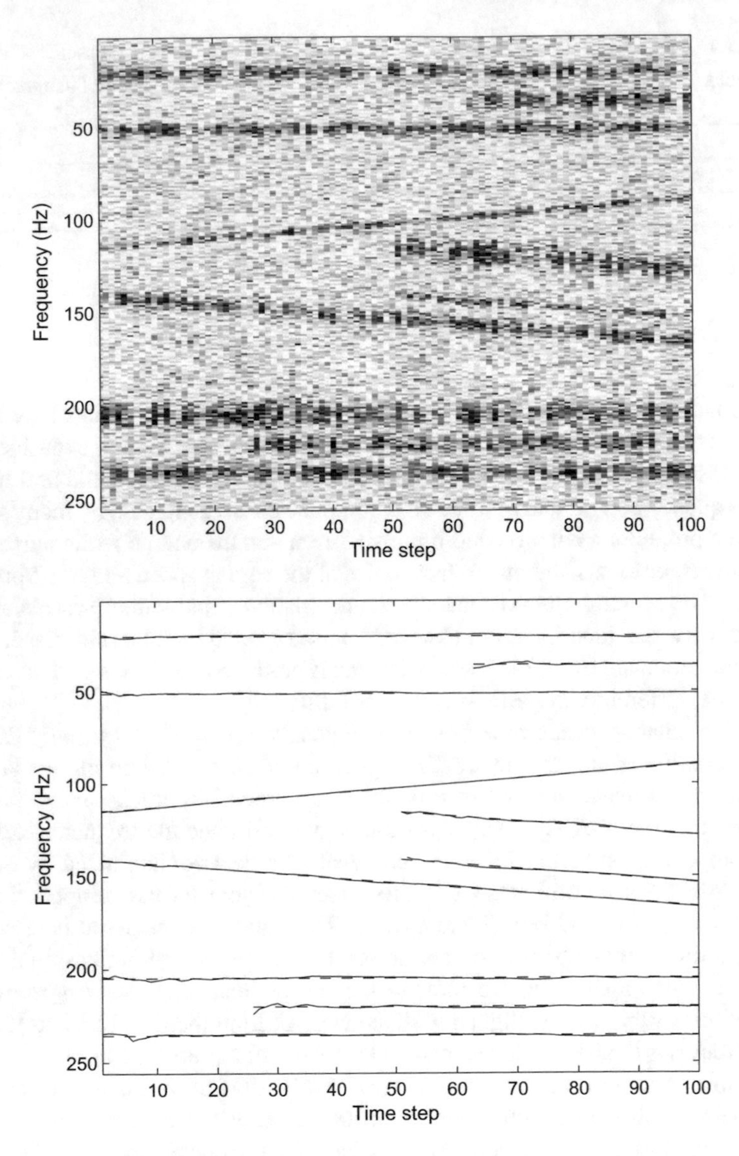

Fig. 11.25 Example 10-component simulated spectrogram (top) and true spectral components overlayed with the H-PMHT-RM estimated tracks

ulated spectrum consisted of 512 bins at 1 Hz spacing with a 1 second frame period. The amplitude of each component randomly fluctuated with a Swerling I model. The assumed target model was a constant velocity model, which in this case means that we model the centre frequency of each component and its rate of change. The clutter pdf was artificially inflated to improve average performance. Figure 11.25 shows an example trial and the H-PMHT-RM tracks generated for that trial.

Table 11.8 Simulated spectra

SNR (dB)	**x** RMS	Σ RMS	Missed %	Duplicate %	Num false
6	0.34	1.23	0.2	0.3	0.03
7	0.29	1.39	0.0	0.2	0.00
8	0.26	1.41	0.2	0.0	0.04
10	0.23	1.66	0.1	0.2	0.02
12	0.20	1.68	0.1	0.2	0.02

The output quality was averaged over 100 Monte Carlo trials using cardinality and accuracy performance measures. Table 11.8 shows the RMS estimation error for the centre frequency of each component $\mathbf{x}$ and the spread variance Σ, which we have referred to as the appearance parameter in earlier chapters. Both of these quantities are scalars because we only present the frequency component of the state and not the frequency rate. The cardinality measures are the average percentage of missed components, the average percentage of duplicate components, and the average number of false components.

As expected, the centre frequency estimation performance improves as the average SNR increases. The cardinality errors are low for all SNRs. However, the spread estimation error actually gets worse as the SNR increases. A possible reason is that the tails of higher SNR components spread farther before they drop below the clutter which causes the association weights for high SNR components to be spread more. This, in turn, leads to a higher estimated spread. The artificial inflation of the clutter pdf mitigates this but it would appear that the inflation was tuned better to the lower SNR components.

11.5.2 Narrowband Experimental Data

Next, we apply the H-PMHT-RM to narrowband passive sonar data collected from an experimental array. The hydrophone data was first transformed into bearing-time using the minimum variance distortionless response beamformer where the beamformed data at each time step is obtained by averaging the discrete Fourier transforms of 8 1-second long signal segments with 50% overlap, for bearings in the range of [0°, 359°] with 1° resolution. Two vessels were then identified by manual inspection and the bearing-time profile for each vessel was manually tracked to extract the underlying time-frequency spectrum. Each spectrum was processed using the 'order-truncate-average' (OTA) algorithm [24] that applies robust processing based on median filtering, and separates the spectrum into two components: a smoothed broadband spectrum, and a narrowband spectrum. We used a median filter of length 15 and set the scaling factor for the sample median to $R = 1$.

With real data, we do not know what the true spectral components are, so it is not possible to compare the track output of H-PMHT-RM with truth. Neither is it possible to compute quantitative performance measures. Instead, our readers will have to be content with a qualitative comparison. To achieve this we construct a *modelled spectrum*, which is the mixture described by the H-PMHT-RM tracks. This is exactly the cell probabilities G_t^i. Figures 11.26 and 11.27 compare the measured sensor time-frequency spectra with the reconstructed model spectra for the two targets. The reconstructed model spectra closely resemble de-noised versions of the sensor image input, implying that the algorithms have captured the significant spectral components. The striation pattern that is visible in Fig. 11.27 is the result of constructive and destructive interference between the direct and the surface-reflected sound profiles, called the Lloyd mirror effect. The suppression or otherwise of these diagonal stripes depends on the assumed dynamics of the target components. The H-PMHT-RM here was tuned to track time-varying components as shown in the simulations. If we had assumed a more constant model then the interference pattern would be suppressed at the expense of true time-varying frequency lines. In a practical situation, the true centre frequency can change as a result of the Doppler effect.

11.5.3 Broadband Experimental Data

H-PMHT-RM was also applied to modelling the broadband components of spectrograms. In this case, the real spectra are smoothed so it can be assumed that the corresponding level of noise is low. Figure 11.28 shows an example broadband spectrogram of a real target, and this spectrogram overlayed with the frequency tracks estimated using H-PMHT-RM. In this case, it is difficult to even qualitatively say much about the algorithm performance. It is clear that H-PMHT-RM has identified four strong components which appear to be an appropriate description of the data.

In principle one could process the narrowband components and the broadband components jointly with a single H-PMHT-RM: the algorithm could learn narrow spreads for some tracks and broad spreads for others. In practice, we doubt that this would robustly work. An alternative would be to first estimate the broadband components and the apply a H-PMHT-RM tuned for narrowband components to the residual background associated images.

11.6 Sonar in Air: Aura

The final sonar example is a Defence Science and Technology Group demonstration system referred to as Aura that uses acoustic signals in the air to illustrate signal processing concepts. It is implemented as an application for PC or IOS devices, such as an iPAD, and is intended to be used by STEM students.

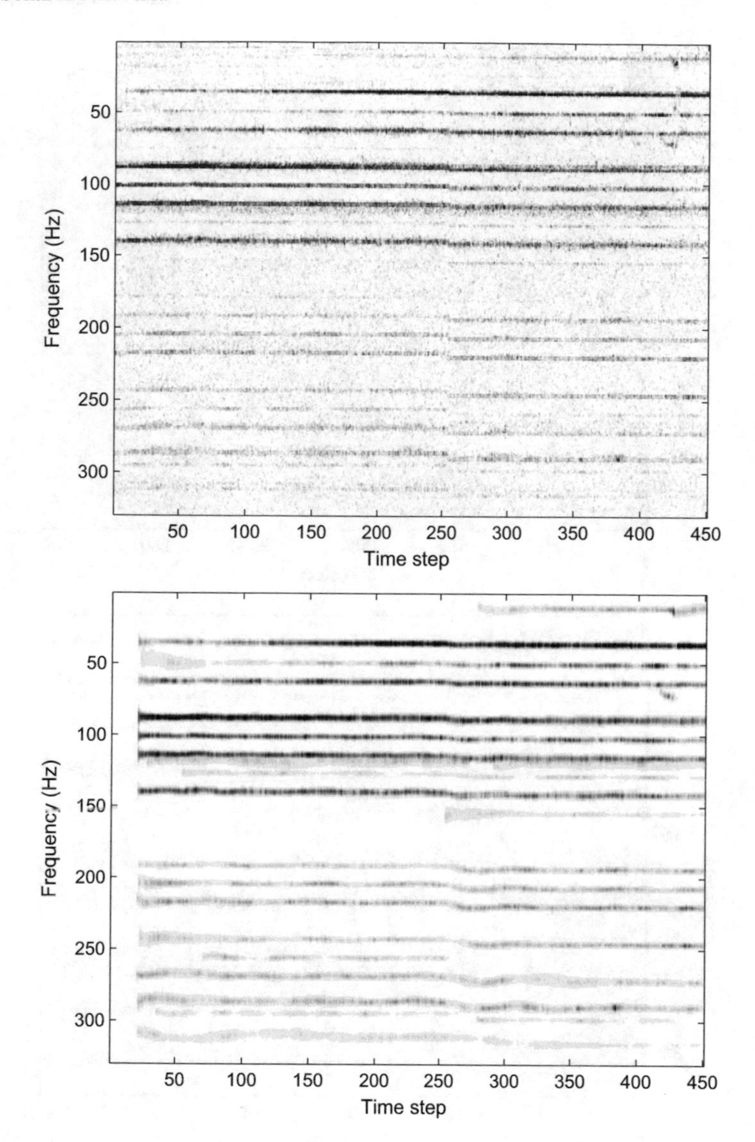

Fig. 11.26 The measured narrowband spectrogram related to target 1 and the H-PMHT-RM modelled spectrum

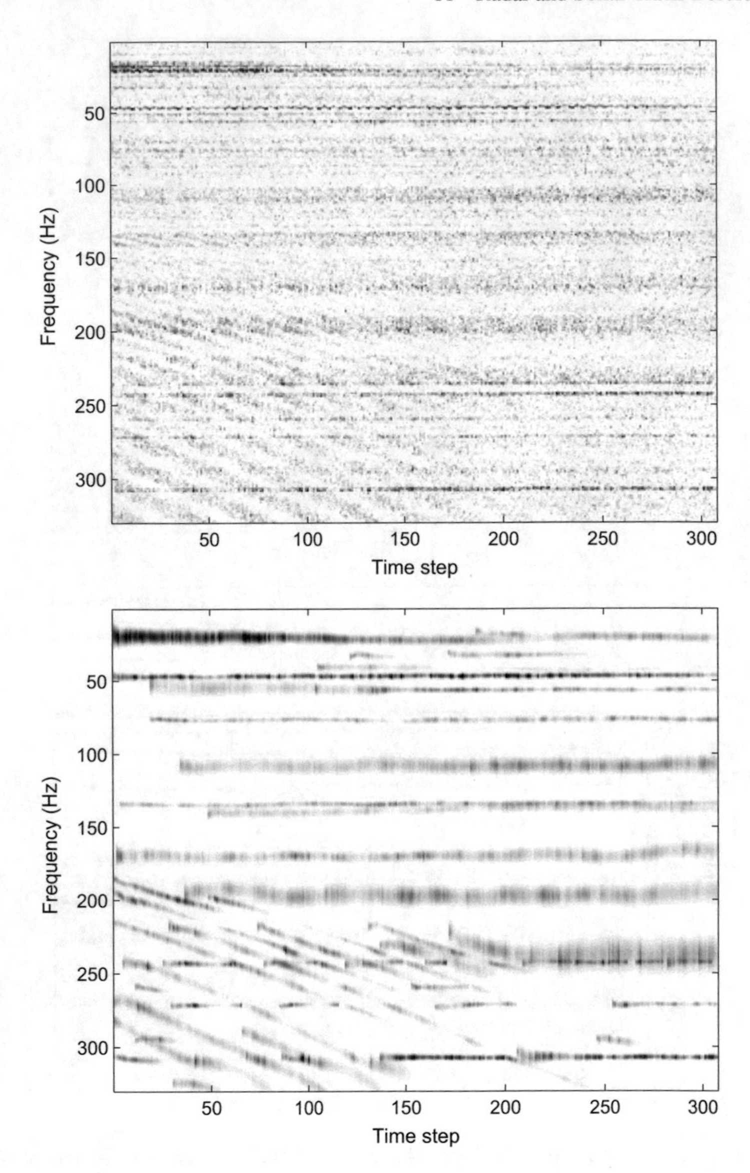

Fig. 11.27 The measured narrowband spectrogram related to target 2 and the H-PMHT-RM modelled spectrum

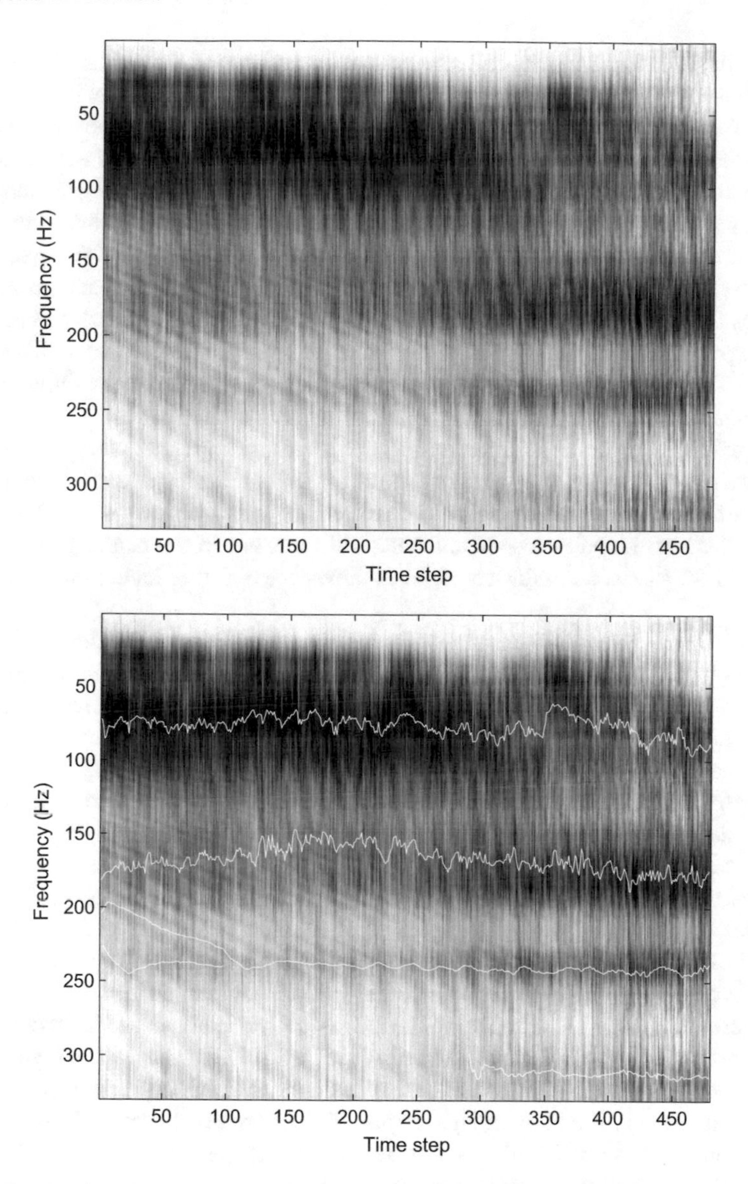

Fig. 11.28 Broadband spectrogram and spectrogram with H-PMHT-RM tracks

We illustrate two data sets collected with a linear frequency modulated continuous wave (FMCW) transmitted signal. The waveform used a 15 kHz carrier frequency and a 6 kHz bandwidth. The waveform repetition frequency for the first data set was 88 Hz and we coherently integrated 32 sweeps to form range-Doppler maps with a coherent integration time of approximately 0.36 s. This results in range-Doppler maps that are very coarse, but there is a trade-off between range and Doppler ambiguity. There is also a trade-off in the coherent integration time (CIT). A longer CIT gives higher SNR but the target can manoeuvre during the integration, which smears the response through Doppler. The second data set used a higher CIT of 0.85 s, which was achieved by integrating 128 sweeps at a repetition frequency of 150 Hz. Figures 11.29, 11.30 and 11.31 show sample frames from the two data sets. The frames show Doppler horizontally and range vertically.

Figure 11.29 shows five frames during the first track drop for the short CIT data set. It shows over the space of five frames the target before, during and after manoeuvre. The frames are very narrow because of the low number of Doppler cells. In each frame, the zero-Doppler response from clutter is clear in the centre of the image. Figure 11.30 shows a similar set of five frames from a little later in the short CIT data set.

Figure 11.31 shows three example frames from the longer CIT data set. The first frame is when the target is non-manoeuvring and the other two show manoeuvres. The improved Doppler resolution is evident from the narrow clutter response in the middle of the Doppler spectrum. The middle and right frames were collected during target manoeuvres and show a high degree of smearing of the target response in Doppler. This has happened because the target changed its radial speed significantly during the coherent integration. The contrast between the frames in Fig. 11.29 and those in Fig. 11.31 shows some of the consequences of the waveform parameter choices. In the shorter CIT case of Fig. 11.29, the target has lower SNR and the Doppler resolution is poor, but it is relatively compact in Doppler. In contrast, the longer CIT data set of Fig. 11.31 has higher SNR and improved Doppler resolution at the expense of a high amount of smearing of the target in Doppler.

Figures 11.32 and 11.33 show the H-PMHT-RM tracks made for the two data sets. The shorter CIT data set 1 dropped track twice during sharp manoeuvres, which was partly due to the loss in SNR when the response smeared in Doppler and partly due to the extreme rate of the manoeuvres. Figure 11.29 shows five frames during the first track drop for the short CIT data set. It shows over the space of five frames the target before, during and after manoeuvre; in this case, the track was dropped because the filter simply failed to respond to the rapid acceleration that is mismatched to the assumed constant velocity target. Figure 11.30 shows frames around the second track drop-out. In this case, the track loss was more due to extremely low target SNR during the manoeuvre. Again, the figure shows frames before, during and after the manoeuvre, but it is not possible to see the target by eye in the during-frames.

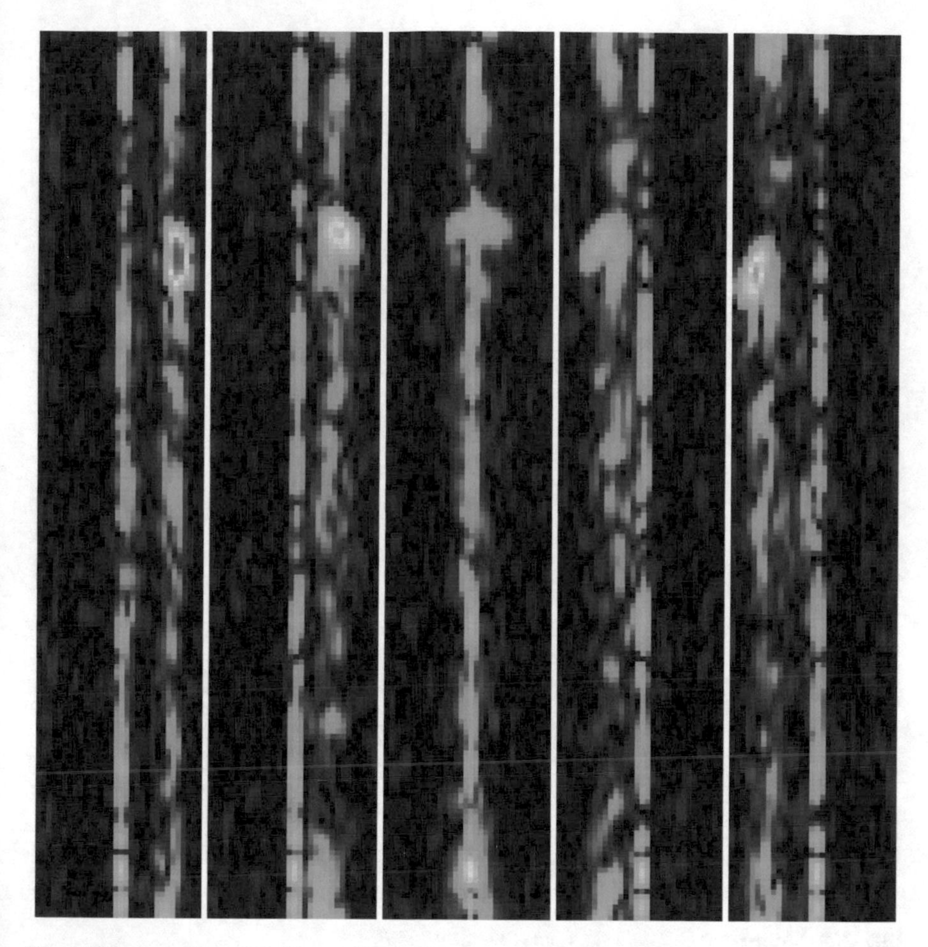

Fig. 11.29 Aura short CIT data set, frames 18–22

11.7 Summary

This chapter has presented three radar and three sonar applications of H-PMHT. In the radar examples, the target response was sufficiently compact to use core H-PMHT with a Gaussian appearance spread matched to the data. In the sonar examples, the target was more spread through the imagery and H-PMHT-RM was generally used. These examples contended with multipath propagation, fluctuating amplitude and non-cooperative clutter distributions.

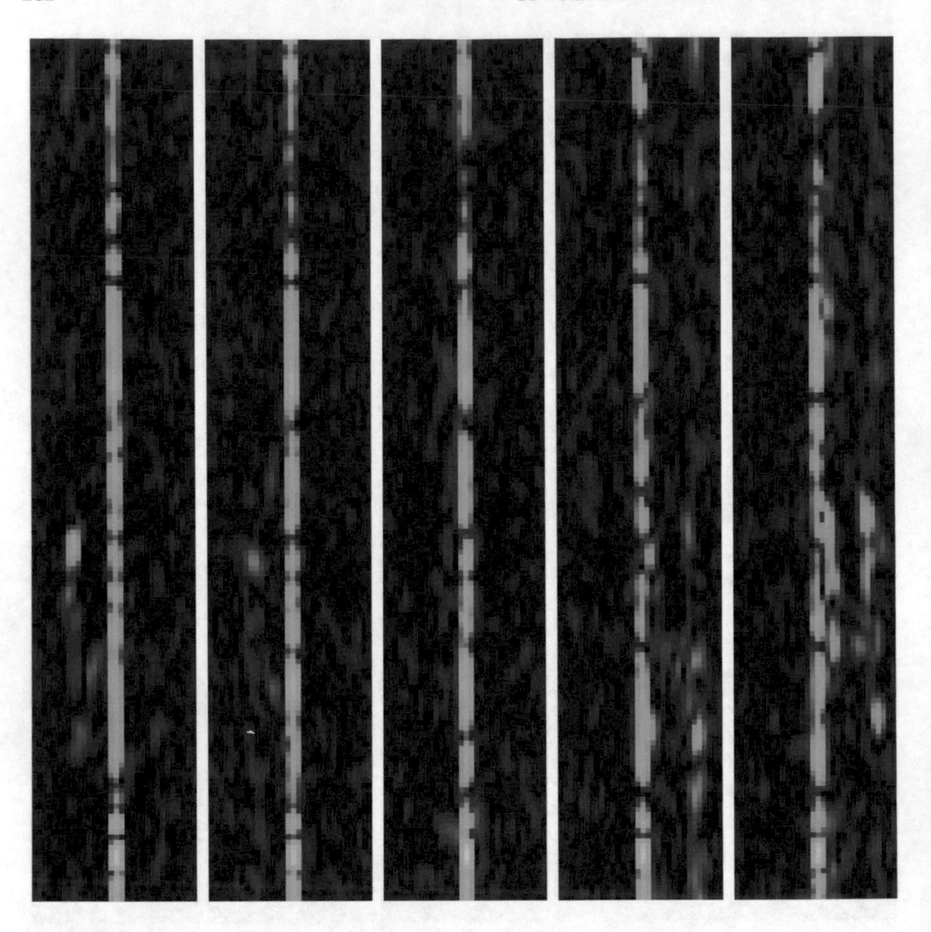

Fig. 11.30 Aura short CIT data set, frames 38–40, 42–43

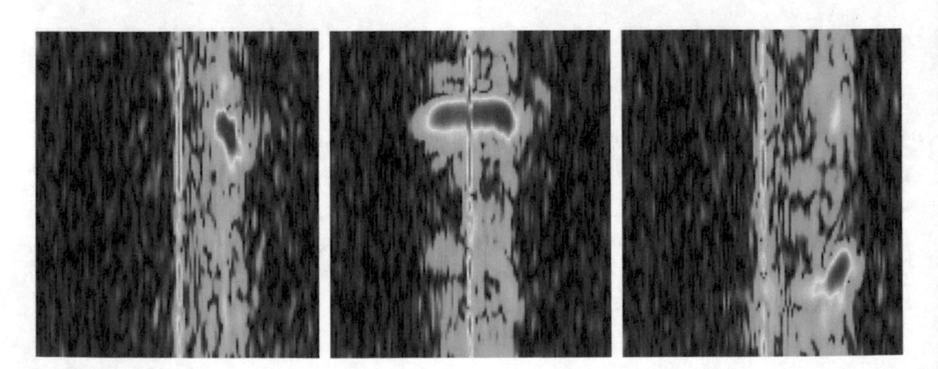

Fig. 11.31 Aura long CIT data set, frames 38, 42, 74

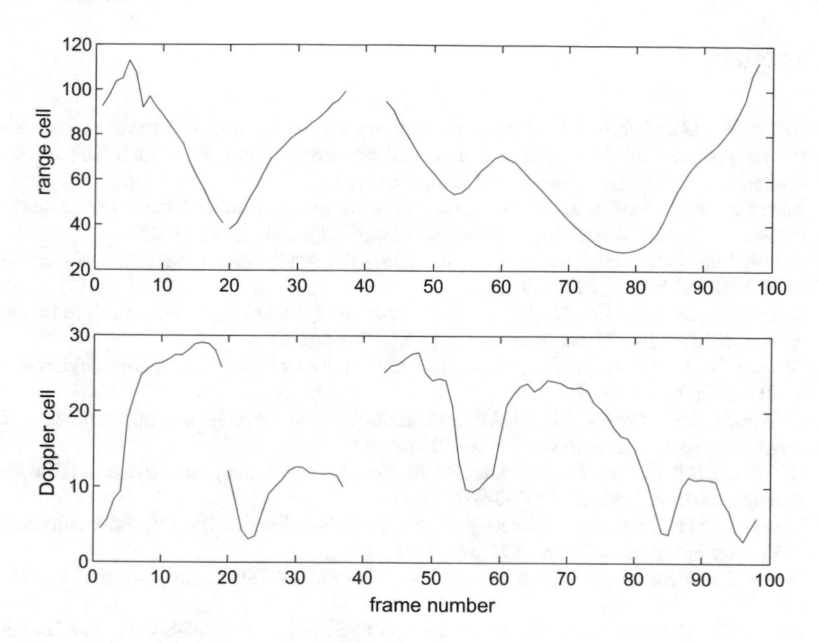

Fig. 11.32 Aura tracks for shorter CIT data set 1

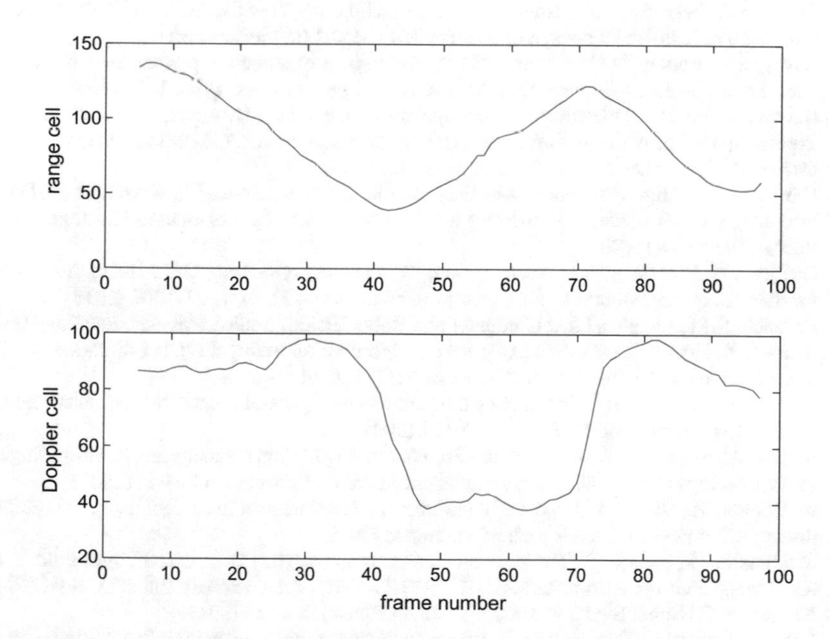

Fig. 11.33 Aura tracks for longer CIT data set 2

References

1. Anderson, R.H., Krolik, J.L.: Multipath track association for over-the-horizon radar using a bootstrapped statistical ionospheric model. In: Conference Record of the Thirty-Third Asilomar Conference on Signals, Systems, and Computers (1999)
2. Anderson, R.H., Krolik, J.L.: Track association for over-the-horizon radar with a statistical ionospheric model. IEEE Trans. Signal Process. **50**(11), 2632–2643 (2002)
3. Bar-Shalom, Y., Willett, P.K., Tian, X.: Tracking and Data Fusion: A Handbook of Algorithms YBS. YBS Publishing, Storrs (2011)
4. Carevic, D., Davey, S.J.: Two algorithms for modeling and tracking of dynamic time-frequency spectra. IEEE Trans. Signal Process. **64**, 6030–6045 (2016)
5. Cervera, M.A., Francis, D.B., Frazer, G.J.: Climatological model of over-the-horizon radar (CMOR). In: URSI (2017)
6. Colegrove, S.B., Davey, S.J.: PDAF with multiple clutter regions and target models. IEEE Trans. Aerosp. Electron. Syst. **39**, 110–124 (2003)
7. Colegrove, S.B., Davey, S.J., Cheung, B.: Clutter rejection using peak curvature. IEEE Trans. Aerosp. Electron. Syst. **42**, 1492–1496 (2006)
8. Davey, S.J.: Histogram PMHT with particles. In: Proceedings of the 14th International Conference on Information Fusion. Chicago, USA (2011)
9. Davey, S.J.: Detecting a small boat with histogram PMHT. ISIF J. Adv. Inf. Fus. **6**, 167–186 (2011)
10. Davey, S.J., Colegrove, S.B.: A unified joint probabilistic data association filter with multiple models. Defence Science and Technology Organisation, Australia, Technical Report DSTO-TR-1184 (2001)
11. Davey, S.J., Fabrizio, G.A., Rutten, M.G.: Detection and Tracking of Multi-path Targets in Over-the-Horizon-Radar using a Group Tracking Model (in preparation)
12. Davey, S.J., Rutten, M.G., Cheung, B.: A comparison of detection performance for several track-before-detect algorithms. EURASIP J. Adv. Signal Process. **2008**, 1–10 (2008)
13. Davey, S.J., Vu, H.X., Fletcher, F., Arulampalam, S., Lim, C.C.: Histogram probabilistic multi-hypothesis tracker with color attributes. IET Radar Sonar Navig. **7**, 178–190 (2015)
14. Davies, K.: Ionospheric Radio Peter Peregrinus Ltd. (1990)
15. Fabrizio, G.A.: High Frequency Over-the-Horizon Radar. McGraw Hill, New York (2013)
16. Fridman, S.V., Nickisch, L.J.: Sifter: Signal inversion for target extraction and registration. Radio Science **39**(1) (2004)
17. Gaetjens, H. X., Davey, S.J., Arulampalam, S., Fletcher, F.K., Lim, C.C.: Histogram-PMHT for fluctuating target models. IET Radar Sonar and Navig. **11**(8), 1292–1301 (2017)
18. Haralick, R.M., Shapiro, L.G.: Computer and Robot Vision. Addison-Wesley, Reading (1992)
19. Liang,L.Y., Pan,Q., Yang,F., Guan,C.: An em algorithm for multipath state estimation in other target tracking. IEEE Trans. Signal Process. **62**(11) (2014)
20. Luginbuhl, T.E., Willett, P.: Estimating the parameters of general frequency modulated signals. IEEE Trans. Signal Process. **52**(1), 117–131 (2004)
21. McDonald, M., Balaji, B.: Continuous-discrete filtering for dim manoeuvring maritime targets. In: Proceedings of the 10th International Conference on Information Fusion (2007)
22. McDonald, M., Balaji, B.: Impact of measurement model mismatch on nonlinear track-before-detect performance. In: IEEE Radar Conference (2008)
23. McDonald, M., Balaji, B.: Track-before-detect using swerling 0, 1, and 3 target models for small manoeuvring maritime targets. EURASIP J. Adv. Signal Process. **2008**(1), 1–9 (2008)
24. Nielsen, R.O.: Sonar Signal Processing. Artech House, Boston (1991)
25. Oron, E., Kumar, A., Bar-Shalom, Y.: Precision tracking with segmentation for imaging sensors. IEEE Trans. Aerosp. Electron. Syst. **29**, 977–987 (1993)
26. Palmer, J., et al.: Illuminator of opportunity bistatic radar research at DSTO. Radar **2008**, 732–736 (2008)
27. Pulford, G.W.: OTHR multipath tracking with uncertain coordinate registration. IEEE Trans. Aerosp. Electron. Syst. **40**(1), 38–56 (2004)

28. Pulford, G.W., Evans, R.J.: A multipath data association tracker for over-the-horizon radar. IEEE Trans. Aerosp. Electron. Syst. **34**(4), 1165–1183 (1998)
29. Rutten, M.G., Percival, D.J.: Joint ionospheric and target state estimation for multipath OTHR track fusion. In: Drummond, O.E. (ed.) Proceedings of SPIE, vol. 4473, pp. 118–129. San Diego, CA, USA (2001)
30. Rutten, M.G., Legg, J., Rivett, C., Davey, S.J.: Multihypothesis target-based multipath track fusion for over-the-horizon radar. Defence Science and Technology Group Technical Report (2016)
31. Skolnik, M.I.: Introduction to Radar Systems. McGraw-Hill, New York (2001)
32. Streit, R.L., Luginbuhl, T.E.: Probabilistic multi-hypothesis tracking. Technical report 10428, NUWC, Newport, Rhode Island, USA (1995)
33. Vu, H.X.: Track-before-detect for active sonar. Ph.D. thesis, The University of Adelaide (2015)
34. Vu, H.X., Davey, S.J., Fletcher, F., Arulampalam, S., Ellem, R., Lim, C.C.: Track-before-detect for an active towed array sonar. In: Acoustics. Victor Harbor, South Australia (2013)
35. White, K.A.B., Percival, D.J.: Multihypothesis fusion of multipath over-the-horizon radar tracks. In: SPIE, vol. 3373 (1998)

Chapter 12
Tracking in Full Motion Video

This chapter presents examples of the application of H-PMHT to problems in video tracking. While the algorithm is application agnostic, to an extent, there are some features of typical video applications that make them different from radar and sonar applications. The first difference is the relative resolution. In radar and sonar it is typical for the physical extent of a target to be no more than a few image pixels, perhaps it can even be assumed to be a point target and spreading in the sensor image is then an artefact of signal processing tapers and a finite aperture. In contrast, in video applications each target is usually spread across a large number of pixels, often hundreds. In this case, the appearance of the target in the sensor image is determined by the target's physical properties, not the sensor blurring function. It is no longer adequate to assume that the target appearance is known and the tracker should estimate it from data. Furthermore, the appearance is usually time varying. Even for rigid bodies, such as vehicles, orientation changes lead to rotations of the appearance. For non-rigid bodies, such as people, the target appearance in the sensor frame is continually changing as parts of the object move, for example, the arms and legs of football players move rapidly as the players run in the video application in Sect. 12.3.

The second feature of video tracking that differentiates it from radar and sonar applications is the update rate. While radar systems may have hundreds of parallel receivers capturing large amounts of data, the sensor processing usually uses significant integration time to improve signal strength. In video applications, the integration time is the exposure time of the frame, which can be of the order of milliseconds. Video sensors often have frame rates in the tens of frames per second and this imposes tight timing restrictions on the tracker. The tracker must have a low computation burden to keep up with the input data rate.

The third feature of video systems is related to the resolution and is occlusion. In radar systems, it is unlikely that an object would be hidden in the shadow of another closer object. In video systems, this is expected. The H-PMHT measurement model

S. J. Davey and H. Gaetjens, *Track-Before-Detect Using Expectation Maximisation*, Signals and Communication Technology,
https://doi.org/10.1007/978-981-10-7593-3_12

assumes superposition of power in the sensor image: when two targets occupy the same pixels then their appearances should add. In video, this is not the case and the closer target will obscure the one behind. The attribute H-PMHT presented in Chap. 9 was developed, in part, to address this issue.

This chapter considers four video tracking examples. The first two are people surveillance problems from the *Performance Evaluation of Tracking and Surveillance* (PETS) community in video tracking. PETS is a video tracking workshop where practitioners are invited to submit algorithms and papers detailing performance on benchmark video sets. Both of the PETS data sets are freely available to download from internet archives along with truth files and detailed descriptions. The first people tracking example is indoor surveillance. The footage is of a foyer in the entrance lobby of the INRIA Labs at Grenoble, France [11], and is referred to as the CAVIAR benchmark. A characteristic of the data is that the scripted scenarios intentionally contain people forming and breaking groups. Also because the camera was mounted close to a doorway there is significant change in the number of pixels occupied by each person as they move towards or away from the camera. The second example is footage of a football game. The team colours are highly contrasting (white and red) and the contrast between players and the field is also high. A challenge in this data is the complicated interactions between a relatively high number of people. The third example is overhead surveillance of an urban traffic scene. In this example there is significant camera motion that needs to be accounted for and cloud between the airborne camera and the road creates a dynamic non-homogeneous clutter background. The final example is space situation awareness. In this application, a telescope fitted with a high-quality digital camera stares at the night sky. As the Earth rotates the stars appear to move across the sensor field of view and satellites are visible through light reflected from the sun. The aim is to detect and track these satellites without prior knowledge of their orbits. In each of the applications, we use H-PMHT-RM to deal with the unknown and varying appearance typical in video tracking.

12.1 Difference Images

In the radar and sonar applications in the previous chapter, we made an implicit assumption that the sensor images were the output of a matched filter and that high-intensity pixels are likely to contain targets. Video imagery is often different to this. For example, a high contrast target could be dark on a light background, or a low contrast target could differ primarily in texture. A common preprocessing step for video tracking is background subtraction, which increases the contrast by building a model of the scene background and then comparing the current frame with the model.

The background model consists of a mean $E\left[\mathbf{z}^i\right]$ and a variance $E\left[\left(\mathbf{z}^i - E\left[\mathbf{z}^i\right]\right)^2\right]$ for each pixel. Numerous approaches exist to derive such a model, for example [2]. A common approach that we will use in this chapter is to use median-based estimates over a large number of frames. Medians are preferred because they are robust to outliers. The background model is used to define a normalised difference image

$$\delta \mathbf{z}_t^i \equiv \frac{\left|\mathbf{z}_t^i - E\left[\mathbf{z}^i\right]\right|}{E\left[\left(\mathbf{z}^i - E\left[\mathbf{z}^i\right]\right)^2\right]^{1/2}}. \tag{12.1}$$

The PETS examples considered here are colour video sequences, whereas the difference image defined above assumes a scalar intensity measurement in each pixel. Colour video generally has three values in each pixel, often red, green and blue channels. The simplest way to treat the colour imagery is to assume that these channels are independent. In practice, the formats used to store images often do not use three independent channels and the red, green and blue values can be derived from two channels. Nevertheless, treating the channels as independent we can combine differences in red, green and blue by forming a Cartesian distance

$$\delta \mathbf{z}_t^i \equiv \left(\left[\delta \mathbf{z}_t^{i,\mathsf{R}}\right]^2 + \left[\delta \mathbf{z}_t^{i,\mathsf{G}}\right]^2 + \left[\delta \mathbf{z}_t^{i,\mathsf{B}}\right]^2\right)^{1/2}, \tag{12.2}$$

where $\delta \mathbf{z}_t^{i,\mathsf{R}}$ is a difference image formed on the red channel, and similarly for green and blue. If we intended to use point measurement tracking, then this difference image could be segmented by applying a threshold that leads to a binary foreground mask [12]. Point measurements can be extracted from the foreground mask by finding connected pixels and then defining a centroid for each collection of connected pixels.

The use of temporal averages explicitly assumes that the statistical distribution of values in each pixel is stationary, which requires not only a fixed scene, but also a fixed camera. In many applications, this is too restricted an assumption. One way to relax the stationary image assumption is to assume that the scene itself is stationary but that the camera moves. Under that model the change in images is due to a transformation that can be estimated. This estimation process is referred to as frame-to-frame registration. The registration method used for the moving camera video in this chapter is the Kanade, Lucas and Tomasi (KLT) feature tracking algorithm [17]. In the KLT algorithm, small features such as corner points are automatically extracted and tracked using an eigenvalue-based corner measure. The tracked features are used as control points to which a frame-to-frame parametric registration model is fitted and used to transform each frame to the common frame of reference, details can be found in [13].

The processes of frame-to-frame registration and background modelling are open research topics in their own right but are beyond the scope of this book. Here we will use KLT registration and a median background model without consideration about whether these are the best available options.

12.2 People Tracking: PETS CAVIAR

The first example uses video sequences from the CAVIAR repository from the 2003 PETS benchmark. These sequences were downloaded from the repository at http://homepages.inf.ed.ac.uk/rbf/CAVIAR/. We acknowledge the EC funded CAVIAR project (IST 2001 37540) for making the benchmark data available for public use. The footage is of a foyer in the entrance lobby of the INRIA Labs at Grenoble, France [11], and is referred to as the CAVIAR benchmark. The frames are 384×288 pixels and collected at a rate of 25 frames per second. A characteristic of the data is that the scripted scenarios intentionally contain people forming and breaking groups. There is also a significant change in the number of pixels occupied by each person as they move towards or away from the camera that was mounted close to a doorway.

We present the results of applying H-PMHT-RM [20, 21] from Chap. 8 to the sequences referred to as *Meet_Crowd* and *Meet_Walk_Split*, which are described in more detail a little later. The H-PMHT-RM tracks are compared with a multi-target particle filter track-before-detect algorithm referred to as PF-JDT (joint detection and tracking). The PF-JDT results were provided by Dr Jamie Sherrah [16]. The CAVIAR results here were first presented in [6, 21].

12.2.1 Splitting and Merging Target Groups

The CAVIAR video sequences have some features that require us to extend the track management framework to achieve good performance. One of the challenges in multiple extended object tracking is correctly identifying the number of objects: when the tracker is able to adapt the target model it becomes possible to incorrectly place two tracks on a single object or one track across multiple objects. This problem is addressed by developing a track manager that incorporates splitting and merging.

The handling of split and merge events in multiple extended object tracking is a special case of the general problem of model order estimation, where the problem is to estimate the number of components in addition to the parameters of each component. A variety of model order estimation approaches for Gaussian mixtures have been proposed: Corduneanu and Bishop presented a method using a *variational Bayesian approximation* [3]. The method starts by initialising a large number of components and then automatically simplifies the mixture by eliminating redundant components [3]. Constantinopoulos and Likas modified this method to instead start with a single component and then progressively split components until subsequent splits no longer improve the model [4, 19]. The methods in [3, 4] use a Wishart prior to estimate the covariance matrices (the extents) of the components, similar to H-PMHT-RM. However, they assumed a stationary distribution with point measurements, whereas here the problem is a dynamic distribution with image measurements.

Figure 12.1 shows an example of object splitting from the CAVIAR database. The images show the original colour video frames stacked with background subtracted

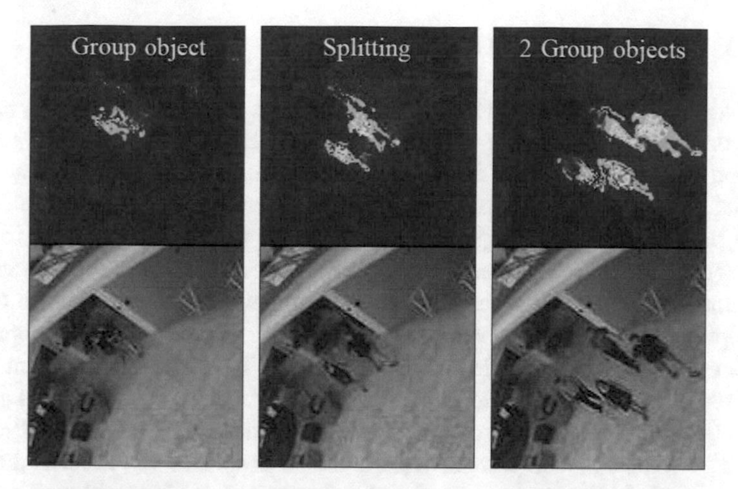

Fig. 12.1 Splitting groups of people

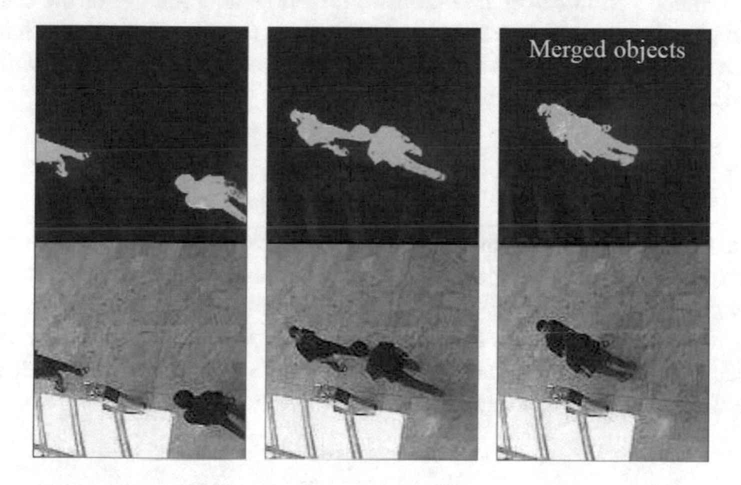

Fig. 12.2 Merging groups of people

difference frames. The scenario is called *Meet_Crowd* and consists of four people successively entering the surveillance area: initially, the first two people are unresolved by the camera; they move apart and appear to be a single target that breaks into two; later another two people enter from behind and are partially occluded, resulting in two groups. Similarly, the image sequence in Fig. 12.2 shows an example of merging targets in a scenario referred to as *Meet_Walk_Split*. Here, two people walk towards each other from the left and from the right. When they meet and shake hands, their associated blobs in the image start merging, which results in an unresolved group target.

12.2.1.1 Merging Components

The merging function is easier to develop than the split function, since merging corresponds to a reduction of information. The method in [3, 4] used the same Gaussian mixture description developed in the beginning of Chap. 3 for a stationary point measurement mixture. It rejected components based on a threshold on the mixing proportion π^m. This approach is essentially the same as the track deletion rule described in Chap. 5. If there is a merge event, then the superfluous components will naturally starve over time. However, if components are rather close or even overlapping, then waiting for vanishing πt^m or Λ_t^m values can take an impractical number of EM iterations: hundreds if not thousands. The point measurement PMHT and H-PMHT are usually limited to ten or fewer EM iterations in practical applications to maintain the real-time computation performance [22], which is sufficient in the case of well-separated targets. What can also happen is that the non-dominant component can drift away from its object and persist on background measurements for an extended period before it peters out. It is more appropriate for the tracker to acknowledge that two tracks have become too close to be resolved and to intervene by removing one of them. This is implemented by measuring the statistical distance between tracks m and j

$$d_t^{m,j} = \left(\mathbf{x}_t^m - \mathbf{x}_t^j\right)^{\mathsf{T}} \left[(\mathsf{P}_t^m)^{-1} + (\mathsf{P}_t^j)^{-1}\right] \left(\mathbf{x}_t^m - \mathbf{x}_t^j\right), \tag{12.3}$$

which are merged if the distance is within a statistical gate. The components are merged using a convex combination [1], which simply means that the merged mean is a weighted combination of the component means. These weights are usually defined in terms of the track covariances, but we found experimentally that the estimated power from each track provides a better weight for H-PMHT-RM. The relative weights to combine tracks m and j are given by

$$\mathsf{w}_t^m = \frac{\Lambda_t^m}{\Lambda_t^m + \Lambda_t^j} \tag{12.4}$$

with $w_t^j = 1 - \mathsf{w}_t^m$.

When merging components it can be desirable to remember the track labels that were combined to form the merged track. This is referred to as the pedigree and enables subsequent analysis to be aware that tracks have been combined, which is important if the user wants to follow a specific object.

12.2.1.2 Splitting Components

An example of splitting group objects is shown in Fig. 12.1; when two formerly unresolved objects move apart and become resolvable, it is desirable for the tracker to create a second track. In the model order estimation method proposed in [4] the

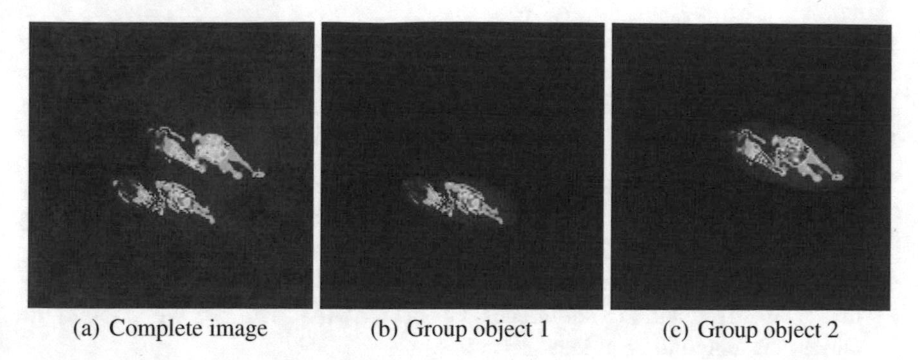

(a) Complete image (b) Group object 1 (c) Group object 2

Fig. 12.3 Associated images for two object groups

mixture of interest is estimated by progressively splitting components. It starts with two initial component candidates on the complete data set and then recursively tries to split each component into two new candidates until no further splits are accepted. EM is used to both estimate the parameters of the components and to identify when a split is rejected. A proposed split is rejected if one of the two component candidates has sufficiently low mixing proportions. The method was applied to a batch of point measurements with fixed components.

The video surveillance application differs from the model order approach in [4] because the components in the scene are dynamic due to movement of the people, and because the video frames should be processed sequentially, not as a batch. The core algorithm of [4] is an EM mixture fitting method and can, therefore, be directly replaced with H-PMHT-RM. A direct adoption of the method in [4] would also imply the fitting of the potential split components *jointly* with all other mixture components. However, in the considered video application it can be assumed that splits in one target are independent of the other targets. The split function proposed for H-PMHT-RM, therefore, isolates the split candidate and holds the remaining components fixed. This decreases the cost of each EM iteration significantly. The isolation is implemented by fitting the split components to the associated image of the original single component. Figure 12.3a shows an intensity image with four people as an example. Assuming that the track database currently contains two group targets, the split function has to calculate two associated images, one for each object. Figure 12.3b comprises the associated image of the lower group object, while Fig. 12.3c refers to the upper group object. Recall that the associated image was defined in (4.67) and are simply the sensor image scaled by the cell assignment weights

$$\mathfrak{z}_t^{i,m} = w_t^{i,m} \bar{\mathbf{z}}_t^i, \tag{12.5}$$

Even with the split candidate isolated, it is too costly to consider all possible splits on every frame. Furthermore, real people do not have a Gaussian shape and could be represented as a complicated mixture of many Gaussian components. In order to keep the processing cost relatively low and to avoid segmenting people or groups into too many components (which has the main penalty of again increasing computation burden), a pre-filter is chosen to identify which targets appear likely to have split before applying H-PMHT-RM to the split components. The pre-filter has three requirements before a component is considered to be a split candidate:

1. The estimated track SNR must be high. It is undesirable to split components that are already of low significance. For the CAVIAR data sets in the following section, we used a 18 dB threshold.
2. The extent of the track must be broad. Again, it is undesirable to split a component that is already small into sub-pieces. For the CAVIAR data sets we required the major axis length to be at least 20 pixels.
3. The associated image should be multimodal. For the CAVIAR data sets, the associated image was segmented using a pixel threshold of 10 dB. At least two segments of greater than 40 pixels had to be found.

For each established track that satisfied the above pre-filter rules, two tracks were initialised on the largest two objects from step 3 above and H-PMHT-RM was performed using the associated image as the measurement frame. On convergence, the split was accepted if $1/\tau \leq \Lambda^1/\Lambda^2 \leq \tau$. For the CAVIAR data sets we used $\tau = 5$. If the split was accepted, then the original track was replaced by the two components of the H-PMHT-RM output. As with merging, a track pedigree record was maintained for analysis.

12.2.2 Tracking Performance

The first scenario is called *Meet_Crowd* and contains four people that entered the scene from a doorway in the North-West and moved across the region, exiting at the South-East corner of the image. The people initially formed two pairs and then one person moved to form a group of three before each of the players left along separate paths. The second scenario is called *Meet_Walk_Split* and contains two main players with numerous other people moving around the edges of the scene. The two people moved from the South-East and West to meet in the middle of the image where they paused and shook hands. Both moved to the West and then they parted as they left the scene. The other people around the edges of the scene were mostly at longer range and so appear as small targets in the video. In the South-East corner of the scene there were two people partially obscured by a window awning. These people did not change position through the video but did make small movements.

The first 25 data frames (1 s) of the *Meet_Crowd* scenario show just the empty scene and were therefore taken to construct a median background model of the image intensity in this scenario. In the *Meet_Walk_Split* data set there are no frames showing the empty scene. In this case, taking the median over the first 300 frames of the data set, led to good results. This model is then used to define difference images, as discussed earlier. We need to use different background models because the background is sensitive to lighting conditions. Figures 12.1 and 12.2 show examples of the original colour imagery on the bottom and the corresponding difference images on the top.

The *Meet_Crowd* and *Meet_Walk_Split* scenario are two sequences of a larger set tested by Sherrah and Ristic with the particle filter for joint detection and tracking (PF-JDT) [16]. They showed that the PF-JDT algorithm gave very good detection performance compared with several other standard methods. The PF-JDT is an imagery version of the PF for track-before-detect (PF-TkBD) which we compared with core H-PMHT and Viterbi TkBD in Chap. 2. There are two main differences between the PF-JDT and the PF-TkBD: first, the PF-JDT uses a stacked state vector to allow the particle filter to follow a number of targets in the scene. Second, it uses a different measurement model to the radar-like model of [7, 15], which is essentially a narrow

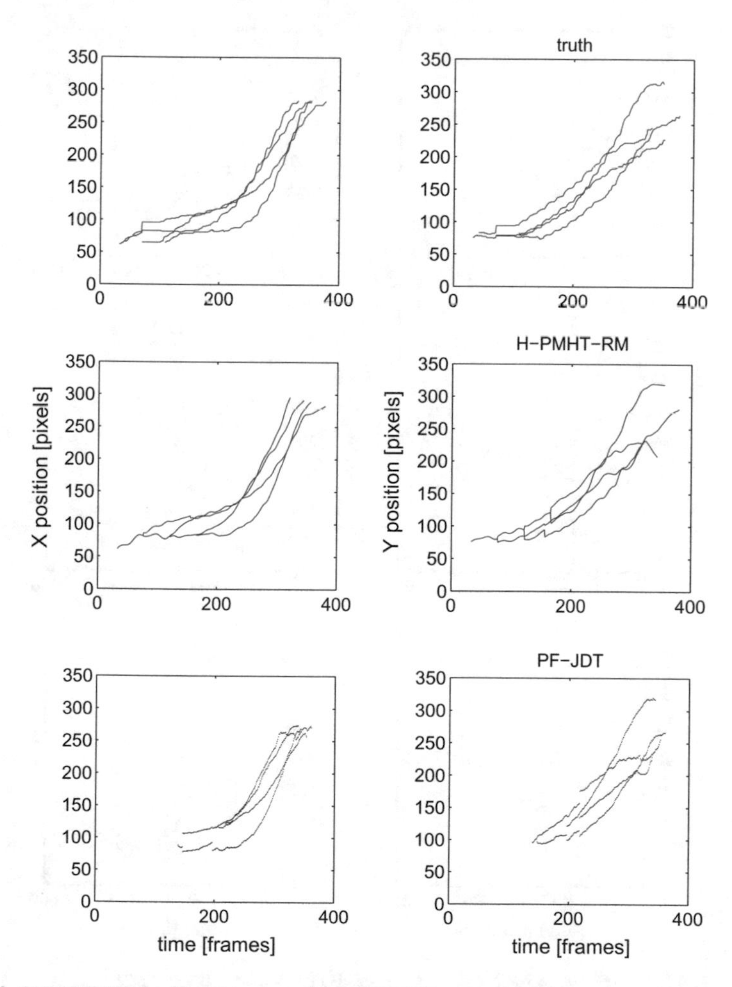

Fig. 12.4 H-PMHT-RM with the truth and the PF-JDT for *Meet_Crowd*

Gaussian appearance model. Instead, each particle has an associated bounding box, which is a rectangular set of pixels within which the target is hypothesised to exist. The track likelihood is a function of the foreground pixels in this bounding box. Since every particle has a different bounding box, PF-JDT estimates the target shape by extending the target state. The method has a potentially different hypothesis for the number and labelling of targets in each particle: post-processing to link the reports across time into contiguous tracks was not performed for the comparison here.

Figure 12.4 compares the H-PMHT-RM tracks for the *Meet_Crowd* scenario with the PF-JDT output and ground truth provided in the CAVIAR database, which was derived from manual labelling. The plots show the x-position and y-position of the targets as a function of frame number (time). The splitting of H-PMHT-RM tracks

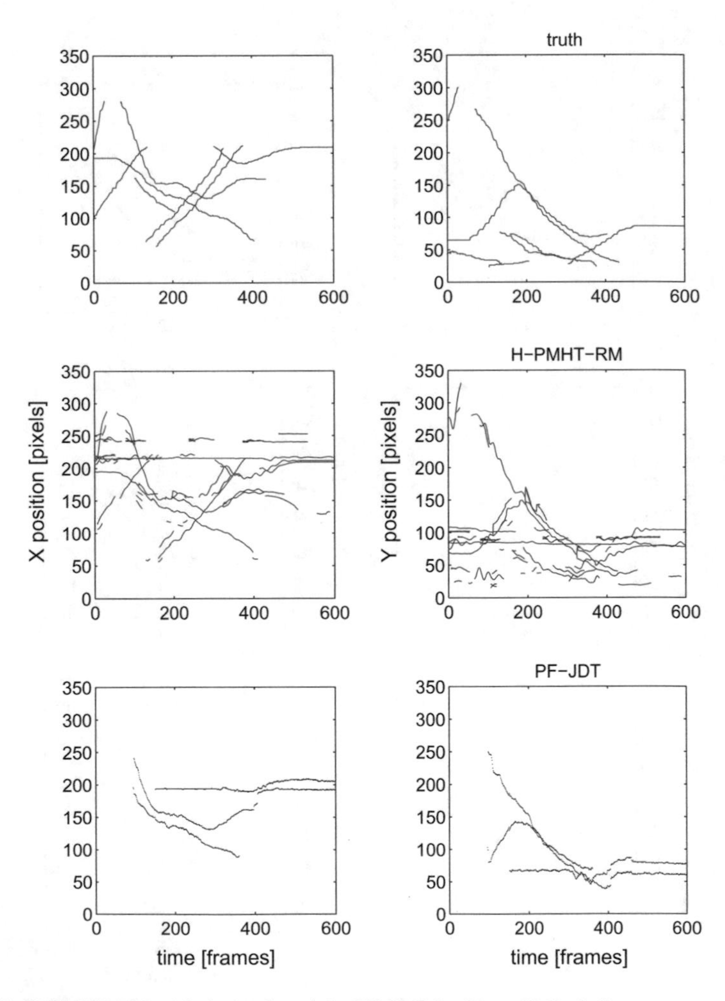

Fig. 12.5 H-PMHT-RM with the truth and the PF-JDT for *Meet_Walk_Split*

can be seen in the y-position plot as the people move apart and become resolvable. Comparing the H-PMHT-RM and PF-JDT outputs, we observe that the H-PMHT-RM tracks start earlier and the target groups split into component tracks in fewer scans. This is because the algorithm is more sensitive to smaller targets: towards the left of the scene where the targets enter the people are much farther from the camera and occupy fewer pixels.

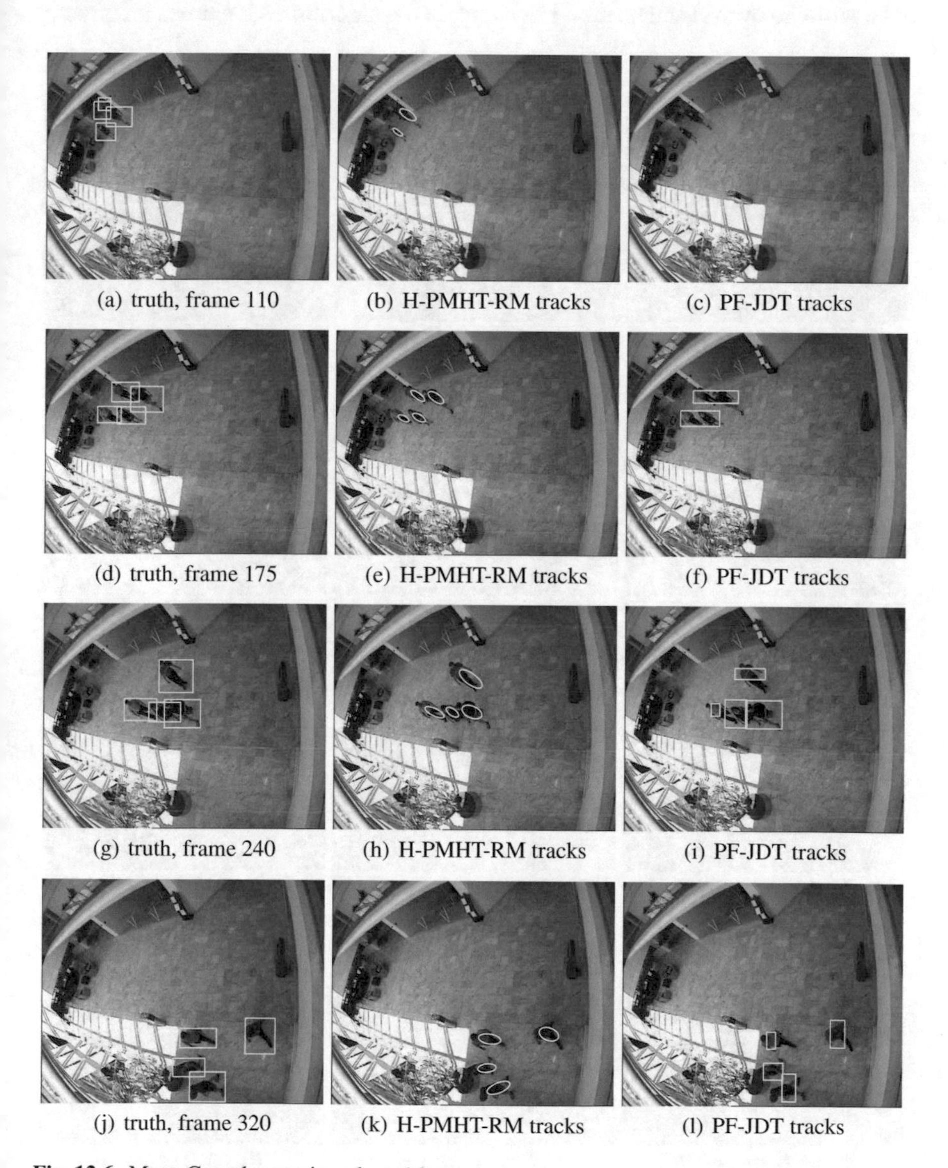

(a) truth, frame 110 (b) H-PMHT-RM tracks (c) PF-JDT tracks

(d) truth, frame 175 (e) H-PMHT-RM tracks (f) PF-JDT tracks

(g) truth, frame 240 (h) H-PMHT-RM tracks (i) PF-JDT tracks

(j) truth, frame 320 (k) H-PMHT-RM tracks (l) PF-JDT tracks

Fig. 12.6 Meet_Crowd scenario, selected frames

Figure 12.5 shows the tracks and ground truth for the *Meet_Walk_Split* scenario. The PF-JDT did a good job of following the two principle players in the scenario, but did not detect the extra people around the scene fringes. This is again most likely due to the implicit size threshold. A single spurious track was formed in the vicinity of coordinate [100, 100] of the scene. In contrast, the H-PMHT-RM was able to detect several of the extra people. During initial testing, the H-PMHT-RM was found to be prone to over-segmentation. For example the algorithm sometimes incorrectly

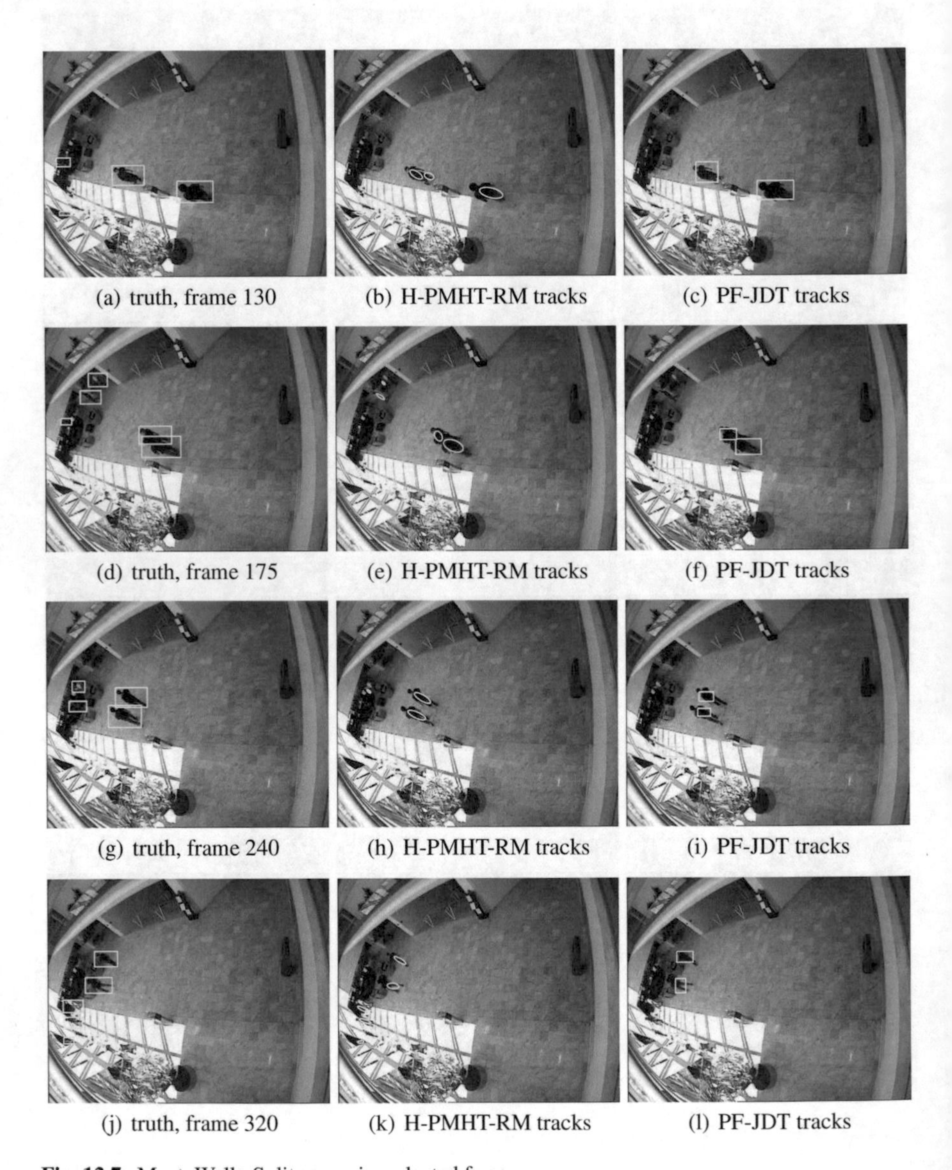

(a) truth, frame 130 (b) H-PMHT-RM tracks (c) PF-JDT tracks

(d) truth, frame 175 (e) H-PMHT-RM tracks (f) PF-JDT tracks

(g) truth, frame 240 (h) H-PMHT-RM tracks (i) PF-JDT tracks

(j) truth, frame 320 (k) H-PMHT-RM tracks (l) PF-JDT tracks

Fig. 12.7 Meet_Walk_Split scenario, selected frames

identified the head and body of a person as two different targets. Intuitively, there is a trade-off between splitting responsiveness and over-segmentation. To mitigate the over-segmentation, a size threshold was introduced for forming new tracks. The CAVIAR repository contains the information required to map the image pixels into a Cartesian spatial frame, so it is possible to apply a consistent minimum size limit across the image. However, a simpler scheme was found to be sufficient. The image is vertically divided into three rectangular areas. These areas correspond to three different levels of depth from the perspective of the camera. The left area is the deepest and the people to be tracked appear very small covering only a small number of pixels. In the middle area, people appear bigger, and in the remaining right area they appear as relatively large objects in front of the camera. This threshold is only set on the level of track seed generation, subsequent track maintenance is carried out by processing the intensity images directly without any size limitation.

The H-PMHT-RM formed intermittent tracks on the people partially obscured by the awning in the *Meet_Walk_Split*, which show up as almost stationary tracks in Fig. 12.5. These people were not included in the CAVIAR truth file, however, it is clear on inspection of the video sequence that there were indeed people there. The detection of these people (who barely move) may be an artefact of the different preprocessing used.

Figures 12.6 and 12.7 show selected frames from the two scenarios. In each figure, the left column shows the image frame with bounding boxes given in the ground truth. The middle column shows the image frame with one-sigma ellipses marked for each H-PMHT-RM track. The right column shows the bounding boxes resulting from the PF-JDT. These snapshots illustrate the difference between the algorithm outputs and are more easily compared with the truth. It is clear that the H-PMHT-RM has generally formed shape covariances with a major axis along the line between each person's feet and head, which is not vertical in the image frame due to the geometry and optics. Frame 130 in Fig. 12.7 illustrates a case where H-PMHT-RM has detected several small targets not detected by PF-JDT. Frames 175 and 240 illustrate the PF-JDT spurious track. The H-PMHT-RM has also done a good job in segmenting closely spaced objects.

12.3 Tracking Football Players: VS-PETS

The second example is from the 2003 Video Surveillance and Performance Evaluation of Tracking and Surveillance (VS-PETS) benchmark and contains footage of a football game. It provides an interesting example of the use of colour because the different teams have strongly contrasting playing strips that are white and red. The results here were first presented in [9]. The particular football sequence tested contains 21 people. There are 17 players in white and red, the referee and 2 officials who wear black, and an additional person who does not wear a uniform. Presumably, this is one of the coaching staff. The frames are at a PAL standard resolution of 720×576 pixels, giving a total image size of a little over 400 kpixels. The VS-PETS archive contains a set of training frames that are manually selected frames from the video that do not contain people. These were used to construct the median background model. The players interact in the video, occluding each other at various times, and enter or leave the field of view. Figure 12.8 shows an example frame from the test data and the corresponding intensity difference image.

The difference image provides strong contrast between the players and the field. It is also sensitive to the ball and to the players' shadows. It is quite effective at suppressing the lines on the field although the flag in the corner of the field does leave an artefact. The difference image is clearly very powerful for detecting the players but it does not contain the colour information that was present in the original frame. The VS-PETS data was chosen to illustrate how the colour information can be used to augment the difference images by different versions of the H-PMHT and improve tracking through occlusions and when players are close.

A difficulty with assessing performance on the VS-PETS data is the truth format, which is not really amenable to analysis. The database contains a compressed version of the video sequence with bounding boxes superimposed in the imagery. The boxes are not available in any other format. The bounding boxes were drawn in different colours for each person. The variation in colours and the compression means that it is not trivial to extract the boxes from the image sequence: we built a line detector for horizontal and vertical lines and used this to construct point measurements. These point measurements were then filtered using point-PMHT. These tracks are the best approximation of truth available: they represent reality better than the H-PMHT outputs but as a truth source they are imperfect. This means that the performance measures derived from them are somewhat compromised.

Three different versions of H-PMHT-RM were tested on the VS-PETS video sequence. First, we use H-PMHT-RM on the difference images with no colour information. This is essentially the same implementation that was used for the previous application. Second, spectral H-PMHT-RM was applied to the individual colour channel difference images. Spectral H-PMHT-RM was described in Chap. 9 and derived in [18]. It treats the colour channels as superposition images, so it does not model occlusion. The input is colour difference images because it models targets as producing high energy, but the colour differences mean that the characteristic colours of individuals can be lost. Third, we illustrate obscuring H-PMHT-RM, which was also described in Chap. 9 and first derived in [9]. The three different options are labelled as greyscale, spectral and attribute.

Figure 12.9 shows the performance metrics derived from 400 frames of the VS-PETS camera-3 sequence. For most of the sequence the RMS error is essentially the same for the three algorithms and after burn-in the number of missed targets is low. However, there is a period around frame 250 where the intensity-only tracking has relatively poor error and it often has slightly worse missed target count. This is because the greyscale tracker has more difficulty resolving close players on different teams. It puts a single track on two players a little more often and this leads to both a missed person and a biased state estimate for the other person. The obscuring H-PMHT-RM has a consistently smaller number of swaps than either the intensity-

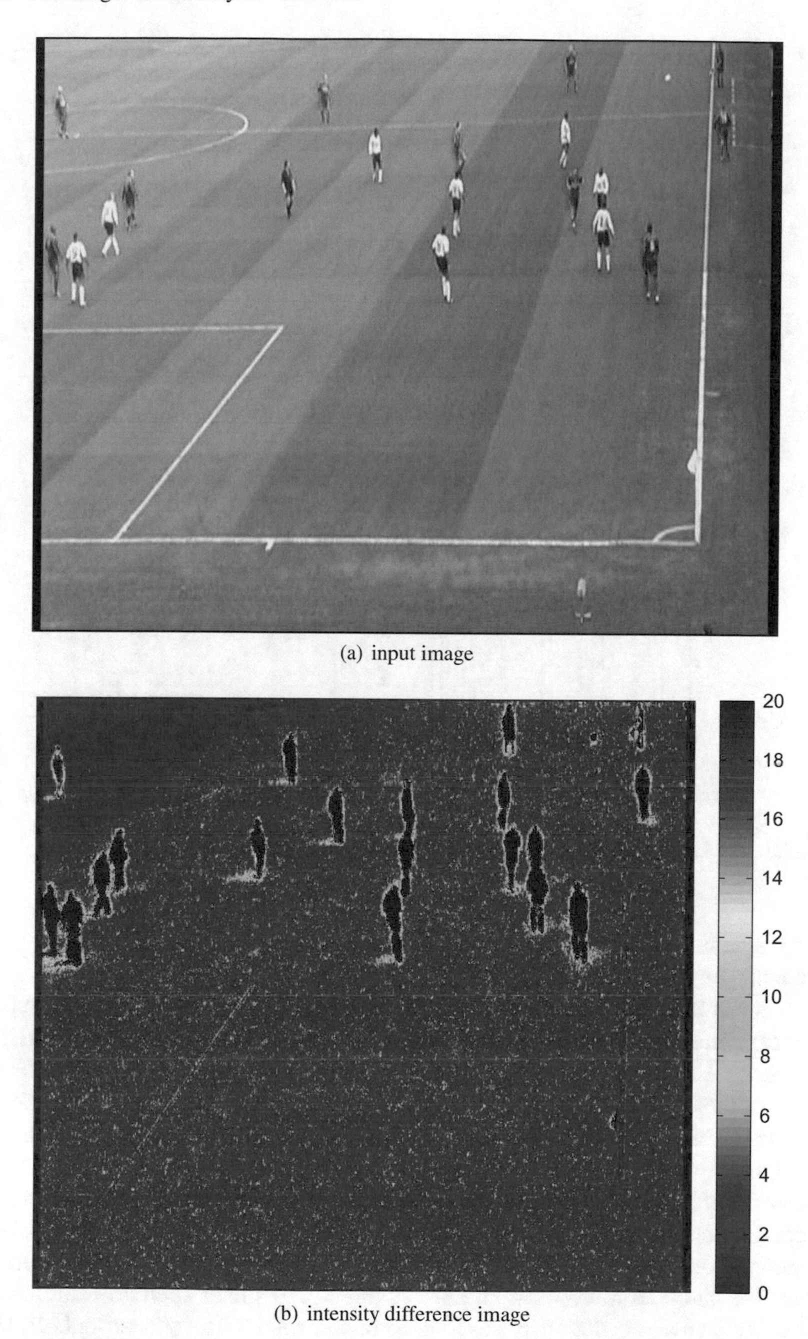

(a) input image

(b) intensity difference image

Fig. 12.8 Example frame from VS-PETS03 soccer archive

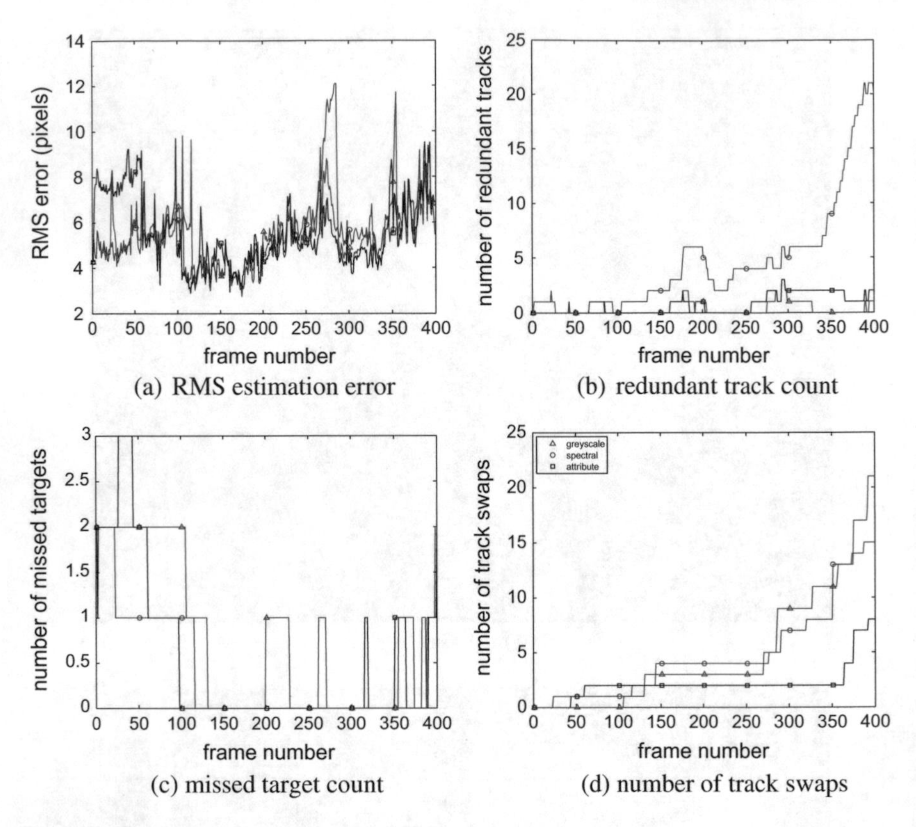

Fig. 12.9 Tracking performance on VS-PETS video

only or spectral tracking. This is because it does a much better job of characterising the colour properties of the players.

The spectral H-PMHT-RM made a large number of redundant tracks because it over-segmented the imagery: in many cases, a single player was associated with multiple tracks. These duplicate tracks created by spectral H-PMHT-RM were generally a result of the tracker not correctly associating all of a target's energy to its track. For example when a player held his arms out then the shape of the player's appearance was highly non-Gaussian and H-PMHT-RM-RM tended to split this into multiple smaller Gaussian-shaped components. The track initiation logic searches the image for energy not associated with an existing track and uses this to start new tracks. If the association of pixels to an existing track leaves residual energy in the difference frame, this can lead to duplicate tracks. Figure 12.10 shows such an example and the background associated image used to form new tracks. The obscuring H-PMHT-RM clearly does the best job of isolating the target energy, which is a reflection of the power of the absolute colour reference model. As the simulations in Chap. 9 showed, the obscuring H-PMHT-RM gives a much lower number of track swaps because it accounts for occlusion. In the experimental data the number of occlusions

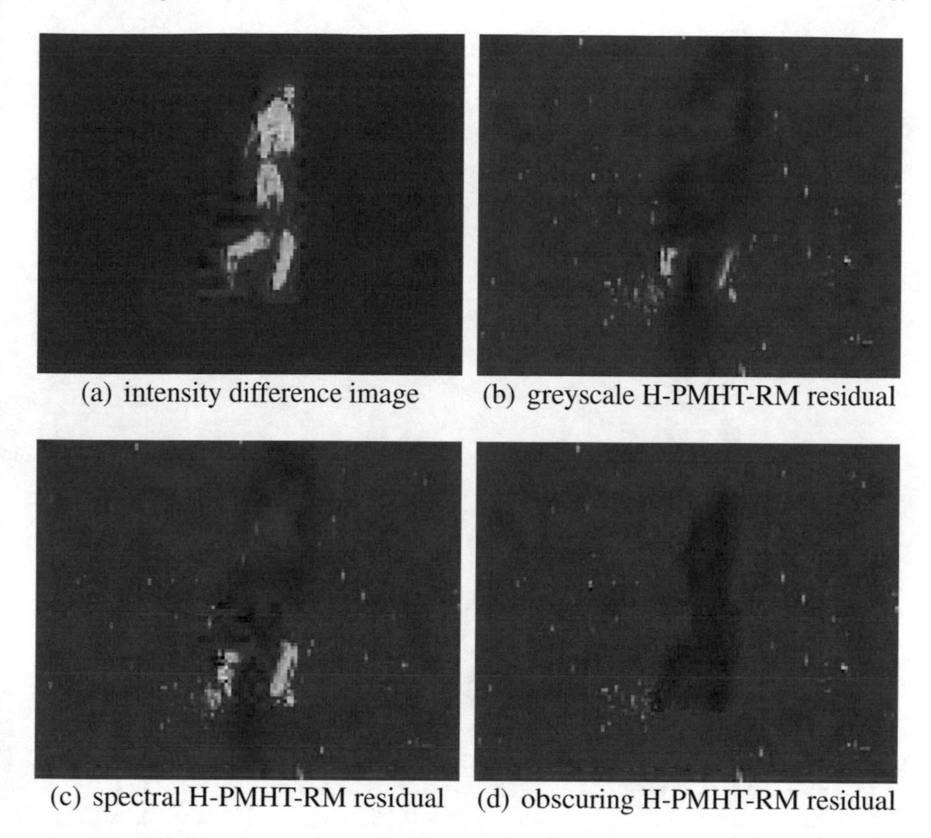

Fig. 12.10 Residual intensity difference images

is much fewer than in the contrived simulation of Chap. 9, so the swap count does not become as high. Towards the end of the sequence, players leave the field of view and this results in track swapping. This could be mitigated by introducing a layer of unobserved cells around the image boundary as discussed in Chap. 4.

Estimating attributes of the people under track improves the tracking performance but it can also create opportunities to estimate non-kinematic properties of the objects. For example, the colour estimates can be used to classify the objects into different types. One way to perform this classification is to use a Gaussian mixture model to estimate the distribution of the attribute data. Both the spectral and obscuring H-PMHT-RM algorithms create a vector of the average intensity in each colour channel. For the spectral case, these are the difference intensities, whereas for the attribute case they are absolute colour values. The mixture modelling here is the same as described in Chap. 3 except that we estimate the number of components by over-modelling and then discarding insignificant components. Here we initialised with nine components. Figure 12.11 shows the mean colour attribute for each track at frame 200. It compares the spectral H-PMHT-RM estimates in Fig. 12.11a with the obscuring H-PMHT-RM estimates in Fig. 12.11b. Each person is colour coded based

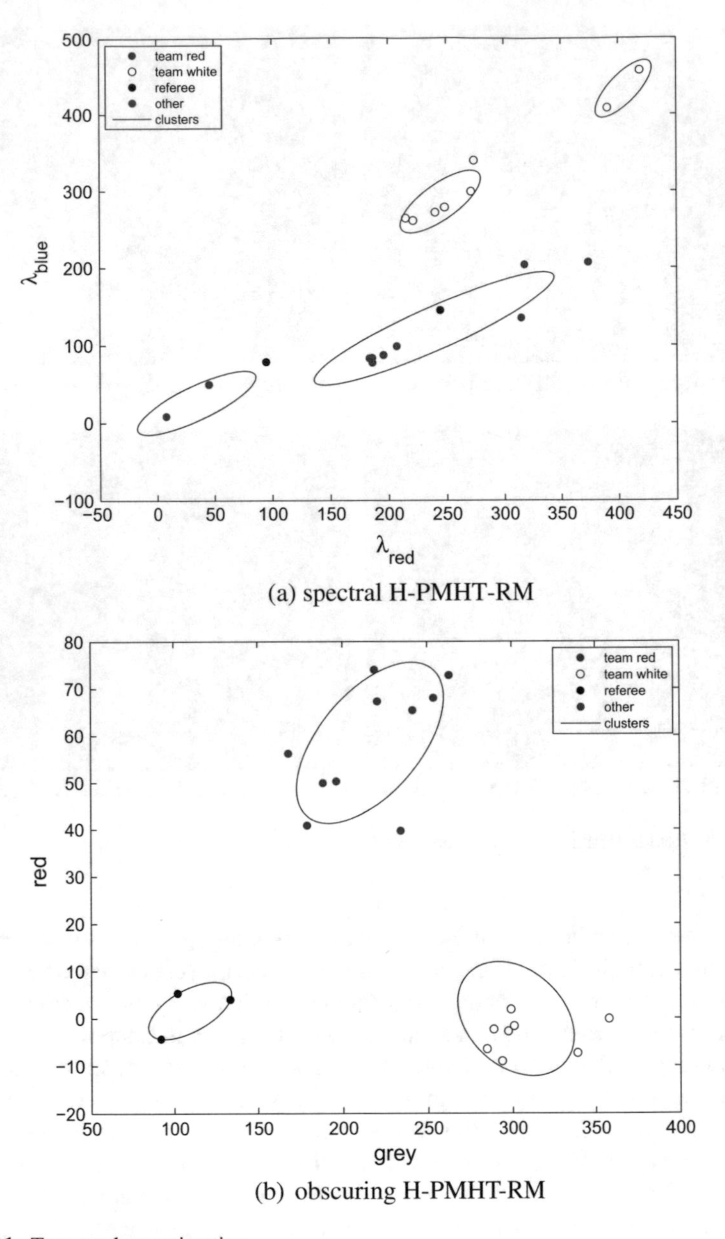

Fig. 12.11 Target colour estimation

on manual classification. For the purposes of visualisation, we collapse the three-dimensional attribute data onto a two-dimensional representation. For the spectral H-PMHT-RM, the axes of the plot correspond to the estimated average difference intensity in the red channel and the blue channel. The green channel was manually determined to be a poor discriminator and is not shown. For the obscuring H-PMHT-RM the absolute colours were projected onto unit vectors in the directions $(1, 1, 1)$ and $(1, -0.5, -0.5)$, which are labelled *grey* and *red*, respectively. These closely align with the two highest eigenvectors of the target colour vector space, so projecting the estimates onto this plane gives the best discrimination. The figure shows ellipses that summarise the colour distribution of the members of each cluster.

The spectral H-PMHT-RM tracks were grouped into four clusters. Two of those corresponded to white-team players and the other two combined red-team players and the other people. The spectral H-PMHT-RM colour estimates did not provide a strong enough feature vector to separate the different groups. In contrast, the obscuring H-PMHT-RM colour vectors were automatically grouped into three clusters that clearly correspond to the two different teams and the officials. There is one outlier that does not fit into any of the clusters: the mixture fitting does not retain clusters with one member. This outlier is the team coach on the sidelines who was dressed differently to the rest of the people. This is a single anecdotal example, so we cannot sensibly define classification measures of performance. Nevertheless, we see that the obscuring H-PMHT-RM has good potential in joint tracking and classification applications.

12.4 Video Moving Target Indication

The PETS examples so far have both been cases of tracking people with a fixed camera. The people occupy a very large number of pixels. The remaining applications are for moving cameras. Next, we consider overhead persistent surveillance of an urban area. This means an aircraft circling with a video camera monitoring traffic on a road network. This application is often referred to as video moving target indication (VMTI). There is a clear compromise in setting the camera magnification between the area of coverage and the number of pixels on a target. Here the magnification is set so that cars on the road are roughly 3–4 pixels across. The preprocessing again uses background subtraction, but in this case, there is the added complication that the camera moves. This means that frame-to-frame registration is required, as described in the beginning of this chapter. For this application we use the background modelling and frame-to-frame registration modules in the Analysts' Detection Support System (ADSS) [14], following the method described in [13]. The video sequence was collected by the Defence Experimentation Airborne Platform (DEAP), a Beech 1900C aircraft configurable with a range of sensors including high-resolution motion imagery sensors, hyperspectral sensors and imaging radar, to trial and experiment with airborne Intelligence Surveillance and Reconnaissance processing, exploitation and dissemination.

Figure 12.12 shows an example frame from the VMTI sequence with three cars relatively close together near the middle of the image. The vehicles are extremely difficult to make out from the raw image frame but they are more clear from the difference image. The figure also shows a local region zoomed in around the cars.

Fig. 12.12 Video input data images: raw image (top); difference image (centre); area containing targets (bottom)

There are several issues with the registration processing that lead to artefacts in the difference images. The first is that the images contain structures like buildings that have sharp edges. These edges create regions of high contrast that give very high differences if there is a small error in registration. This is made worse by the effect of parallax: the registration assumes that the camera observes features in a plane whereas the buildings have significant height. Parallax means that objects closer to the camera appear to move more than those farther away when the camera moves. Although the difference in range is relatively small there is an observable parallax effect on the tops of buildings which is again exacerbated by these being the areas of high contrast in the background. The third effect is that the camera itself has nonlinear distortions near the edges of the frame, as is usual in optical systems. These distortions mean that the transform applied for registration has small errors near the frame edges. In addition to these effects, the environment also introduces other problems. Clouds pass between the camera and the ground that not only obscure the vehicles but also present themselves as large moving objects that create time-varying inhomogeneous clutter. Figure 12.13 shows an example frame where clouds obscure a large portion of the image and give rise to strong spatially correlated difference patches. Again a zoomed local region illustrates the structure of the cloud response. The task of the tracking algorithm is to detect the vehicles without making too many false tracks. A point measurement tracker for this application based on PMHT was described in [5]. The results here are based on H-PMHT and were first presented in [8].

The clouds in the VMTI imagery lead to a tracking problem where the clutter is spatially non uniform and evolves with time. The solution is to build a clutter map, as described in Chap. 8. We illustrate here a map based on a Gaussian mixture representation. H-PMHT-RM is applied directly to the difference images and the targets are selected from the mixture based on size and shape as well as amplitude. The size and shape of the components were determined by finding the eigenvectors and eigenvalues of the estimated appearance covariance matrix $\hat{\Sigma}$. The eigenvalues quantify the length of the major and minor axes of the shape ellipse. The size of promoted tracks was limited by placing a minimum on the smallest eigenvalue and a maximum on the largest eigenvalue. We know from the sensor geometry that a vehicle should occupy 3–4 pixels in width. The shape was limited by putting a threshold on the ratio of the eigenvalues, since the vehicles of interest are roughly square.

Figure 12.14 shows an example of localised high clutter due to registration errors. In the figure, the clutter map is shown as a greyscale intensity image. There is no requirement to accurately describe the clutter, instead the map is a means to improve the tracking quality, so there is no need to count the number of clutter mixture components. The promoted tracks are shown as solid lines. The top image in the figure shows the difference image that is the input to the trackers; the middle image shows the H-PMHT-RM clutter map and overlaid tracks; and the bottom image shows the uniform background core H-PMHT tracks. In this example, there is a single target visible in the scene as marked with a red box in the difference image. The track associated with this target is shown in red in the tracker output images and the remaining blue tracks are all false. It is clear that the H-PMHT-RM clutter map has detected the target without forming any false tracks. There is a strong clutter

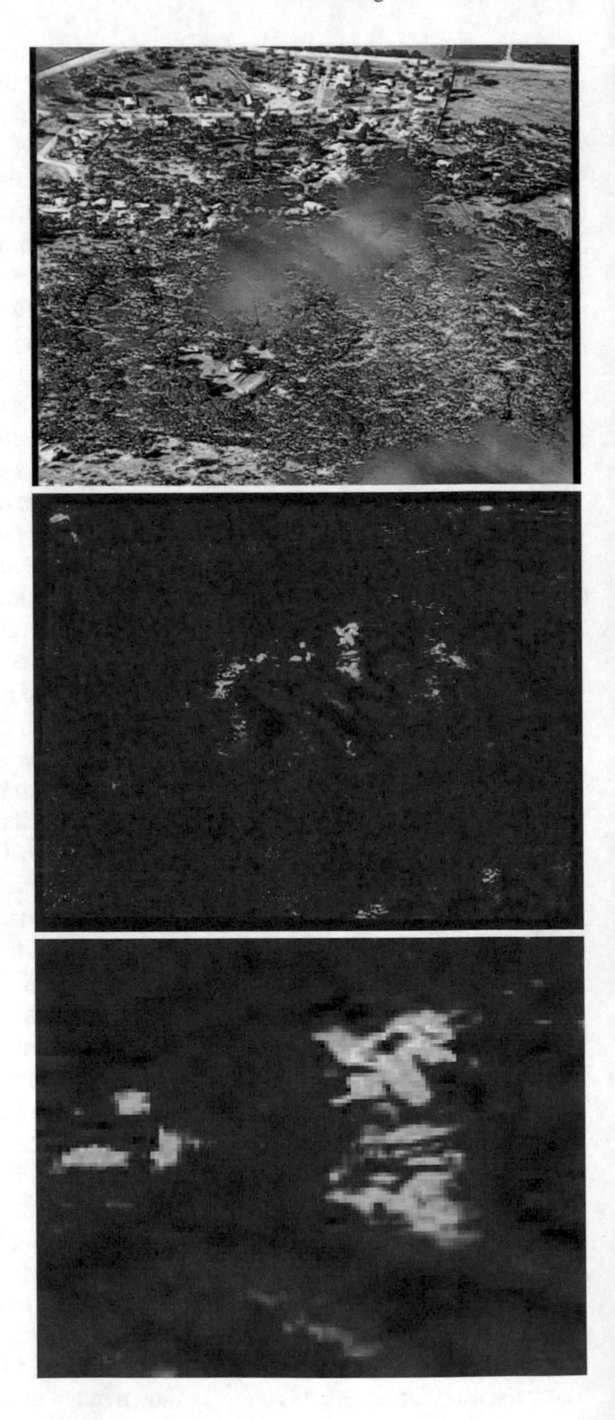

Fig. 12.13 Video images with clouds: raw image (top); difference image (centre); area containing clouds (bottom)

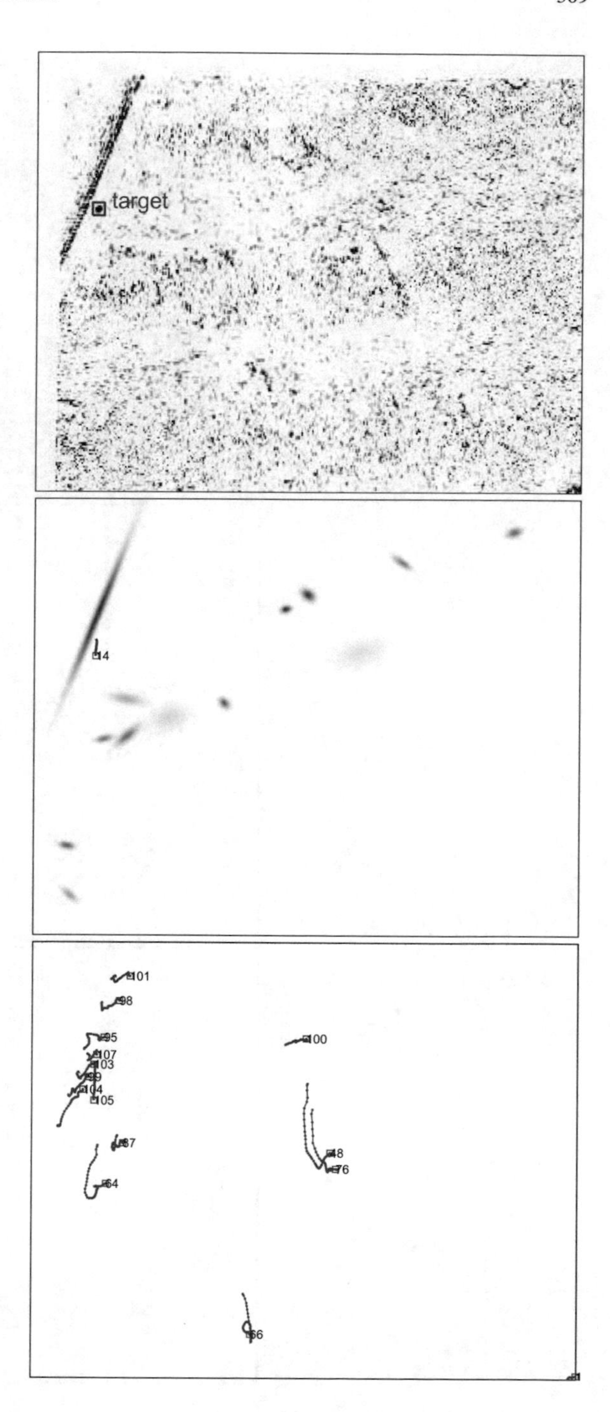

Fig. 12.14 Tracking performance comparison: input difference image (top); H-PMHT-RM with clutter map (middle); H-PMHT-RM with uniform background (bottom)

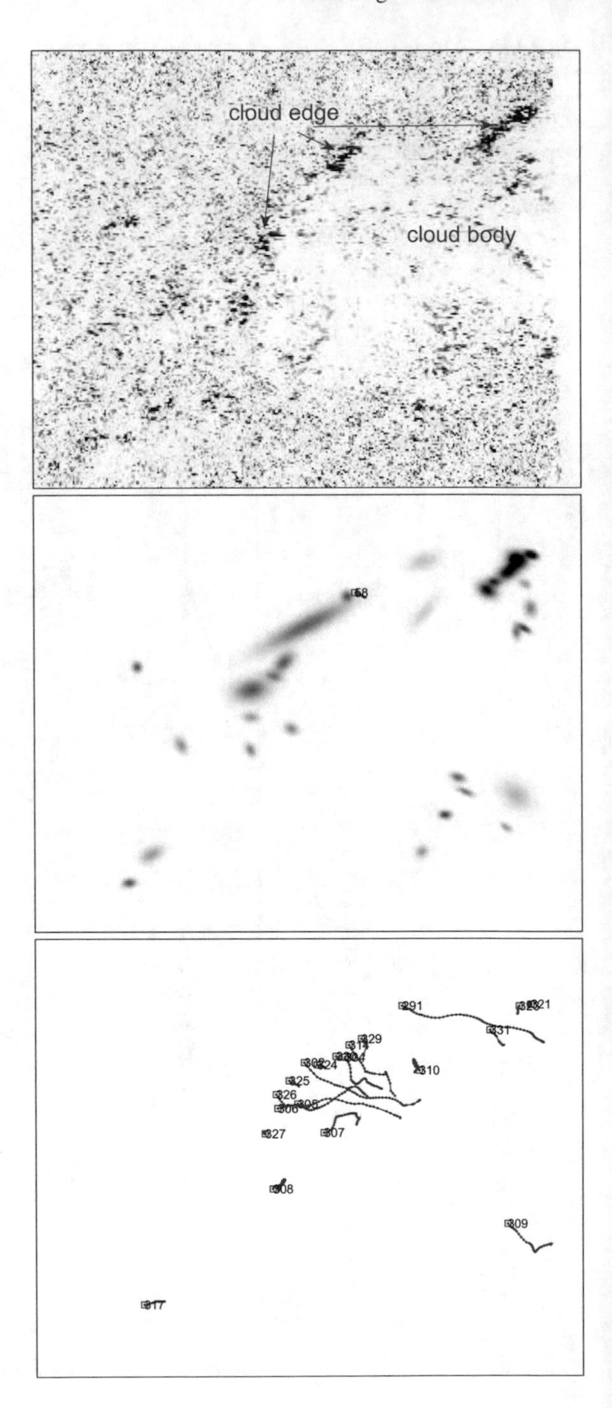

Fig. 12.15 Cloud performance comparison: input difference image (top); H-PMHT-RM with clutter map (middle); H-PMHT-RM with uniform background (bottom)

component corresponding to the localised dense clutter and a collection of other minor components. In contrast, the uniform background H-PMHT has produced a swarm of tracks on the high clutter patch and without careful inspection of the video data, it is difficult to recognise that there is a target amongst them.

Figure 12.15 shows an example with clouds: around a third of the image is obscured by cloud that moves quickly from right to left. There are several patches of high clutter along the cloud leading edge and a large region of very low clutter in the cloud body. In this case, there are no targets visible because they are obscured. The H-PMHT-RM clutter map output shows a single false track formed on one of the high clutter patches. The uniform background again shows numerous false tracks. In the H-PMHT-RM clutter map there are a number of close components in the top right that could have been merged into one large component, which would reduce the computation cost somewhat. However, it was found that looser merging rules caused the target track in Fig. 12.14 to merge with the strong clutter component near it and so were not used.

12.5 Space Situation Awareness

The final application considered in this chapter is the use of optical imagery in support of space situation awareness (SSA). This is the task of maintaining accurate estimates of the positions of objects in orbit with the main objective of avoiding collisions. Due to the reliance on space-based assets, SSA has become a major concern to both military and commercial systems. There are a variety of different sensors that are currently used to compile a catalogue of orbital objects; the results here are for visible wavelength images collected from a telescope held at a fixed azimuth and elevation. Due to the Earth's rotation, the camera slowly pans across the background star-field. The camera detects satellites using light reflected from the sun. We observe geosynchronous satellites that are close to stationary in the imagery.

The basic strategy used in this section is to simply track everything in the image frame, stars and satellites alike. In the application context, this is a fairly dumb brute force method. There are many ways that one might exploit knowledge of the geometry and sensor characteristics to automatically pre-screen the star field before it reaches the tracker. The stars form a background appearance function that looks very much like a large collection of targets. One variation of the approach would be to treat the stars as a group target and learn the camera pan-rate by estimating the velocity of the group. However, here we simply treat the imagery as an opportunity to test the algorithm on genuine sensor data that contains an extremely high number of targets. This is interesting from the perspectives of testing robustness and scaling. The number of tracks generated on the video is extremely high. Whereas typical multi-target studies in the literature consider five or six targets we will be tracking thousands of stars! An early version of the results in this section was presented in [10].

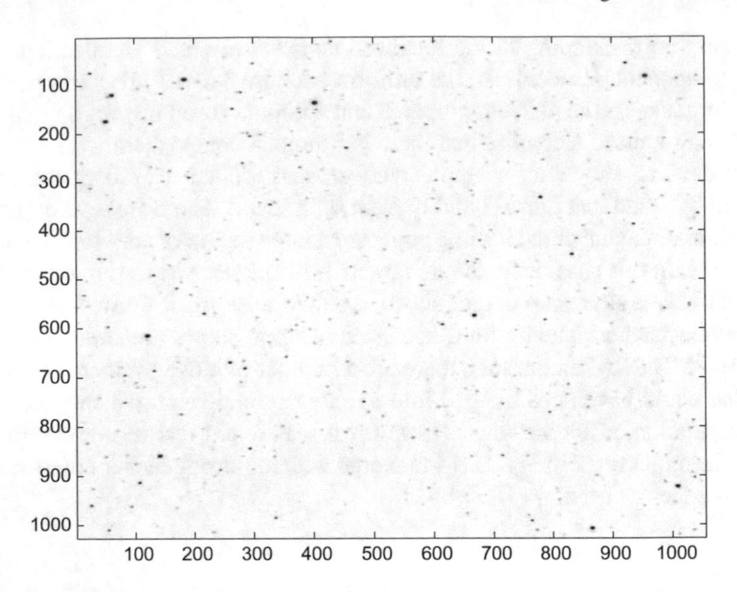

Fig. 12.16 Example frame

12.5.1 Experimental Data

We illustrate SSA using data sets from an experimental optical system developed by the Defence Science and Technology Group. This optical system provides a test bed to prototype tracking and data fusion algorithms as applied to space surveillance.

In the first data set, the telescope was steered to point towards satellite GORIZONT1 (NORAD ID 11158), which is in an eccentric, inclined orbit close to geostationary. The imagery was collected using a 1024 × 1024 pixel camera with a field of view of 76 × 76 arc-minutes. The mount followed the satellite for 17 frames each with a 1 second exposure time. After that point, the telescope was held at a fixed azimuth and elevation where a further 82 frames were collected, after which the satellite GORIZONT1 left the field of view. A 1 second exposure is short enough that the stars are only slightly elongated when the telescope is following the satellite and both the stars and the satellite are only slightly elongated at a fixed azimuth and elevation. At this exposure length, it is challenging to differentiate between the satellite and the stars based on a single image frame; both are spread over only a few pixels. Figure 12.16 shows a single example frame of the telescope imagery when the telescope was steered to a fixed azimuth and elevation. Numerous stars of varying brightness are spread over the image and the location of the satellite within this frame is far from obvious.

The other data sets were collected using a larger 2048 × 2048 focal plane that gave four times the coverage area. In this case the telescope was held at a fixed azimuth and elevation for the entire collection. The scene contains several satellites.

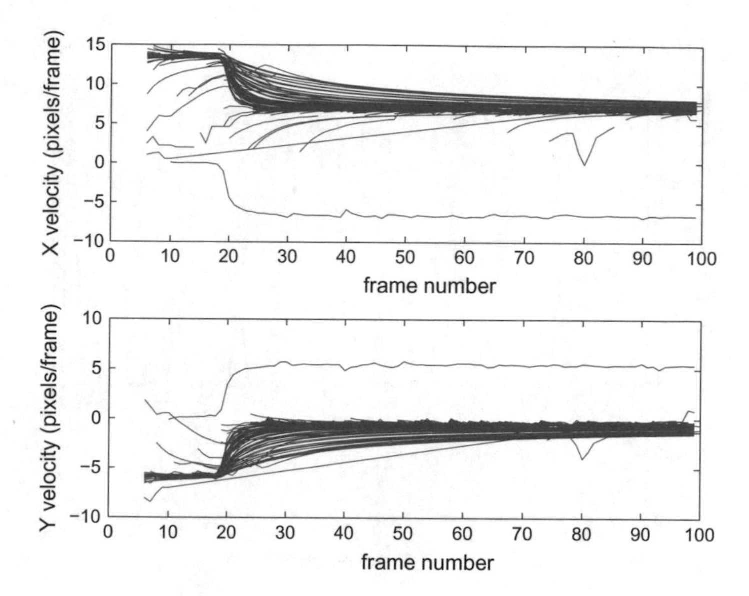

Fig. 12.17 Velocities of H-PMHT tracks

The magnification was chosen to give the same resolution as the first data set, but the exposure time was increased to 5 s.

Tracks were created by providing normalised image frames as the input to H-PMHT-RM. The normalisation was performed by dividing the raw image by its median. Note that the target appearance in this application is dominated by smearing caused by movement of the camera while the aperture is open. This means that the major axis of an ellipse approximating the shape is along the velocity vector. This feature was used to improve initialisation of the track state. Figures 12.19 and 12.20 show example small regions around satellites. The stars are significantly smeared but the satellite is a relative compact point because it is geosynchronous.

12.5.2 Performance on Experimental Data

For the first data set, H-PMHT-RM created 436 tracks, mostly on stars. One of these tracks that on satellite GORIZONT1 and was manually identified by inspecting the data. Figure 12.17 shows the X and Y velocities of the tracks. The satellite track is marked in red: clearly, its motion is inconsistent with the star-field. The first five frames for each track are a nominal burn-in period where the track acquires a velocity estimate and these are not shown in the figure. The track position is not plotted because such plots are very congested.

For the first 17 frames the telescope was steered to point at the satellite and so the satellite velocity is approximately zero. The star-field velocity at this time is

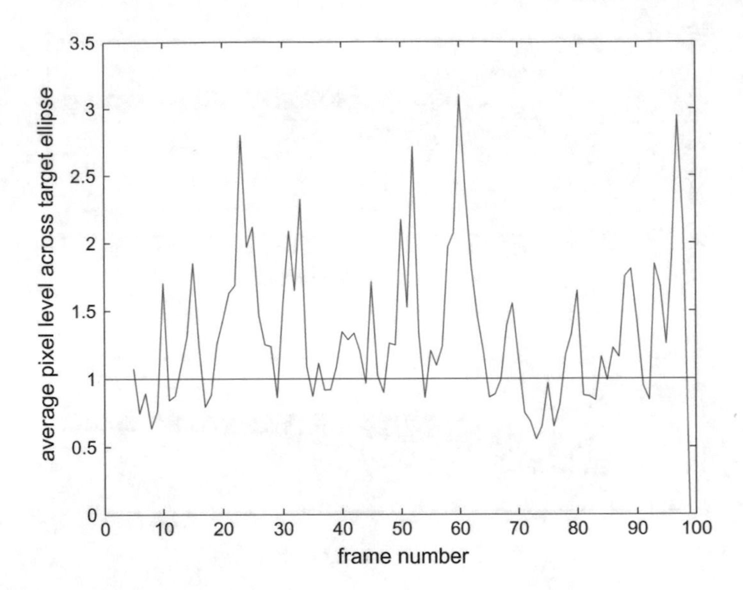

Fig. 12.18 Frame-by-frame image power

approximately [13.8, −5.9]. After this, the telescope stopped moving and stared at a fixed angle. This produced an apparent manoeuvre for all of the targets. The satellite track changed to a velocity of [−6.6, 5.5] and the star-field changed to a velocity of approximately [7, −0.5].

Figure 12.18 shows the instantaneous image power associated with the satellite track averaged over the estimated extent of the image response. This extent is given by the area of the 1-sigma ellipse defined by the spread matrix $\hat{\Sigma}_t^m$, i.e. $A = \pi \left| \hat{\Sigma}_t^m \right|^{\frac{1}{2}}$. A horizontal red line marks an average pixel level of unity, which is the average background level because the images have been normalised. The response is particularly high in frames 52 and 60. There is no particular change in geometry here, the increased brightness is purely a random fluctuation.

Figure 12.19 shows the telescope image cropped to a local region around the H-PMHT-RM satellite track. An ellipse marks the track location and shows the estimated appearance spread from the algorithm. The first frame shown is at the time when the telescope was automatically steered to follow the satellite and as a result, the estimated shape is approximately circular. After frame 17, the satellite was moving relative to the telescope steer direction and this caused a smearing of the response: the H-PMHT-RM estimated shape becomes an ellipse with a major axis along the direction of motion. Around frames 30 and 46 the satellite path through the image plane passes very close to strong stars in the background: these are the two rightmost frames of the top row in Fig. 12.19. The H-PMHT track was not distracted by the presence of a much stronger nearby object because it had a track on the star is each case. The greatest dynamic range is between the satellite and the star in frames 31

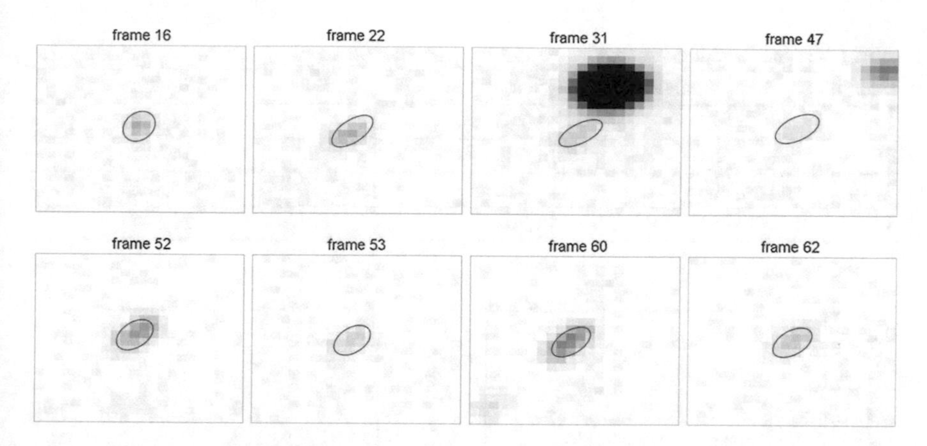

Fig. 12.19 Sample frames

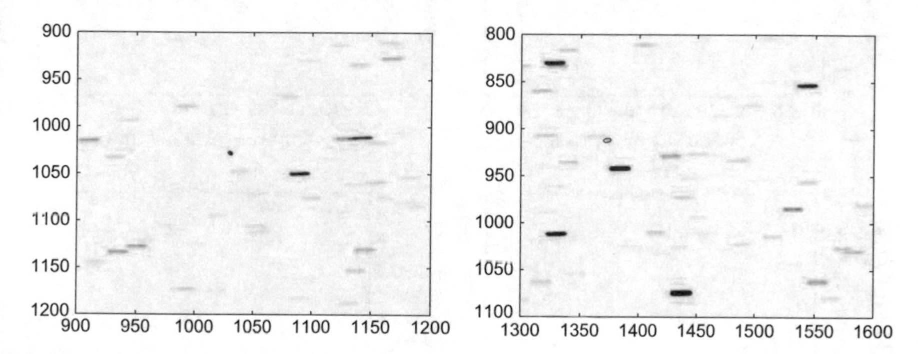

Fig. 12.20 Frame-by-frame image power

with a relative brightness difference of approximately 30 dB, this is the frame second from the right on the top row. The first two frames in the lower row of Fig. 12.19 shows a relatively strong frame 52 and the more typical frame immediately following. These highlight the fluctuations in amplitude and how it is often difficult to see. The right two frames in the lower row show a similar example from frames 60 and 62.

Three further data sets were collected using a larger focal plane camera at lower magnification to give the same spatial resolution. Multiple satellites were visible in each of the data sets. Figure 12.20 shows two of these satellites from the first data set and typical star smears around them. The satellite shown on the left is quite strong, but the satellite on the right is much weaker. These images also show the multitude of stars that are present at varying intensities. The strong stars would easily be tracked by a point-measurement method but there are many more that would likely not be detected. The tracks from H-PMHT-RM were automatically classified by comparing the track velocity with the median velocity over all the tracks and placing a threshold on the difference. Table 12.1 shows the number of frames in each sequence and the numbers of tracks for stars and satellites created for each. We list the number of

Table 12.1 Star tracking

Data set	Frames	Stars	Satellites	Tentative
1	29	1,705	4	28,006
2	55	1,682	3	15,671
3	26	1,412	2	13,862
4	45	1,701	2	21,547

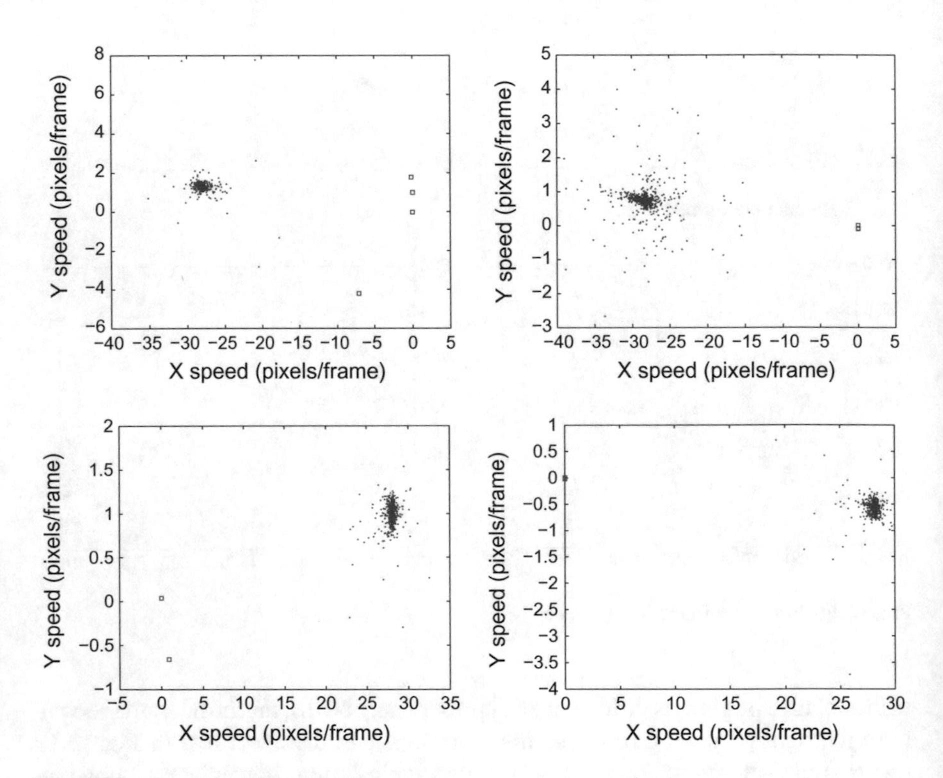

Fig. 12.21 Star and satellite velocities

tentative tracks even though the majority of these do not survive to become established because it reflects how hard the tracker is working. In each case, there are at least 10,000 tentative tracks and in the most dense data set, the tracker formed around a thousand tentative tracks per frame! This illustrates that the H-PMHT-RM has been able to scale to an extremely large tracking problem.

Figure 12.21 shows the velocity estimates of all of the established tracks formed for the sequences. For each data set, the stars are marked as dots and the satellites with red boxes. Clearly, there is substantial separation between the two sets of tracks.

12.6 Summary

This chapter presented four video applications of H-PMHT. The random matrix H-PMHT-RM was used for each of them because the flexibility it provides is required for video applications where the true appearance is usually unknown and time varying. We also illustrated the use of attribute data in H-PMHT through colour video. The contrast between the different types of target in this chapter and the previous show the variety of applications that can be handled by H-PMHT and the extreme number of tracks maintained in the star-tracking application illustrates its excellent scalability.

References

1. Bar-Shalom, Y., Willett, P.K., Tian, X.: Tracking and Data Fusion: A Handbook of Algorithms. YBS (2011)
2. Bouwmans, T.: Statistical background modeling for foreground detection: a survey. Handbook of Pattern Recognition and Computer Vision. World Scientific Publishing, Singapore (2010)
3. Corduneanu, A., Bishop, C.M.: Variational Bayesian model selection for mixture distributions. In: Jaakkola, T., Richardson, T. (eds.) Artificial Intelligence and Statistics 2001, pp. 27–34. Morgan Kaufmann, Burlington (2001)
4. Constantinopoulos, C., Likas, A.: Unsupervised learning of Gaussian mixtures based on variational component splitting. IEEE Trans. Neural Netw. **18**, 745–755 (2007)
5. Davey, S.J.: Video moving target indication using PMHT. In: IEE Seminar on Target Tracking (2006)
6. Davey, S.J., Wieneke, M.: H-PMHT with an unknown arbitrary target. In: Proceedings of ISSNIP (2011)
7. Davey, S.J., Rutten, M.G., Cheung, B.: A comparison of detection performance for several track-before-detect algorithms. EURASIP J. Adv. Signal Process. **2008** (2008)
8. Davey, S.J., Vu, H.X., Arulampalam, S., Fletcher, F., Lim, C.C.: Clutter mapping for histogram PMHT. Statistical Signal Processing, pp. 153–156. Gold Coast, Queensland (2014)
9. Davey, S.J., Vu, H.X., Fletcher, F., Arulampalam, S., Lim, C.C.: Histogram probabilistic multi-hypothesis tracker with color attributes. IET Radar Sonar and Navig. (2015)
10. Davey, S.J., Bessell, T., Cheung, B., Rutten, M.G.: Track before detect for space situation awareness. In: Digital Image Computing: Techniques and Applications (2015)
11. Fisher, R.: PETS04 surveillance ground truth data set. In: Proceedings of the Sixth IEEE International Workshop on Performance Evaluation of Tracking and Surveillance, pp. 1–5 (2004)
12. Haritaoglu, I., Harwood, D., Davis, L.: W4: real-time surveillance of people and their activities. IEEE Trans. Pattern Anal. Mach. Intell. **22** (2000)
13. Jones, R., Ristic, B., Redding, N.J., Booth, D.M.: Moving target indication and tracking from moving sensors. In: Digital Image Computing: Techniques and Applications (2005)
14. Redding, N.J.: Design of the analysts' detection support system for broad area aerial surveillance. Technical report TR-0746, Defence Science and Technology Organisation, Australia (1998)
15. Ristic, B., Arulampalam, S., Gordon, N.J.: Beyond the Kalman Filter: Particle Filters for Tracking Applications. Artech House, Boston (2004)
16. Sherrah, J., Ristic, B., Redding, N.J.: Evaluation of a particle filter to track people for visual surveillance. In: Digital Image Computing: Techniques and Applications (DICTA) 2009, Melbourne, Australia, pp. 96–102 (2009)

17. Shi, J., Tomasi, C.: Good features to track. In: Computer Vision and Pattern Recognition, CVPR 94, pp. 593–600 (1994)
18. Streit, R.L., Graham, M.L., Walsh, M.J.: Multitarget tracking of distributed targets using histogram-PMHT. Digit. Signal Process. **12**(2), 394–404 (2002)
19. Tzikas, D.G., Likas, A.C., Galatsanos, N.P.: The variational approximation for Bayesian inference. IEEE Signal Process. Mag. **25**, 131–146 (2008)
20. Wieneke, M., Davey, S.J.: Histogram PMHT with target extent estimates based on random matrices. In: Proceedings of the 14th International Conference on Information Fusion. Chicago, USA (2011)
21. Wieneke, M., Davey, S.J.: Histogram-PMHT for extended targets and target groups in images. IEEE Trans. Aerosp. Electron. Syst. **50**(3) (2014)
22. Willett, P.K., Ruan, Y., Streit, R.L.: PMHT: problems and some solutions. IEEE Trans. Aerosp. Electron. Syst. **38**, 738–754 (2002)

Chapter 13
Related Methods and Future Directions

There are a number of other tracking algorithms in the literature that are closely related to PMHT and H-PMHT. In this chapter, we briefly review the most closely related methods and discuss the differences between them. We also look at a couple of promising areas of future development with H-PMHT.

The first method is referred to as the Competitive Attention Tracker (CAT) and is a grid-based track-before-detect method. The CAT approach is motivated by efficient implementation and uses an image representation of the state pdfs. It has a heuristic data association stage that is modelled on PMHT.

The Adaptive Dirichlet process mixture model-based Rao-Blackwellised particle filter is a track-before-detect that borrows concepts from H-PMHT by constructing shot measurements to enable the soft association of energy in an image pixel. It blends particle filtering concepts such as resampling and sample-based data association.

The interpolated-Poisson PMHT is a close cousin of H-PMHT that uses a heuristic extension of the Poisson distribution to the real numbers to directly describe the measurement likelihood without using the intermediate step of point measurements.

Finally, we consider an alternative EM strategy where the maximisation is performed over the assignments instead of the states. This has been considered already for point measurement sensors but appears to have much more potential in the context of H-PMHT.

13.1 Competitive Attention Tracker and CACTuS

The Competitive Attention Tracker (CAT) is a grid-based track-before-detect method that uses a marginal representation of the kinematic pdf by separating the position and velocity components into image pdf representations [6]. The position pdf is stored as an image of the same size as the sensor measurement and we denote it as

$$p_{t|t}^{m,i} = p(h(\mathbf{x}_t^m) \in W^i | \mathbf{Z}_1, \dots \mathbf{Z}_t), \tag{13.1}$$

S. J. Davey and H. Gaetjens, *Track-Before-Detect Using Expectation Maximisation*, Signals and Communication Technology,
https://doi.org/10.1007/978-981-10-7593-3_13

and the velocity pdf is stored as an image where the resolution corresponds to a speed increment of 1 pixel per frame. The velocity pdf is denoted by $v_{t|t}^{m,i}$. The CAT was developed under a computer-vision philosophy and is generally presented in terms of image functions. For example, the predicted position and velocity distributions are written as convolutions

$$v_{t|t-1}^{m} = v_{t-1|t-1} \otimes a, \tag{13.2}$$

$$p_{t|t-1}^{m} = p_{t-1|t-1} \otimes v_{t|t-1}^{m}, \tag{13.3}$$

where a is the acceleration pdf. In this book, we would have written $v_{t|t-1} = \int p(v_t|v_{t-1})v_{t-1|t-1}\mathrm{d}v_{t-1}$ and the two are equivalent. Rather than dynamically selecting the number of components, the CAT tiles the image plane with a large fixed number of these marginalised grid-trackers. It uses a heuristic pixel association weight defined by the ratio

$$c_t^{m,i} = \frac{p_{t|t-1}^{m,i}}{\sum_{s=1}^{M} p_{t|t-1}^{s,i}}. \tag{13.4}$$

It also calculates a localised position prediction pdf defined as

$$\gamma_t^{m,i} = \frac{p_{t|t-1}^{m,i} G_t^{m,i}}{\sum_{i=1}^{I} p_{t|t-1}^{m,i} G_t^{m,i}}. \tag{13.5}$$

where the pixel-target cell probabilities are determined assuming a Gaussian appearance function with a circular covariance matrix Σ. The diagonals of Σ are set to $16r$ where r is the equivalent radius of the mean pixel area per track. Since we have I pixels and M tracks, this is $r = \sqrt{I/(\pi M)}$. Note that CAT does not include a clutter model.

The position pdf is updated using the pixel association weight, the localised position prediction pdf and a single-pixel likelihood ratio $\mathscr{L}(\mathbf{z}_t^i, \mathbf{a}_t^i)$. The updated pdf is given by

$$p_{t|t}^{m,i} \propto \mathscr{L}(\mathbf{z}_t^i, \mathbf{a}_t^i) c_t^{m,i} \gamma_t^{m,i}, \tag{13.6}$$

The 'observed velocity' is given by $\mathscr{L}(v_t^m) = \mathscr{L}(\mathbf{z}_t^i, \mathbf{a}_t^i) \otimes p_{t|t-1}^m$ and the velocity pdf update is

$$v_{t|t}^{m,i} \propto \mathscr{L}(v_t^{m,i}) v_{t|t-1}. \tag{13.7}$$

The track quality is quantified by the quantity $\sum_i \left(p_{t|t}^{m,i}\right)^2$, which has a maximum value of unity if the position pdf has collapsed to a Kronecker delta. That is, quality is measured in terms of the concentration of the pdf.

The CAT looks quite similar to core H-PMHT for a Gaussian appearance function because the heuristic pixel association weight was reportedly inspired by the point measurement PMHT [6]. However, there are several differences. The use of a fixed number of tracks is minor: one could certainly choose to implement H-PMHT in this way and similarly, one could fairly simply build a clutter component into CAT. Similarly, the use of an image-based marginalised pdf representation could be applied to H-PMHT: in a sense, we can view H-PMHT as describing a general framework and CAT as jumping directly to a particular implementation choice. The bigger difference is that CAT is a heuristic method whereas H-PMHT uses a fixed mathematical framework. In practice, this means that CAT has been extended by further heuristic modifications to the pixel weights or localised prediction. In contrast, modifications to H-PMHT have required more detailed mathematics but also deliver general results rather than application-specific methods that cannot be known to generalise. The most important difference is that the associated images used by H-PMHT are based on the current MAP point estimate whereas, the CAT association weights vary depending on the application: they can include uncertainty.

The most significant extension to CAT is referred to as Competitive Attentional Correlation Tracker using Shape (CACTuS) and extends CAT to include an appearance model [10]. Like the image-based position and velocity pdfs, CACTuS uses an image-based appearance model, that is a template. As such, it bears a strong resemblance to the Dirichlet appearance learning H-PMHT described in Chap. 8. The difference is that the CACTuS model is binary: each pixel is either foreground, and hence belongs to the target, or background. The shape estimate is a probability of each pixel in the template being foreground. This influences the single-pixel likelihood ratio which becomes $\mathscr{L}\left(\mathbf{z}_t^i, \mathbf{a}_t^i | s_{t|t-1}^m\right)$ where $s_{t|t-1}^m$ is the shape pdf.

13.2 Adaptive Dirichlet Process Mixture Model-Based Rao-Blackwellised Particle Filter

The Adaptive Dirichlet process mixture model-based Rao-Blackwellised particle filter (ADPM-RBPF) is a TkBD algorithm that blends ideas from particle filtering and H-PMHT [1]. Similar to the H-PMHT, it creates a set of shot measurements and then assigns these to individual components in a mixture model. Whereas H-PMHT uses quantisation to create these measurements and then using a limit to recover the original sensor data, ADPM-RBPF uses the sensor data as a probability mass function and then draws n random samples from this probability. The number of shot measurements drawn is a parameter of the algorithm.

The ADPM-RBPF models the assignment of shots using a multinomial distribution and it uses a Dirichlet process to model the assignment priors π_t^m. As discussed in Chap. 8, the Dirichlet distribution is a conjugate prior to the hyperparameter of a multinomial distribution. In English, this means that if we have a Dirichlet distribution on π_t^m and then we collect some multinomial measurements, the posterior is also Dirichlet. This is essentially why we used a Dirichlet representation of the appearance template. A Dirichlet process is a prior for the case where we do not know how many targets there are, that is $\pi_t^{1:M}$ forms a probability vector, but M is unknown. In the context of ADPM-RBPF, this allows the filter to learn how many targets are present in the scene.

In particle filtering, the term *Rao-Blackwellise* refers to a mathematical arrangement that makes the filter more efficient. Particles are useful for nonlinear non-Gaussian systems, but in many applications only part of the system is like this and the rest can be modelled as linear and Gaussian. For example, if I know a sequence of positions then the velocity is straightforward to infer. This can be used to separate the filtering problem into estimating position given measurements and then estimating velocity given position. It does not matter how difficult the position estimation is, the velocity inference is still straightforward. In the context of a particle filter, this means that we sample position but run a conditional Kalman filter over velocity, which reduces the dimensions of the space we need to sample. For ADPM-RBPF, the sampled space is the assignment of shot measurements, $\mathbb{K}$. Conditioned on this assignment, that is if we know which measurement goes with which target, then the kinematic state $\mathbb{X}$ can be estimated using a Kalman filter. This allows ADPM-RBPF to use a relatively small number of particles to explore the assignment space without needing to jointly explore the kinematic space.

The ADPM-RBPF assumes that the mixture components are Gaussian and uses an inverse Wishart prior to estimate the spreading variance, which is the same approach as H-PMHT-RM. Initial comparisons between ADPM-RBPF and H-PMHT-RM show that H-PMHT-RM gives better performance results at a fraction of the computation cost [1] but this method is very new and further development is expected.

13.3 Interpolated Poisson

The H-PMHT makes a soft assignment of energy in each image pixel by creating a collection of shot measurements and then marginalising over hard assignments of the shots. Using a set of shot measurements allows the use of the Poisson distribution to describe the measurement likelihood. This is helpful because of the superposition properties that it brings. However, the variance of the Poisson distribution is the same as its mean. As the quantisation step size is reduced the mean and variance of the measurement model increase at the same rate. We can reconstruct the power associated with a single track by simply multiplying the number of associated shots with the step size. Because the mean of the Poisson distribution increases linearly with the inverse step size, this reconstructed power does not vary with step size. However, the variance of the Poisson distribution also increases linearly with the inverse step size whereas the variance of the reconstructed power is the Poisson variance scaled by the step size squared. This means that reducing the step size reduces variability. In the limit, there is no measurement variance, so the H-PMHT cannot be used as a generative model to simulate data.

For any particular quantisation step size $\hbar^2$ and assignment $\mathbf{K}_t$ the proportion of the pixel energy associated with target m is a ratio of natural numbers that is a rational number. In the limit as $\hbar^2 \rightarrow 0$ then the set of possible energy proportions approaches an interval between 0 and 1 on the rationals. So one interpretation of the H-PMHT mechanics is that it provides a means to approximate the relative proportion of target energy in each pixel as a rational number. If we had a distribution directly on this proportion that obeyed superposition we would not need quantisation and it would provide a generative model.

The interpolated Poisson PMHT [7, 8] is a TkBD approach based on this idea. The Poisson distribution is a probability mass over natural numbers and is given by

$$p(n; \lambda) = \exp\{-\lambda\}\frac{\lambda^n}{n!}. \tag{13.8}$$

The factorial term in the denominator can be replaced by the gamma function $\Gamma(n+1)$, which interpolates the factorial function over the complex plane. The interpolated Poisson density is then

$$p(z; \lambda) = \exp\{-\lambda\}\frac{\lambda^z}{\Gamma(z+1)}, \tag{13.9}$$

where z is now a non-negative real number. We can show that this function is not a pdf because it doesn't integrate to unity [7]. However, if we ignore this inconvenience, an algorithm that looks a lot like H-PMHT can be derived.

In many respects, the interpolated Poisson PMHT is a different way of deriving Poisson PMHT. The main difference is that by using this interpolated function there is no need for shot measurements and their corresponding limits and consequently there is no need for a resampled target prior. Unfortunately, it ultimately does not solve the problems associated with taking the limit of a quantised measurement because it requires an incorrect assumption, that the interpolated function is a pdf.

Another option along a similar theme is to model the pixel energy as gamma distributed [9], that is

$$p_\gamma(\mathbf{z}; \theta, \lambda) = \exp\left\{-\frac{\mathbf{z}}{\theta}\right\}\frac{\mathbf{z}^{\lambda-1}}{\Gamma(\lambda)\theta^\lambda}, \tag{13.10}$$

where θ is a scale parameter and the mean is $\theta\lambda$. For a fixed θ, the gamma distribution admits superposition, that is if $\mathbf{z}^1 \sim p_\gamma(\mathbf{z}; \theta, \lambda^1)$ and $\mathbf{z}^2 \sim p_\gamma(\mathbf{z}; \theta, \lambda^2)$ then $\mathbf{z}^1+\mathbf{z}^2 \sim p_\gamma(\mathbf{z}; \theta, \lambda^1+\lambda^2)$. Further, the proportion due to component m, that is $\mathbf{z}^1/(\mathbf{z}^1+\mathbf{z}^2)$ follows a Dirichlet distribution. This is analogous to the relationship between a Poisson count and a multinomial assignment. At the time of writing this book, this idea had not been explored in detail.

13.4 Association Maximisation

A characteristic feature of H-PMHT is that it uses EM to marginalise over assignments and then finds MAP estimates for the states. This is the strategy adopted by the PMHT association method. There are two problems with this. First, the output of the algorithm is a point estimate, whereas in many applications we would prefer a probability density, or at least a covariance. This could be addressed by extending the point measurement PMHT covariance estimation method discussed in Chap. 5. The second problem is that the association uses this state point estimate and does not account for uncertainty.

An alternative strategy is to marginalise over the states and then find the MAP estimates for the assignments. This has been used for point measurement problems, for example [4, 5]. For point measurements, it is not a very appealing approach because it leads to global nearest neighbour association. However, for H-PMHT the assignment is over shot measurements. This means that making a hard assignment of each shot still leads to a soft assignment of the total energy in a pixel. In fact, we can easily show that the assignment weights are almost the same as H-PMHT except that they use the state distribution, not a point estimate! These weights then lead to associated images and the expectation step amounts to independently finding the state distribution for each target [2]. We briefly outline how this works.

Consider the H-PMHT development at the quantisation stage. Pixel i contains n_t^i. To simplify the notation, assume that the measurement rates $\mathbb{L}$ are known. As derived in Chap. 6, the complete data likelihood is

$$p_{\mathsf{comp}}\left(\mathbf{X}_t, \mathbf{K}_t, \mathbf{N}_t\right) = \frac{\exp\{-\Lambda_t\}}{n_t!} \prod_{m=0}^{M} \left[p\left(\mathbf{x}_t^m|\mathbf{x}_{t-1}^m\right) p\left(\Lambda_t^m|\Lambda_{t-1}^m\right)\right] \prod_{i=1}^{I} \prod_{r=1}^{n_t^i} \Lambda_t^{k_t^r} G_t^{k_t^r,i}, \quad (13.11)$$

which can equivalently be written in terms of the number of measurements assigned to each target $n_t^{m,i}$

$$p_{\mathsf{comp}}\left(\mathbf{X}_t, \mathbf{K}_t, \mathbf{N}_t\right) = \frac{\exp\{-\Lambda_t\}}{n_t!} \prod_{m=0}^{M} \left[p\left(\mathbf{x}_t^m|\mathbf{x}_{t-1}^m\right) p\left(\Lambda_t^m|\Lambda_{t-1}^m\right) \prod_{i=1}^{I} \left(\Lambda_t^m G_t^{m,i}\right)^{n_t^{m,i}}\right], \quad (13.12)$$

The EM auxiliary function under this alternative formulation is

$$\begin{aligned} \mathcal{Q}\left(\mathbf{K}_t|\hat{\mathbf{K}}_t\right) &= E_{\mathbf{X}_t|\hat{\mathbf{K}}_t,\mathbf{N}_t}\Big[\log\left\{p_{\mathsf{comp}}\left(\mathbf{X}_t, \mathbf{K}_t, \mathbf{N}_t\right)\right\}\Big], \\ &= C + E_{\mathbf{X}_t|\hat{\mathbf{K}}_t,\mathbf{N}_t}\Big[\log\left\{p\left(\mathbf{K}_t, |\mathbf{X}_t, \mathbf{N}_t\right)\right\}\Big]. \end{aligned} \quad (13.13)$$

The conditional probability above is the combinatorial scaling term and the pixel product from (13.12). Assigning more measurements to the component with the highest $\Lambda_t^m G_t^{m,i}$ increases this but the factorial scaling term penalises assignments that depart from an even distribution of measurements: the two terms balance each other. It is not obvious that this expression has a single maximum or where it is, but this is the case. An intuitive way to recognise this is to approximate the multinomial expression in (13.12) with a Gaussian. This is a common method for large counts, which is exactly what we have for small quantisation steps $\hbar^2$ [3]. The mean of the approximate Gaussian is given by

$$\bar{n}_t^{m,i} = \frac{\Lambda_t^m G_t^{m,i}}{\sum_{s=0}^{M} \Lambda_t^s G_t^{s,i}} n_t^i. \tag{13.14}$$

The EM auxiliary function is the expectation over the log of these Gaussians. With a little chicanery of the usual flavour, the reader may be able to show that the result of this is equivalent to the log-likelihood of a single Gaussian with a mean given by

$$\tilde{n}_t^{m,i} = \frac{\Lambda_t^m E_{\mathbf{X}_t|\hat{\mathbf{K}}_t,\mathbf{N}_t}\left[G_t^{m,i}\right]}{\sum_{s=0}^{M} \Lambda_t^s E_{\mathbf{X}_t|\hat{\mathbf{K}}_t,\mathbf{N}_t}\left[G_t^{s,i}\right]} n_t^i. \tag{13.15}$$

This expression is almost the same as the associated image that we derived all the way back in Chap. 4 except that the weights use the expected value of the pixel probability, not a value calculated using a point estimate for the state. This is a fantastic thing because it means that the weights account for uncertainty.

The expectation step of this variation is decoupled because it is conditioned on the current estimate of the assignments and consists of a bank of M independent filters that produce pdf estimates, not point estimates. The development of this version of H-PMHT is still enfolding but it could provide something revolutionary.

References

1. Carevic, D., Davey, S.J.: Two algorithms for modeling and tracking of dynamic time-frequency spectra. IEEE Trans. Signal Process. **64** (2016)
2. Davey, S.J., Williams, J.L.: Marginal multi-target track-before-detect (in preparation)
3. Gelman, A., Carlin, J.B., Stern, H.S., Rubin, D.B.: Bayesian Data Analysis. Chapman and Hall/CRC, Boca Raton (2004)
4. Li, S., Chen, S., Leung, H., Bosse, E.: Joint data association, registration, and fusion using EM-KF. IEEE Trans. Aerosp. Electron. Syst. **46** (2010)
5. Pulford, G.W., La Scala, B.F.: MAP estimation of target manoeuvre sequence with the expectation-maximisation algorithm. IEEE Trans. Aerosp. Electron. Syst. **38**, 367–377 (2002)
6. Strens, M.J.A., Gregory, I.N.: Tracking in cluttered images. Image Vis. Comput. **21**, 891–911 (2003)
7. Vu, H.X.: Track-before-detect for active sonar. Ph.D. thesis, The University of Adelaide (2015)

8. Vu, H.X., Davey, S.J., Arulampalam, S., Fletcher, F., Lim, C.C.: H-PMHT with a Poisson measurement model. In: Radar, pp. 446–451. Adelaide, South Australia (2013)
9. Williams, J.L.: Personal communications (2016)
10. Wong, S.C.: Algorithms and architectures for visual tracking. Ph.D. thesis, The University of South Australia (2010)

Chapter 14
Summary

This brings us to the end of our journey together. If you've made it this far without skipping too many chapters, then thanks for your time! This book has aimed to be a complete resource for the Histogram Probabilistic Multi-Hypothesis Tracker, it can be fundamentally divided into three parts dealing with the background of H-PMHT, extensions to the method and applications. The final chapter discusses how H-PMHT relates to other tracking algorithms, particularly in the field of track-before-detect. It also describes some of our ongoing research.

This book also links the mathematical descriptions with a software toolbox that realises these algorithm variations. The toolbox can be used as a benchmark by other researchers or as a starting point for further development.

When all of the extensions described in this book are included, the H-PMHT can be described as a tracker for image-based sensor measurements that:

- Has been demonstrated to give comparable output quality to numerical implementations of the Bayesian filter at a fraction of the computation cost;
- Can scale to thousands of targets in millions of sensor pixels;
- Can handle nonlinear target dynamics;
- Can handle any known target appearance function (in several ways);
- Or estimate an unknown appearance using parametric or non-parametric models;
- Can automatically and dynamically determine the number of targets in a scene;
- Can make use of attribute information, like colour;
- Has an intrinsic model for interaction of overlapping targets that support either superposition or occlusion;
- Has been demonstrated in radar, sonar and video sensor applications.

S. J. Davey and H. Gaetjens, *Track-Before-Detect Using Expectation Maximisation*, Signals and Communication Technology,
https://doi.org/10.1007/978-981-10-7593-3_14

Appendix A
Comparison Algorithms

This appendix provides an overview of the models and algorithms used for comparisons in this book. These are generally standard textbook methods and so we will not try to derive or motivate them, rather this appendix simply serves to be explicit about what we have compared with.

The target and measurement models used to produce the comparison results in Chap. 2 are reviewed first. These models are quite widely used and were the basis of the series [6–8] in which H-PMHT was shown to be a strong contender for multi-target TkBD. The statistical models for noise-only and target-plus-noise described here are consistent with the way that image measurements are simulated throughout the book. That is, we generally use complex Gaussian noise in each pixel and add a non-zero mean when a target is present.

The remainder of the appendix describes the point measurement and TkBD algorithms that we compare against H-PMHT through the book. Readers with interest in understanding these methods better should refer to other tracking texts, such as [3, 4, 10]

A.1 Target and Measurement Models

The difference between TkBD and detect-then-track is the measurement process, not the target model. The particle filter implementation used in this book uses the almost constant velocity target model, as described in Chap. 1. The Viterbi algorithm uses a discretised version of the same. Both methods are easily capable of incorporating more complex models but as TkBD is more about measurements than the target process we avoid unnecessary complications.

S. J. Davey and H. Gaetjens, *Track-Before-Detect Using Expectation Maximisation*, Signals and Communication Technology,
https://doi.org/10.1007/978-981-10-7593-3

A.1.1 Target Model

The almost constant velocity target model was discussed in Chap. 1, here we show a simple discretised version of it. Whereas the continuous state space represents the set of allowable target positions and velocities as a four-dimensional volume, the discretised state space samples this volume and constrains the state to that set of samples. The advantage of this is that difficult integrals become summations which can be evaluated numerically even if they cannot be analytically simplified.

The simplest form of discrete state space is a uniform grid. For two-dimensional position and velocity it is intuitive to index the state space using a vector $\boldsymbol{j} \equiv [j_x, j_{\dot{x}}, j_y, j_{\dot{y}}]^\mathsf{T}$, such that

$$\mathbf{x}(\boldsymbol{j}) \equiv \left[\Delta_x j_x, \tfrac{\Delta_x}{\Delta_t} j_{\dot{x}}, \Delta_y j_y, \tfrac{\Delta_y}{\Delta_t} j_{\dot{y}}\right]^\mathsf{T}, \tag{A.1}$$

where Δ_x and Δ_y are the grid spacing values.

In this case, the state evolution equation is equivalent to an evolution over the index

$$\boldsymbol{j}_{t+1}^m = \mathsf{F}_{\text{grid}} \boldsymbol{j}_t^m + \tilde{\boldsymbol{v}}_t^m, \tag{A.2}$$

where

$$\mathsf{F}_{\text{grid}} = \begin{bmatrix} \mathsf{F2}_{\text{grid}} & 0 \\ 0 & \mathsf{F2}_{\text{grid}} \end{bmatrix}, \quad \mathsf{F2}_{\text{grid}} = \begin{bmatrix} 1 & 1 \\ 0 & 1 \end{bmatrix}, \tag{A.3}$$

and $\tilde{\boldsymbol{v}}_t^m$ is a discrete noise process with a known probability mass function (pmf). It is advantageous to limit the support of the pmf of $\tilde{\boldsymbol{v}}_t^m$ because it reduces the required computation effort. In other words, we will choose $P(\tilde{\boldsymbol{v}}_t^m)$ to be identically zero everywhere except within a small distance of $[0, 0, 0, 0]^\mathsf{T}$.

A simple example for $\tilde{\boldsymbol{v}} = \{v_j\}$ is the pmf

$$P(\tilde{\boldsymbol{v}}) = \begin{cases} \alpha & \tilde{\boldsymbol{v}} = [0, 0, 0, 0]^\mathsf{T}, \\ \frac{1-\alpha}{80} & \tilde{\boldsymbol{v}} \neq [0, 0, 0, 0]^\mathsf{T} \text{ and } |v_j| \leq 1, \\ 0 & \text{otherwise,} \end{cases} \tag{A.4}$$

where $0 \leq \alpha \leq 1$. For $\alpha > 0.5$, this assigns the majority of the probability mass to the event that the target moves deterministically, with the remainder equally spread across the 80 elements of a $3 \times 3 \times 3 \times 3$ hypercube around it.

A.1.2 Sensor Model

We now briefly describe the sensor model commonly used for TkBD. This is not the same model that is used by H-PMHT but it is statistically consistent with the way that images have been generated for the examples in this book. Both of the Chap. 2 alternative TkBD approaches use it. Recall that the tracking input is an image. In this appendix, we stick with the most common example, which is a two-dimensional image.

Let $\mathbf{z}_t^i$ denote the ith pixel in the sensor image at time t, and let $\mathbf{Z}_t = \{\mathbf{z}_t^i\}$ denote a stacked vector of all the pixels of the image. The contribution of the target depends on the physical shape and orientation of the target, the point spread function (psf) of the sensor, and the amplitude of the signal, A_t. The mathematical function that defines this contribution is referred to as the appearance function and we assume that it is known within the appendix. Extensions to learning the appearance function are described in Chap. 8. The contribution of the target to pixel i is denoted by $G^i(\mathbf{x})$. This function is discussed in detail in Chaps. 4, 7 and 8. The target can often be treated as a point scatterer in radar applications, in which case the appearance function is a property of the sensor and is defined by the spectral response of the windows used during Fourier Transforms for forming range and Doppler cells and during beam-forming.

For a single target, the measurement function is

$$\mathbf{z}_t^i = A_t G^i(\mathbf{x}_t) + w_t, \tag{A.5}$$

where w_t is additive sensor noise. The H-PMHT naturally allows for multiple targets but the comparison methods do not. We restrict ourselves to the single target case here because that is what is used for the comparison in Chap. 2. The statistics of the sensor noise may vary depending on the physical characteristics of the sensor. In some applications, Gaussian noise can be assumed, but in radar-based applications, it is common to assume complex Gaussian noise. Here we present only the case where the sensor output is the magnitude of an image with complex Gaussian noise. Other examples are presented in [7].

The comparison is chiefly interested in detection, so it is important to quantify outputs as a function of target strength. Define the peak signal to noise ratio (peak-SNR) as

$$\gamma_t = \frac{\left(A_t \max_i G^i(\mathbf{x}_t)\right)^2}{\sigma_w^2}, \tag{A.6}$$

where σ_w^2 is the variance of the sensor noise and $\max_i G^i(\mathbf{x}_t)$ is the highest value of the psf. The psf is scaled so that it sums to unity. The peak SNR is often expressed in a decibel scale, i.e. as $10\log_{10}\{\gamma_t\}$.

When the phase of the target response is unknown then the standard approach [10] is to assume that the noise is spatially uniform and uncorrelated. This means that the envelope response in each sensor pixel is conditionally independent given the target state, or the knowledge that there is no target.

When no target is present, the pdf of the envelope at pixel i at time t, $\mathbf{z}_t^i$, follows a Rayleigh distribution:

$$p^0\left(\mathbf{z}_t^i\right) = \frac{\mathbf{z}_t^i}{\sigma_w^2} \exp\left\{-\frac{\left(\mathbf{z}_t^i\right)^2}{2\sigma_w^2}\right\}, \tag{A.7}$$

and when a target is present, it follows a Rician distribution:

$$p^1\left(\mathbf{z}_t^i|\mathbf{x}_t\right) = \frac{\mathbf{z}_t^i}{\sigma_w^2} \exp\left\{-\frac{\left(\mathbf{z}_t^i\right)^2 + A_t^2\left(G^i(\mathbf{x}_t)\right)^2}{2\sigma_w^2}\right\} \mathsf{I}_0\left(\frac{A_t G^i(\mathbf{x}_t)\mathbf{z}_t^i}{\sigma_w^2}\right), \tag{A.8}$$

where $\mathsf{I}_0(\cdot)$ is the modified Bessel function of order 0, given by

$$\mathsf{I}_0(x) = \sum_{m=0}^{\infty} \frac{(x/2)^{2m}}{m!\,\Gamma(m+1)}. \tag{A.9}$$

The likelihood ratio for pixel i is then

$$\mathscr{L}\left(\mathbf{z}_t^i|\mathbf{x}_t\right) = \exp\left\{-\frac{A_t^2\left(G^i(\mathbf{x}_t)\right)^2}{2\sigma_w^2}\right\} \mathsf{I}_0\left(\frac{A_t G^i(\mathbf{x}_t)\mathbf{z}_t^i}{\sigma_w^2}\right). \tag{A.10}$$

Since the pixels are assumed to be conditionally independent, the likelihood of the envelope of the whole image is the product over the pixels

$$\begin{aligned}\mathscr{L}\left(\mathbf{Z}_t|\mathbf{x}_t\right) &= \prod_{i=1}^{I} \mathscr{L}\left(\mathbf{z}_t^i|\mathbf{x}_t\right) \\ &= \exp\left\{-\frac{A_t^2\sum_{i=1}^{I}\left(G^i(\mathbf{x}_t)\right)^2}{2\sigma_w^2}\right\} \prod_{i=1}^{I} \mathsf{I}_0\left(\frac{A_t G^i(\mathbf{x}_t)\mathbf{z}_t^i}{\sigma_w^2}\right).\end{aligned} \tag{A.11}$$

The first term in this expression is a function of the total target power whereas the second term is a function of the correlation between the modelled target power and the measurements. The Bessel function grows monotonically and more quickly than the exponential, so the likelihood ratio is unbounded. For targets that spread over a large number of pixels, the implementation should be careful to avoid numerical issues from the very large possible likelihood values.

A.2 Probabilistic Data Association Filter

In this manuscript, we follow the *visibility* based formulation of PDAF due to Colegrove [5]. More general forms can be found in texts such as [3]. Various authors have argued the merits of models referred to as *existence*, *perceivability* and *null-state*.

From our perspective the practical results of these methods are the same, so we prefer to use the terminology of the original authors.

The fundamental idea behind probabilistic data association (PDA) is to treat the true source of each measurement k_t^r as a random variable. PMHT assumes a Poisson distributed number of measurements from each target whereas PDA assumes a binomial number of measurements, that is either zero or one. It only considers one target at a time, so it is more convenient to define an assignment variable that points from the target to its measurement rather than the assignment k_t^r for PMHT that points from a measurement to its source. Let θ denote the measurement caused by the target, $\theta = 0$ denotes a missed detection and $\theta = -1$ denotes the possibility that there is no target at all. The posterior distribution of the kinematic state can then be written as

$$\begin{aligned} p\left(\mathbf{x}_t|\mathbf{Y}_{1:t}\right) &= \sum_{\theta_t=-1}^{n_t} p\left(\mathbf{x}_t, \theta_t|\mathbf{Y}_{1:t}\right), \\ &= \sum_{\theta_t=-1}^{n_t} p\left(\mathbf{x}_t, |\theta_t, \mathbf{Y}_{1:t}\right) p\left(\theta_t|\mathbf{Y}_{1:t}\right). \end{aligned} \tag{A.12}$$

The density $p\left(\mathbf{x}_t, |\theta_t, \mathbf{Y}_{1:t}\right)$ is the output of a Kalman filter for a known measurement. The assignment probabilities are typically denoted by $\beta_t^r = p\left(\theta_t|\mathbf{Y}_{1:t}\right)$ and are given by [5]

$$\beta_t^r = \frac{b_t^r}{\sum_{r=-1}^{n_t} b_t^r}, \tag{A.13}$$

$$b_t^{-1} = \frac{1 - Pv_{t|t-1}}{Pv_{t|t-1}}, \tag{A.14}$$

$$b_t^0 = 1 - P_{\text{D}}, \tag{A.15}$$

$$b_t^r = P_{\text{D}} g(\mathbf{y}_t^r)\gamma(\mathbf{a}_t^r) n_t |W|, \tag{A.16}$$

where $\gamma(\mathbf{a}_t^r)$ is the likelihood ratio of the amplitude of measurement r and

$$Pv_{t|t-1} = (1 - \beta_{t-1}^{-1})(1 - p_{\text{death}}) + \beta_{t-1}^{-1} p_{\text{birth}}, \tag{A.17}$$

is the predicted visibility probability, that is the prior probability that there is actually a target. Track maintenance decisions are made based on the posterior visibility probability, $Pv_{t|t} = 1 - \beta_t^{-1}$.

A.3 Particle Filter TkBD

As introduced in Sect. 1.2, the fundamental Bayesian filtering relationship is

$$p\left(\mathbf{x}_t|\mathbf{Z}_{1:t}\right) \propto \mathscr{L}\left(\mathbf{Z}_t|\mathbf{x}_t\right) \int p\left(\mathbf{x}_t|\mathbf{x}_{t-1}\right) p\left(\mathbf{x}_{t-1}|\mathbf{Z}_{1:t-1}\right) d\mathbf{x}_{t-1}, \tag{A.18}$$

where the measurement likelihood has been replaced with a likelihood ratio. The ratio is proportional to the likelihood but has the advantage of being unity for most of the state space.

For point measurement tracking, the densities are often assumed to be Gaussian and linear and the integral can be analytically solved, leading to the Kalman Filter. For a general nonlinear problem, such as TkBD, the integral cannot be solved analytically. The most popular method to solve these nonlinear problems is sequential Monte Carlo sampling, that is, the particle filter.

The particle filter uses a random sample approximation to the densities in the Bayesian filter,

$$p(\mathbf{x}) \approx \sum_{n=1}^{N_p} w^n \delta(\mathbf{x} - \boldsymbol{p}^n), \tag{A.19}$$

where N_p is the number of particles, $\delta(\cdot)$ is the Dirac delta function, $\boldsymbol{p}^n$ is the nth particle, which is a sample from the state space, and w^n is the weight of the nth particle. The weights sum to unity.

There are several different types of particle filter: the form usually used for TkBD is referred to as the Sequential-Importance-Resampling (SIR) particle filter. It consists of the following steps: propagate the particles forwards in time from $t-1$ to t by drawing an independent sample from the state process for each particle; weight the particles by the likelihood (ratio) of the measurement at t; and resample the particle population. The resampling step does not need to be performed on every time iteration. Its purpose is to prevent the particle population from degenerating: a new particle set is created by making N_p draws on the existing population with replacement, using the weights as the selection probability. Detailed discussions of particle filtering and applications can be found in [10]. The particle filter was first applied to TkBD by Salmond in [12] and Boers in [2]. This was extended by Ristic and Rutten in [10, 11].

A direct application of SIR to TkBD provides a way of updating a state estimate using image measurements, but it does not incorporate a mechanism to detect the presence of a target. This is done by introducing an existence variable, which is a binary Markov chain that denotes whether or not there is a target present in the scene; the particle filter approach here deals only with the single target case. The particles are then used to describe the conditional distribution of the state given that there is a target. When there is no target then we define $\mathbf{x}_t \equiv \emptyset$.

Let E_t be an indicator variable such that $E_t = 1 \rightarrow \mathbf{x}_t \neq \emptyset$ and $E_t = 0 \rightarrow \mathbf{x}_t = \emptyset$. E_t is often referred to as the target existence variable and $p_t^{\emptyset}$ denotes the probability that the target does not exist, that is the null-state probability,

$$p_t^{\emptyset} := p\left(\mathbf{x}_t = \emptyset | \mathbf{Z}_{1:t}\right) = p(E_t = 0 | \mathbf{Z}_{1:t}). \tag{A.20}$$

The transition probabilities $p(E_t|E_{t-1})$ are specified by the priors P_{birth} and P_{death}

$$p(E_t|E_{t-1}) = \begin{cases} 1 - P_{\text{birth}} & E_t = 0, \quad E_{t-1} = 0, \\ P_{\text{death}} & E_t = 0, \quad E_{t-1} = 1, \\ P_{\text{birth}} & E_t = 1, \quad E_{t-1} = 0, \\ 1 - P_{\text{death}} & E_t = 1, \quad E_{t-1} = 1. \end{cases} \tag{A.21}$$

The non-null marginal state distribution is approximated with particles

$$p(\mathbf{x}_t|\mathbf{x}_t \neq \emptyset, \mathbf{Z}_{1:t}) = p(\mathbf{x}_t|E_t = 1, \mathbf{Z}_{1:t}) \approx \sum_{n=1}^{N_p} w_t^n \delta(\mathbf{x}_t - \boldsymbol{p}_t^n). \tag{A.22}$$

The null-state probability recursion is simply the Bayesian filter equation

$$p\left(E_t|\mathbf{Z}_{1:t}\right) \propto p(\mathbf{Z}_t|E_t) \sum_{E_{t-1}=0}^{1} p\left(E_t|E_{t-1}\right) p\left(E_{t-1}|\mathbf{Z}_{1:t-1}\right). \tag{A.23}$$

The likelihood $p(\mathbf{Z}_t|E_t)$ can be expressed as a function of the particle weights.

Using Bayes' rule and the law of total probability, the state distribution can be written as the mixture of two components, one representing the distribution of new (birth) targets appearing at time t, and the other representing continuing targets,

$$\begin{aligned} p\left(\mathbf{x}_t|E_t = 1, \mathbf{Z}_{1:t}\right) &\propto \mathscr{L}(\mathbf{Z}_t|\mathbf{x}_t) P_{\text{birth}} p_{t-1}^{\emptyset} p(\mathbf{x}_t|E_{t-1} = 0) \\ &+ \mathscr{L}(\mathbf{Z}_t|\mathbf{x}_t)(1 - P_{\text{death}})(1 - p_{t-1}^{\emptyset}) \int p(\mathbf{x}_t|\mathbf{x}_{t-1}) p(\mathbf{x}_{t-1}|\mathbf{Z}_{1:t-1}) d\mathbf{x}_{t-1}. \end{aligned} \tag{A.24}$$

The particle filter TkBD algorithm, therefore, constructs two sets of particles. The first set, $\{w_b^n, \boldsymbol{p}_b^n\}$, represents the density of new (birth) targets. The second set, $\{w_c^n, \boldsymbol{p}_c^n\}$, represents the density of targets that continue from $t-1$. The mixing proportions between these two sets are determined by the prior existence probability, $p_{t-1}^{\emptyset}$, and by the birth and death probabilities, P_{birth} and P_{death}.

The updated existence probability is calculated based on the particle weights, and then the two sets of particles are combined. A resampling stage reduces the total number of particles to a prescribed level and the resulting posterior particle set has uniform weights.

The algorithm declares a target detected when the null-state probability, $p_t^{\emptyset}$, is below a tunable threshold. The state estimate is then found by taking the mean of the

Algorithm 4 Particle Filter

1: Create a set of N_b birth particles by placing the particles in the highest intensity cells

$$\boldsymbol{p}_b^n \sim q(\mathbf{x}|E_{t-1} = 0, \mathbf{Z}_t). \tag{A.25}$$

Note that the placement of birth particles is independent from the set of continuing particles. The un-normalized birth particle weights are calculated using the likelihood ratio and proposal density (A.25)

$$\tilde{w}_b^n = \frac{\mathscr{L}(\mathbf{Z}_t|\boldsymbol{p}_b^n)\ p(\mathbf{x}_t = \boldsymbol{p}_b^n|E_{t-1} = 0)}{N_b\ q(\mathbf{x}|E_{t-1} = 0, \mathbf{Z}_t)}, \tag{A.26}$$

where the particular form of the likelihood depends on the measurement model assumed for the problem.

2: Create a set of N_c continuing particles using the system dynamics for the proposal function, with un-normalized weights

$$\tilde{w}_c^n = \frac{1}{N_c} \mathscr{L}\left(\mathbf{Z}_t|\boldsymbol{p}_c^n\right). \tag{A.27}$$

3: Determine mixing probabilities using sums of un-normalized weights

$$\tilde{M}_b = P_{\text{birth}}\ p_{t-1}^{\emptyset} \sum_{n=1}^{N_b} \tilde{w}_b^n, \tag{A.28}$$

$$\tilde{M}_c = (1 - P_{\text{death}})(1 - p_{t-1}^{\emptyset}) \sum_{n=1}^{N_c} \tilde{w}_c^n. \tag{A.29}$$

4: Update the null-state probability using the un-normalized weights

$$p_t^{\emptyset} = \frac{P_{\text{death}}(1 - p_{t-1}^{\emptyset}) + (1 - P_{\text{birth}}) p_{t-1}^{\emptyset}}{\tilde{M}_b + \tilde{M}_c + P_{\text{death}}(1 - p_{t-1}^{\emptyset}) + (1 - P_{\text{birth}}) p_{t-1}^{\emptyset}}. \tag{A.30}$$

5: Normalize the particle weights

$$\hat{w}_b^n = \frac{P_{\text{birth}}\ p_{t-1}^{\emptyset}}{\tilde{M}_b + \tilde{M}_c}\ \tilde{w}_b^n, \tag{A.31}$$

$$\hat{w}_c^n = \frac{(1 - P_{\text{death}})(1 - p_{t-1}^{\emptyset})}{\tilde{M}_b + \tilde{M}_c}\ \tilde{w}_c^n. \tag{A.32}$$

6: Combine the two sets of particles into one large set and resample from $N_b + N_c$ down to N_c particles with uniform weights.

---

state vectors of all particles. The particle filter based TkBD[1] method is summarised in Algorithm 4.

[1] *ParticleTracker*
The H-PMHT toolbox contains the function ParticleTracker that applies a single target particle filter implementation of Bayesian TkBD to a sequence of images.

Algorithm 5 Grid Bayes Filter

1: Pre-compute data likelihood ratios, $\mathscr{L}(\mathbf{Z}_t|\mathbf{x}_t^j)$
2: Initialize $\alpha_0(\emptyset) = 1$ and $\alpha_0(j_0 \neq \emptyset) = 0$
3: **for** $k = 1$ to T **do**
4: Forwards filter:

$$\alpha_t(\mathbf{x}_t^j) = \mathscr{L}(\mathbf{Z}_t|\mathbf{x}_t^j) \sum_{\mathbf{x}_{t-1}^j} p(\mathbf{x}_t^j|\mathbf{x}_{t-1}^j)\alpha_{t-1}(\mathbf{x}_{t-1}^j)$$

5: **end for**
6: Initialize $\beta_t(\mathbf{x}_t^j) = (N_x N_{\dot{x}} + 1)^{-1}, \forall \mathbf{x}_t^j$
7: **for** $k = T - 1$ to 1 **do**
8: Backwards filter:

$$\beta_t(\mathbf{x}_t^j) = \sum_{\mathbf{x}_{t+1}^j} \mathscr{L}(\mathbf{Z}_{t+1}|\mathbf{x}_{t+1}^j)p(\mathbf{x}_t^j|\mathbf{x}_{t+1}^j)\beta_{k+1}(\mathbf{x}_{t+1}^j)$$

9: **end for**
10: Determine marginal posterior pmf

$$p(\mathbf{x}_t^j|\mathbb{Z}) = \frac{\alpha_t(\mathbf{x}_t^j)\beta_t(\mathbf{x}_t^j)}{\sum_{\mathbf{x}_t^j} \alpha_t(\mathbf{x}_t^j)\beta_t(\mathbf{x}_t^j)}$$

11: At each scan, choose the state with the highest probability mass

A.4 Grid Bayes

The particle filter uses a random sampling approximation to the difficult integrals in the Bayesian filter. Another option is to numerically evaluate the filter over a discrete grid and approximate the various functions as piecewise constant over that grid [13]. This amounts to a Reimann sum approximation to the required integrals and will converge to the Bayesian filter provided the Reimann integral exists. We are happy to assume that the underlying pdfs are continuous and smooth, so Reimann integration will suffice. In the particle filter, it is inefficient to create samples that correspond to a non-existent target, so it carries a separate probability $p(E)$. When we use a finite state representation, it is more convenient to achieve the same effect by augmenting the state space with an additional null-state that implies there is no target. Mathematically, we use the shorthand $p(\emptyset)$, which is equivalent to $p(E = 0)$ in the particle context.

Algorithm 5 presents a forwards–backwards formulation of track-before-detect using the birth and death model described above. Details about the general forwards–backwards method can be found in [9]. This method is appropriate for batch processing problems. If a recursive algorithm is preferred, such as for real-time processing, then the backwards filter is omitted and the forwards filter is the recursive pdf estimator.

The main shortcoming of this method is that the computation cost required to estimate the target location to a high degree of accuracy may be excessive.

A.5 Viterbi TkBD

The Viterbi algorithm is an optimal sequence estimation method for discrete state systems [1, 15]. Whereas the particle filter produces an estimate of the state pdf at each frame, the Viterbi algorithm directly produces a single best state at each frame. The definition of best depends on the cost function used: if the measurement likelihood is used then the estimate is a maximum likelihood estimate, if the state transition probabilities are used then the estimate is a maximum a posteriori estimate. The sequence nature of the Viterbi algorithm means that it is a batch method that optimises over all the states at once. In contrast, the particle filter described in the previous section is a time-recursive method. The advantage of sequence estimation is that the output is guaranteed to not contain zero-probability transitions. The disadvantage is that it increases computation cost.

The Viterbi algorithm was the most popular grid method in the 1990s [1, 14] and research interest still continues [16]. The way it is applied to TkBD where the state space is continuous is to approximate the state as a grid of discrete points and then to solve this approximate problem. Since the sample locations are fixed, the resolution of the grid is a fundamental limit on estimation accuracy. The total number of sequences in a batch of length T is $||\mathbb{X}||^T$ with $||\mathbb{X}||$ the number of samples in the discrete-space grid. The main trick of the Viterbi algorithm is to exploit the Markov nature of the state process to avoid enumerating all of the state sequences. This reduces the complexity to $(T-1)||\mathbb{X}||^2$, which is further reduced to $(T-1)||\mathbb{X}||N_{t|t-1}$ where $N_{t|t-1} \ll ||\mathbb{X}||$ is the number of states that each particular $\mathbf{x}$ is allowed to transition into. Clearly, there is an advantage in not choosing a transition probability function with infinite support, like the Gaussian used everywhere else. In many Viterbi TkBD papers a uniform probability is applied over a local transition region.

The Viterbi algorithm works by factorising the sequence through j_t into the sequence leading into j_t and the sequence following it. From the Markov chain property of the state, we know that the probability of sequences after j_t does not depend on how we got here, just where we are now. In other words, the best path from A to C via B is the same as the best path from A to B followed by the best path from B to C. Mathematically, this is conditioning on j_t

$$
\begin{aligned}
p(j_{0:t-1}, j_{t+1:T}|\mathbf{Z}_{1:T}, j_t) &\propto p(j_{0:t-1}, j_{t+1:T}, \mathbf{Z}_{1:T}|j_t), \\
&= p(\mathbf{Z}_t|j_t)p(\mathbf{Z}_{1:t-1}|j_{0:t-1})p(\mathbf{Z}_{t+1:T}|j_{t+1:T}) \qquad \text{(A.33)} \\
&\quad \times p(j_{0:t-1}|j_t)p(j_{t+1:T}|j_t), \\
&\propto p(j_{0:t-1}, \mathbf{Z}_{1:t-1}|j_t)p(j_{t+1:T}, \mathbf{Z}_{t+1:T}|j_t). \qquad \text{(A.34)}
\end{aligned}
$$

Suppose that at time k, we know the most probable path through each grid point and we know the associated path likelihood. Denote the most probable ancestor of j_t as $\theta_t(j_t)$ and let the cost of the path through j_t be the negative path log-likelihood, $C_t(j_t)$. The Viterbi algorithm is a recursion for $\theta_t(j_t)$ and $C_t(j_t)$.

The cost of a path through j_t and j_{t+1} is

$$C_t(j_t) - \log\left\{p\left(j_{t+1}|j_t\right)\right\} - \log\left\{p\left(\mathbf{Z}_{t+1}|j_{t+1}\right)\right\}, \tag{A.35}$$

so the cost of the best path through j_{t+1} is the lowest of these, namely

$$C_{t+1}(j_{t+1}) = \min_{j_t}\left\langle C_t(j_t) - \log\left\{p\left(j_{t+1}|j_t\right)\right\}\right\rangle - \log\left\{p\left(\mathbf{Z}_{t+1}|j_{t+1}\right)\right\}. \tag{A.36}$$

The most probable ancestor of j_{t+1} is the value of j_t that results in the lowest cost,

$$\theta_{t+1}(j_{t+1}) = \arg\min_{j_t}\left\langle C_t(j_t) - \log\left\{p\left(j_{t+1}|j_t\right)\right\}\right\rangle. \tag{A.37}$$

At the end of the batch, the estimated state is the one with the lowest final cost, $C_t(j_t)$, and the estimated path is found by stepping backwards through the ancestors starting with $\theta_t(j_t)$.

The costs of all paths will grow with time since they are the accumulated log-likelihood. However, the optimisation is invariant under shifts, so we can arbitrarily choose to make $C_t(\emptyset) = 0$ by shifting the costs at each time step. We can also replace the data likelihood with the likelihood ratio, which is equivalent to shifting the costs by $\log\{P(\mathbf{Z}_{t+1}|\emptyset)\}$.

The Viterbi[2] algorithm as applied to TkBD is summarised in algorithm 6. The key feature of the algorithm is that it implicitly optimises over all possible state sequences by making locally optimal decisions. We use a null-state representation again, as with the forwards–backwards approach. This is not common, but is a natural extension and gives an integrated track score statistic. See [8] for a more detailed discussion.

[2] *ViterbiTracker*
The H-PMHT toolbox contains the function ViterbiTracker that applies a single target Viterbi filter implementation of Bayesian TkBD to a sequence of images.

Algorithm 6 Fixed-grid Viterbi

1: Initialize $C_0(\emptyset) = 0$ and $C_0(j_0) = \infty$ for all other states
2: **for** each frame $t = 1 \ldots T$ **do**
3: Calculate the un-normalized cost of the null state

$$c_t^0 = \min_{j_{t-1}} \left\langle C_{t-1}(j_{t-1}) - \log \{p\,(j_t = \emptyset | j_{t-1})\} \right\rangle$$

4: Calculate the normalized state costs

$$C_t(j_t) = -\log \mathscr{L}(\mathbf{Z}_t | j_t) + \min_{j_{t-1}} \left\langle C_{t-1}(j_{t-1}) - \log p\,(j_t | j_{t-1}) \right\rangle - c_t^0$$

5: The previous state in the most likely sequence leading to j_t is given by

$$\theta_t(j_t) = \arg\min_{j_{t-1}} \left\langle C_{t-1}(j_{t-1}) - \log p\,(j_t | j_{t-1}) \right\rangle$$

6: **end for**
7: The estimated state at time T is

$$\hat{j}_T = \arg\min_{j_T} C_T(j_T)$$

8: **for** each frame $t = T - 1 \ldots 1$ **do**
9: The estimated state at time k is found by backtracking

$$\hat{j}_t = \theta_{t+1}\left(\hat{j}_{t+1}\right)$$

10: **end for**

References

1. Barniv, Y.: Dynamic Programming Algorithm for Detecting Dim Moving Targets. In: Multitarget-Multisensor Tracking: Advanced Applications, Artech House (1990)
2. Boers, Y., Driessen, J.N.: Particle filter based detection for tracking. In: Proceedings of the American Control Conference (2001)
3. Bar-Shalom, Y., Willett, P.K., Tian, X.: Tracking and Data Fusion: A Handbook of Algorithms YBS (2011)
4. Blackman, S.S., Popoli, R.: Design and Analysis of Modern Tracking Systems. Artech House, Norwood, MA (1999)
5. Colegrove, S.B., Davey, S.J.: PDAF with multiple clutter regions and target models. IEEE Trans. Aerosp. Electron. Syst. **39**, 110–124 (2003)
6. Davey, S.J., Rutten, M.G., Cheung, B.: A comparison of detection performance for several track-before-detect algorithms. EURASIP J. Adv. Signal Process. **2008** (2008)
7. Davey, S.J., Rutten, M.G., Cheung, B.: Using phase to improve track-before-detect. IEEE Trans. Aerosp. Electron. Syst. **48**(1), 832–849 (2012)
8. Davey, S.J., Rutten, M.G., Gordon, N.J.: Integrated Tracking, Classification and Sensor Management: Theory and Applications, Chapter. Track Before Detect Techniques

9. Jang, B., Rabiner, L.: An Introduction to Hidden Markov Models, IEEE ASSP Magazine (1986)
10. Ristic, B., Arulampalam, S., Gordon, N.J.: Beyond the Kalman Filter: Particle Filters for Tracking Applications Artech House (2004)
11. Rutten, M.G., Gordon, N.J., Maskell, S.: Recursive track-before-detect with target amplitude fluctuations. IEE Proc. Radar Sonar Navig. **152(5)** (2005)
12. Salmond, D.J., Birch, H.: A particle filter for track-before-detect. In: Proceedings of the American Control Conference (2001)
13. Stone, L.D., Streit, R.L., Corwin, T.L., Bell K.L.: Bayesian Multiple Target Tracking. Artech House (2013)
14. Tonissen, S.M., Evans, R.J.: Performance of dynamic programming techniques for track-before-detect. IEEE Trans. Aerosp. Electron. Syst. **32**, 1440–1451 (1996)
15. Viterbi, A.J.: Error bounds for convolutional codes and an asymptotically optimum decoding algorithm. IEEE Trans. Inf. Theory **13**, 260–269 (1967)
16. Wie, Y., Morelande, M.R., Kong, L., Yang, J.: An efficient multi-frame track-before-detect algorithm for multi-target tracking. IEEE J. Select. Top. Signal Process. **7**(3), 421–434 (2013)

Appendix B
H-PMHT Toolbox

We briefly describe here the important components of the H-PMHT toolbox. A more detailed manual is available online.

B.1 Core Data Structures

The toolbox uses two core data structures: *hpmht_parameters* and *hpmht_data*. hpmht_parameters stores the static input parameters, such as the process noise covariance and track manager decision thresholds. Some of these are common across the different incarnations of H-PMHT in the toolbox, such as image dimensions. Others are only used by particular variants; the H-PMHT-RM has a rate parameter that defines the time correlation scale of the shape matrix, see Chap. 8. hpmht_parameters also contains function pointers that tell the track manager how to create new tracks, update tracks with new image data and make status decisions. These are described a little later in Sect. B.3.

hpmht_data stores the dynamic outputs of the tracker. It contains three track lists: candidate tracks; established tracks; and old tracks. The track manager in Sect. B.3 creates candidate tracks on potential targets and these are promoted to established tracks if they satisfy the track promotion rules. Poor candidates are discarded but established tracks that are no longer associating measurement data are stored as old tracks. Only the established tracks and old tracks should be treated as output data; candidates would not be shown to a system user. hpmht_data also stores counters for the track identifier numbers. We use two separate counters for candidate tracks, and established and old tracks.

S. J. Davey and H. Gaetjens, *Track-Before-Detect Using Expectation Maximisation*, Signals and Communication Technology,
https://doi.org/10.1007/978-981-10-7593-3

B.2 Scenario Generation

Throughout this book, several variations of the H-PMHT are compared on a common simulation scenario, referred to as the canonical multi-target scenario. This scenario acts as a benchmark for the different methods and is provided in the toolbox, which enables users to reproduce most of the performance graphs shown in the book. We also suggest that this scenario could be the starting point for a broader benchmarking conversation across the tracking community for multi-target track-before-detect. The scenario is described in detail in Chap. 1.

Two versions of the scenario are contained in the toolbox: *GenerateCanon* produces the standard scenario described above; *GenerateCanonColour* produces a variation of the scenario used for colour video tracking and discussed in more detail in Chap. 9.

B.3 Track Manager

The different variants of H-PMHT are built on a common track management framework. This can be described in terms of a Poisson birth density and a hierarchical Gaussian mixture, but we leave this nicety to other places, try Chap. 5. The framework of tracks is illustrated in Fig. 5.5 which we repeat here; we retain two living track lists for candidate and established tracks, and a third track list for old tracks, which are dead established tracks. The candidates are automatically formed by the manager and are hypothesised new components of the mixture. Candidates become established once they pass a status test.

The function *HPMHTTracker* contains the spine of the track manager. It receives a sequence of image frames and uses the parameters in hpmht_data to select which H-PMHT variant to use. hpmht_data contains function pointers to implement the blocks in the standardised track manager in Fig. 5.5. These functional blocks are now briefly described (Fig. B.1).

B.3.1 Hierarchical Data Association

Figure 5.5 illustrates the hierarchical access to data implemented by the manager. Established tracks are first associated with the sensor image which results in a set of associated images: one for each track and another for the background Chap. 5. The background associated image is the input to a second round of H-PMHT that updates the candidate tracks and the associated image from this stage is used to form new candidate tracks. The layering prevents the tracker from creating multiple tracks on the same object response. The H-PMHT measurement model is a mixture, so it is

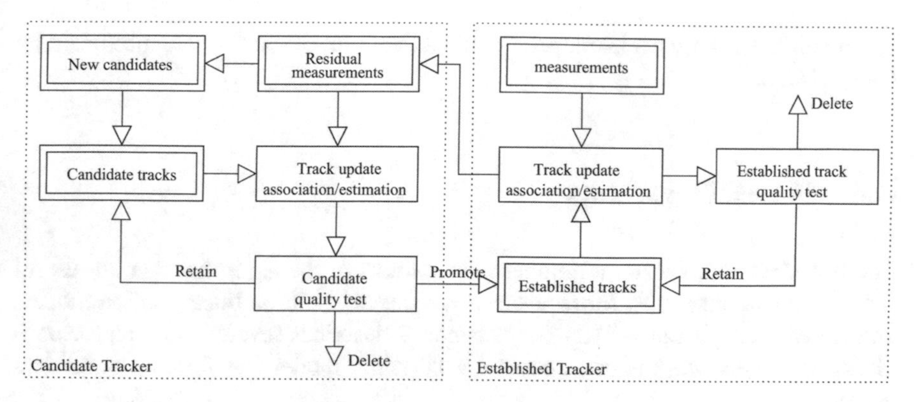

Fig. B.1 Track management flow diagram

always possible to stack two tracks on top of each other and share the target power between them.

B.3.2 Candidate Formation

The track manager forms new candidate tracks by finding local maxima in the background associated image. A threshold can be applied to reduce computation overheads if required. The manager uses the function *CreateTrack* to form a new instance of the track structure and initialise it using each local peak.

B.3.3 Track Update

The track manager calls the *UpdateTrack* function and provides it with an input image. This input is either the sensor measurement or a background associated image. The update function predicts tracks forwards in time, performs data association and corrects state estimates. It returns a collection of modified tracks and a residual background associated image.

B.3.4 Status Decisions

Status functions are used to promote candidates to established tracks, to discard poor candidates, and to terminate poor established tracks. There are two status functions: *CandStatus* defines the statistical tests and rules used to promote and discard candidate tracks; *EstStatus* defines the tests used to terminate established tracks.

The majority of these decisions are based on the track power estimate produced by *UpdateTrack*.

B.4 H-PMHT variants

The H-PMHT toolbox has a number of variations of the algorithm that are useful for different applications. Most of these variants are built on linear motion models and Gaussian appearance functions. Chapter 7 describes several methods for non-Gaussian targets, the toolbox contains a Dirichlet model that forms a numerical estimate.

B.4.1 Core H-PMHT

The core H-PMHT uses a multinomial measurement prior. The default tracking parameters for the core H-PMHT are created by *MakeHPMHTParams('core')* or *MakeHPMHTParams()*.

B.4.2 Single Target Chip H-PMHT

This is a variant of core H-PMHT that uses efficient implementation to deal with very large images and target counts. It combines strategies described in Chap. 5 and applies only to two-dimensional input images with non-skewed Gaussian appearance. The default tracking parameters are created by *MakeHPMHTParams('STC')*.

B.4.3 Poisson H-PMHT

The multinomial assignment model in core H-PMHT requires a normalising constraint across all targets that prohibits the introduction of a dynamic model (Chap. 6). The alternative approach is to use an independent Poisson prior on each individual target. The Poisson assignment prior improves track maintenance decisions, especially for targets with fluctuating amplitude. The default Poisson H-PMHT tracking parameters are created by *MakeHPMHTParams('Poisson')*. The resulting tracks contain parameters for a gamma distributed posterior over the Poisson rate.

B.4.4 H-PMHT with Random Matrices

The above H-PMHTs all assume a known Gaussian target appearance function, that is the shape of the blob of energy caused by each target is the same, is Gaussian, and the covariance matrix of this Gaussian is known. H-PMHT-RM relaxes this assumption by allowing each target to have an unknown Gaussian spreading matrix, Chap. 9. It learns the shape spreading matrix by applying an inverse Wishart prior updating this with data. The original H-PMHT-RM was built on core H-PMHT and uses the multinomial model. This can be setup using *MakeHPMHTParams('core H-PMHT-RM')*. A version incorporating the Poisson assignment prior is initialised with *MakeHPMHTParams('Poisson H-PMHT-RM')*.

B.4.5 Dirichlet Appearance H-PMHT

The Dirichlet H-PMHT is an extension of the Poisson H-PMHT that uses a grid approximation to the appearance function instead of a known Gaussian. It is capable of learning arbitrary appearance functions as in Chap. 9 but is quite costly to run. It is initialised with *MakeHPMHTParams('Dirichlet')*.

B.5 Other Tracking Algorithms

For completeness, the toolbox also contains some elementary implementations of point measurement tracking algorithms. These include a single frame detector, point measurement PMHT and Probabilistic Data Association.

Index

S. J. Davey and H. Gaetjens, *Track-Before-Detect Using Expectation Maximisation*, Signals and Communication Technology,
https://doi.org/10.1007/978-981-10-7593-3

Printed in the United States
By Bookmasters